Turbulent Flow

Turbulent Flow

R.J. GARDE

Professor Emeritus
Civil Engineering Department
University of Roorkee
Roorkee, India

JOHN WILEY & SONS
NEW YORK • CHICHESTER • BRISBANE • TORONTO • SINGAPORE

First Published in 1994 by
WILEY EASTERN LIMITED
4835/24 Ansari Road, Daryaganj
New Delhi 110 002, India

Distributors:

Australia and New Zealand :
JACARANDA WILEY LIMITED
PO Box 1226, Milton Old 4046, Australia

Canada :
JOHN WILEY & SONS CANADA LIMITED
22 Worcester Road, Rexdale, Ontario, Canada

Europe and Africa :
JOHN WILEY & SONS LIMITED
Baffins Lane, Chichester, West Sussex, England

South East Asia :
JOHN WILEY & SONS (PTE) LIMITED
05-04, Block B, Union Industrial Building
37 Jalan Pemimpin, Singapore 2057

Africa and South Asia :
WILEY EASTERN LIMITED
4835/24 Ansari Road, Daryaganj
New Delhi 110 002, India

North and South America and rest of the world :
JOHN WILEY & SONS INC.
605, Third Avenue, New York, NY 10158, USA

Library of Congress Cataloging-in-Publication Data

ISBN 0-470-23340-0 John Wiley & Sons Inc.
ISBN 81-224-0562-2 Wiley Eastern Limited

Printed in India at S.P. Printers, Noida.

To

Prof. M.L. Albertson

and

Dr. A.S. Apte

Preface

Turbulence is a phenomenon which occurs in a very wide range of situations both in natural context, e.g., in streams, oceans and atmosphere, as well as in technological context, namely, in hydraulic, chemical, aeronautical and naval engineering. In all these cases Reynolds number of the gross flow characterising its state is very large, and therefore, the turbulent state rather than the laminar state is regarded as natural and unavoidable.

Since the pioneering contribution made by Osborne Reynolds at the end of the nineteenth century, considerable developments have taken place in the understanding of the nature of turbulence, measurement of parameters of turbulence, mathematical formulations and numerical solutions of problems in turbulent flow. These studies have also revealed that the problem of turbulence, always regarded as difficult, is, in fact, extremely difficult. Hence, the time scale, representing the time required for significant advance in understanding the subject, has to be enlarged accordingly. Several books, such as those written by Hinze, Rodi, Lumley et al., Bradshaw, Townsend, Batchelor, and Monin and Yaglom, are available for those who wish to study turbulence.

In spite of these developments and availability of the above mentioned books, relatively little new information on turbulence has permeated to Civil Engineers dealing with hydraulic analysis and design, building aerodynamics, and environmental analysis and design.

This is so because there has developed communication barrier between those conducting research in turbulence and those who would like to use the essential findings in understanding and solving practical problems. This is mainly due to the inherent complexities of the turbulence phenomenon, the large number of scientists with different backgrounds working in this area, and the limited ability of majority of engineers to comprehend the complexities of mathematics of turbulence.

I am convinced that the Civil Engineers involved in hydraulic analysis and design, as well as the environmental engineers, will be greatly benefitted by getting familiar with the recent advances in turbulence. The present book is written to fulfil this purpose.

The book gives the basic analytical framework for description of turbulent flows and discusses various types of turbulent flows commonly encountered by the Civil Engineer. The book can be easily followed by engineering graduates who have taken a couple of courses in fluid mechanics. The material presented is arranged in ten chapters. The first chapter deals with Laminar Flow discussing briefly the characteristics of N.S. equations and typical solutions thereof. The second chapter deals with Transition from Laminar to Turbulent Flow. The third chapter briefly highlights the nature of Turbulent Flow, while the fourth chapter deals with the Equations of Motion, namely, continuity, momentum and energy equations. The fifth chapter gives the basic notions of the Statistical

Theory of Turbulence, while the sixth deals with Turbulence Models. The seventh chapter is devoted to the Measurement of Turbulence in air and water. The eighth, ninth and tenth chapters deal with description of Wall Turbulent Shear Flows, Free Surface Flows, and Free Turbulece Shear Flows. The choice of material presented in this book is governed by my experience, its relevance to the civil engineer, and by the ease with which it can be presented in a simpler form. From this point of view the effect of compressibility has been omitted throughout the book. In the preparation of this book, I have heavily relied on the works of those who have written before me on this subject and whose works are considered classics in this field. I am indebted to all these authors.

I wish to take this opportunity to express my sincere appreciation to Dr. S. Narasimhan, former Professor of Civil Engineering at Indian Institute of Technology, Bombay for meticulously going through the manuscript and making useful comments and suggestions most of which have been incorporated. I am also thankful to Ms. Jaya Pahvalkar, CWPRS, Pune for suggesting some additions in the chapter on Measurement of Turbulence which have been included.

I wish to express my thanks to Mr. P.P. Rao and Ms. Neelima Vadnere for typing the manuscript, Mr. Mehta for preparing the art work, and Mr. Praveen Bell for designing the cover.

The book is dedicated to Prof M.L. Albertson and Dr. A.S. Apte who have influenced me greatly throught out my active life, and to whom I am greatly indebted.

R. J. GARDE

CWPRS
P.O. Khadakwasla
Pune 411024

List of Symbols

a_o	distance of virtual origin of wake from centre of the body
A	cross-sectional area
b	half width of the jet or the wake at any distance x
b_o	half width of jet at the beginning
B	channel or two dimensional conduit width
C	concentration of the solute
C_f	local skin friction coefficient
C_D	drag coefficient
C_{DO}	corrected drag coefficient
C_f	average skin friction coefficient
C_L	lift coefficient
C_p	specific heat at constant pressure
C_v	absolute volume concentration
d	width of grid bar
D_o	initial diameter of round jet
D_L	longitudinal dispersion coefficient
D_m	molecular diffusion coefficient
E	kinetic energy
$E(K)$	three dimensional energy spectrum function
E_c	Eckert number
f	Darcy Weisbach friction factor
f_m	natural frequency
f_p	friction factor with polymer
f_s	friction factor without polymer
$f(r)$	longitudinal correlation coefficient
F_D	drag force
F_r	Froude number
$F_1(n)$	normalised one dimensional energy spectrum function
g	gravitational acceleration
$g(r)$	lateral correlation coefficient
h	height of plate or object; triple velocity correlation coefficient
h_f	energy loss
H	ratio of displacement thickness δ^* to momentum thickness θ of the boundary layer
I	current
I_u	turbulence intensity
k	average height of roughness; kinetic energy of turbulence per unit volume; also coefficient of conductivity

l, l_k, l_m, l_w	mixing lengths
L	eddy size, macroscale of turbulence, length
L_f, L_g	longitudinal and lateral integaral macroscales of turbulence
L_j	length of hydraulic jump
L_m	dispersion length or mixing length of dispersion
L_{mo}	Monin–Obukhov length scale
M	size of grid opening; also momentum flux
n	frequency
N	rotational speed in rpm
p	pressure at a point
p_o	reference pressure
p_w	wall pressure
P	power
Pr	Prandtl number $(= \mu_g C_p/K)$
Q	dishcarge
Q_0	discharge at the beginning
r	radical distance
R	radius of sphere or pipe, also resistance
Ro	resistance at reference temperature
R_b	Rossby number $(= U_o/L_o\, w)$
Re	Reynolds number $(= U_o\, L_o\, \rho/\mu)$
Ri	Richardson number $\left(= -\frac{g}{\rho}\frac{\partial \rho}{\partial y} \Big/ \left(\frac{\partial u}{\partial y}\right)^2 \right)$
$R_E(t)$	Eulerian correlation coefficient
$R_L(t)$	Lagrangian correlation coefficient
R_w	wire resistance
S	channel slope; Strouhal number; surface of the control volume
t	time, or time interval
T	sampling time, return period
T_e	Eulerian microtime scale
T_E	Eulerian macrotime scale
u	local velocity in x direction
u_d	velocity difference in wake $(= U_o - \bar{u})$
u_m	maximum velocity, or centreline velocity
u_r	velocity of Karman vortices
u_*	shear velocity $(= \sqrt{\tau_o/\rho})$
U_o	free stream or reference velocity
U_d	maximum vlaue of u_d
U_g	geostrophic wind velocity
U_m	average velocity in a section
v	local velocity in y direction
v_m	Kolmogorov's velocity scale

v_t	characteristic turbulent velocity
$\forall$	control volume
V_x, V_r, V_θ	velocities in x, r, θ direction, respectively
w	local velocity in z direction
$W(K, t)$	energy transfer function of Kolmogorov
x	variable distance in x direction
X	component of body force per unit mass, or per unit volume, in x direction
y	variable distance in y direction
y'	distance from the boundary where $u = 0$ according to logarithmic law
y_1, y_2	depths of sections 1, 2, respectively
y_o	depth of flow in a channel
Y	component of body force per unit mass or volume, in y direction
Y_g	height above the ground at which geostrophic velocity occurs
z	variable distance in z direction
Z	component of body force per unit mass or volume, in z direction; also a variable in turbulence modelling ($Z = k^m l^n$)
$'$	fluctuating component as in u'
$-$	time averaged quantity as in $\bar{u}$
o	reference quantity as in U_o
(3*)	as quoted in reference 3

Greek Notations

α	wave number ($= 2\pi/\lambda$); also energy correction factor
β	momentum correction factor: also circular frequency ($= \beta_r + i\beta_i$)
γ	specific weight of fluid ($= \rho g$); also intermittency factor of turbulence
δ	nominal boundary layer thickness
δ^*	displacement thickness of boundary layer
Γ	diffisivity coefficient; also circulation
ε	energy loss per unit volume; also eddy kinematic viscosity
ε_m	momentum transfer coefficient
ε_s	sediment transfer coefficient
η	dimensionless y distance e.g. $\frac{y}{\delta}$ or $\frac{y}{b}$; also eddy dynamic viscosity
θ	momentum thickness of boundary layer
κ	Karman constant
λ	wave length; mean free path of molecules; Pohlhausen number $\left(= \frac{\delta^2}{\mu\rho U_o}\frac{\partial p}{\partial x}\right)$
λ_f, λ_g	longitudinal and lateral microlength scales of turbulence
μ	dynamic viscosity of fluid
μ_t	dynamic eddy viscosity of fluid
ν	kinematic viscosity of fluid

ν_t	kinematic eddy viscosity of fluid
ξ	non-dimensional distance; vorticity $= \left(\frac{\partial v}{\partial x} - \frac{\partial u}{\partial y}\right)$
ρ	mass density of fluid
σ	standard deviation
τ	shear stress
τ_t	turbulent shear stress
τ_o	bed shear stress
ϕ	scalar quantity, such as temperature or concentration; also a dimensionless variable $\left(\phi = \frac{y}{ax}\right)$
χ	Rouse stability parameter
ψ	stream function; also drag reduction parameter
ω	angular velocity $\omega = \frac{1}{2}\left(\frac{\partial v}{\partial x} - \frac{\partial u}{\partial y}\right)$; also fall velocity
ω_o	fall velocity in quiet fluid
$\pi(x)$	profile parameter of Cole

Contents

CHAPTER I

Laminar Flow

1.1 INTRODUCTION

In classical Hydrodynamics the fluid is assumed to be inviscid (i.e., having zero viscosity) and incompressible (i.e. having constant mass density). Such a fluid is known as *ideal fluid*, and in the case of flow of such a fluid there is no tangential shear stress between the adjacent fluid layers. Further, at the surface of contact of an ideal fluid and the solid body, continuity requires that the normal velocity of the fluid and the boundary should be the same; but there is a relative tangential velocity, or velocity of slip.

Valuable information on characteristics of flow of a real fluid, under certain conditions, such as rapidly converging flow, can be obtained by assuming the fluid to be ideal and applying the theory of classical Hydrodynamics. However, this cannot be done as a general rule. The existence of tangential stresses, and the impossibility of slip at the boundary constitute the main differences between the flow of real fluid and the ideal one. In the case of class of real fluids known as *Newtonian Fluids*, for one dimensional flow in x direction, shear stress in x direction on the xz plane is given by Newton's formula $\tau = \mu\,(\partial u/\partial y)$, where μ is the coefficient of dynamic viscosity having dimensions of $ML^{-1}\,T^{-1}$, and $\partial u/\partial y$ is the velocity gradient. The ratio of coefficient of dynamic viscosity μ to the mass density of fluid ρ is known as the coefficient of kinematic viscosity ν, having the dimensions of $L^2\,T^{-1}$. Under ordinary circumstances, μ for liquids depends on temperature only, μ decreasing with increase in temperature. On the other hand μ for gases increases as the temperature rises. This difference in the variation of viscosity of liquids and gases with change in temperature is because of the relative importance of cohesive force and molecular activity in liquids and gases. In liquids, the molecules are very closely spaced and, consequently, cohesive forces are large. Cohesion decreases with increase in temperature, and hence the viscosity of liquids decreases with the increase in temperature. In gases, the cohesive force is quite small because gas molecules are far apart, hence existence of viscosity is more due to transfer of molecular momentum. Since the molecular activity increases with the increase in temperature, viscosity of gases increases with the increase in temperature.

Theoretical explanation of Newton's formula for shear stress in a viscous fluid was first given by Maxwell in 1860 for gases using kinetic theory. This is briefly discussed below. Consider an imaginary surface AB in xz plane in which velocity u varies with y; let $\partial u/\partial y$ represent the velocity gradient there. If one considers a unit volume of gas there with N molecules of mass m each, it can be assumed that these molecules are in random motion with velocity

c, and, further, that $N/3$ are moving in x direction, $N/3$ in y direction and $N/3$ in z direction. Of the $N/3$ molecules moving in y direction, assume that $N/6$ move upwards and $N/6$ move downwards across AB. Hence, the mass of gas flowing upwards and downwards per second per unit area is $Nmc/6$. If λ is the mean free path of molecules, the net rate of momentum transfer per unit area is

$$\frac{Nmc}{6}\left(u + \frac{\partial u}{\partial y}\lambda - \left(u - \frac{\partial u}{\partial y}\lambda\right)\right) = \frac{Nmc\lambda}{3}\frac{\partial u}{\partial y}$$

Since according to Newton's Second Law of Motion, the rate of change of momentum must be equal to force, shear stress on plane AB is $\tau = \frac{Nmc\lambda}{3}\frac{\partial u}{\partial y}$.

However, Nm = mass per unit volume i.e., mass density ρ. Hence

$$\tau = \frac{\rho c\lambda}{3}\frac{\partial u}{\partial y}$$

Comparing this with $\tau = \mu \frac{\partial u}{\partial y}$, the Newton's formula, one gets

$$\mu = \frac{1}{3}\rho c\lambda$$

This derivation is extremely simplified. More accurate calculations have given an expression for μ as $0.499\rho c\lambda$ in place of $\rho c\lambda/3$. It can be seen from this formula that the usual ideas of kinetic theory would make μ independent of pressure and directly proportional to the square root of absolute temperature. However, experimentally it has been found that μ increases with increase in temperature more rapidly than the theoretical results.

In engineering problems one comes across fluid flows involving Newtonian fluids flowing at low velocities and with small dimensions. Lubrication problems, flow of water through porous material, flow past immersed bodies at very small velocities (or its counterpart of motion of tiny particles in still fluids), flow of viscous fluids through small diameter tubes and sheet flow of viscous fluid, are some of the numerous important problems falling in this category. This category of problems is characterised by the fact that, in all these cases, viscous force per unit volume of fluid is very large compared to the inertial force per unit volume. These flows are called *Laminar Flows.* The equations governing laminar flow and some typical solutions for specific cases are discussed below to focus attention on the characteristics of laminar flows which are quite different from those of turbulent flows.

1.2 EQUATIONS OF MOTION

As in the case of ideal fluid flow, the laminar flow is analysed by writing the equations of motion which are a mathematical statement of Newton's

Second Law of Motion. According to Newton's Second Law, the summation of the components of all external forces acting on a fluid in a given direction is equal to mass of fluid multiplied by the total acceleration in that direction. Considering a unit volume of incompressible fluid, the equations of motion in cartesian coordinate system take the following form, when pressure, gravity and viscous forces control the motion.

$$\rho\left[\frac{\partial u}{\partial t}+u\frac{\partial u}{\partial x}+v\frac{\partial u}{\partial y}+w\frac{\partial u}{\partial z}\right]=X-\frac{\partial p}{\partial x}+\mu\left[\frac{\partial^2 u}{\partial x^2}+\frac{\partial^2 u}{\partial y^2}+\frac{\partial^2 u}{\partial z^2}\right]$$

$$\rho\left[\frac{\partial v}{\partial t}+u\frac{\partial v}{\partial x}+v\frac{\partial v}{\partial y}+w\frac{\partial v}{\partial z}\right]=Y-\frac{\partial p}{\partial y}+\mu\left[\frac{\partial^2 v}{\partial x^2}+\frac{\partial^2 v}{\partial y^2}+\frac{\partial^2 v}{\partial z^2}\right]$$

$$\rho\left[\frac{\partial w}{\partial t}+u\frac{\partial w}{\partial x}+v\frac{\partial w}{\partial y}+w\frac{\partial w}{\partial z}\right]=Z-\frac{\partial p}{\partial z}+\mu\left[\frac{\partial^2 w}{\partial x^2}+\frac{\partial^2 w}{\partial y^2}+\frac{\partial^2 w}{\partial z^2}\right] \quad \ldots (1.1)$$

These equations were first derived independently by Navier and Stokes; hence they are commonly known as *Navier–Stokes equations.* Here ρ and μ are the mass density and the dynamic viscosity of the fluid, respectively, u, v and w are the components of velocity in the x, y and z directions, respectively; p is the pressure and X, Y and Z are the components of body force per unit volume in the three directions. The terms in the bracket on the left hand side represent the total acceleration in a given direction, which consists of local acceleration (e.g. $\partial u/\partial t$, etc.) due to unsteadiness of the flow, and the convective acceleration $\left(\text{e.g. } u\frac{\partial u}{\partial x}+v\frac{\partial u}{\partial y}+w\frac{\partial u}{\partial z}, \text{ etc.}\right)$ which is due to the nonuniformity of flow. On the right hand side, the last term in the bracket represents the viscous forces per unit volume arising out of the components of shear stresses and additional normal stresses induced due to the presence of viscosity. It can be seen that when μ is equal to zero, Equations 1.1 reduce to the well known Euler's equations of motion.

Similarly, equations for laminar flow can be derived in cylindrical polar coordinate system. If r, θ and x denote the radial, azimuthal and axial coordinates respectively, and V_r, V_θ and V_x denote the respective velocities in these directions, then for incompressible Newtonian fluids, N.S. equations are

$$\rho\left(\frac{\partial V_r}{\partial t}+V_r\frac{\partial V_r}{\partial r}+\frac{V_\theta}{r}\frac{\partial V_r}{\partial \theta}-\frac{V_\theta^2}{r}\right)$$

$$=F_r-\frac{\partial p}{\partial r}+\mu\left(\frac{\partial^2 V_r}{\partial r^2}+\frac{1}{r}\frac{\partial V_r}{\partial r}+\frac{1}{r^2}\frac{\partial^2 V_r}{\partial \theta^2}+\frac{\partial^2 V_r}{\partial x^2}-\frac{V_r}{r^2}-\frac{2}{r^2}\frac{\partial V_\theta}{\partial \theta}\right)$$

$$\rho\left(\frac{\partial V_\theta}{\partial t}+V_r\frac{\partial V_\theta}{\partial r}+\frac{V_\theta}{r}\frac{\partial V_\theta}{\partial \theta}+V_x\frac{\partial V_\theta}{\partial x}+\frac{V_\theta V_r}{r}\right)$$

$$=F_\theta-\frac{1}{r}\frac{\partial p}{\partial \theta}+\mu\left(\frac{\partial^2 V_\theta}{\partial r^2}+\frac{1}{r}\frac{\partial V_\theta}{\partial r}+\frac{1}{r^2}\frac{\partial^2 V_\theta}{\partial \theta^2}+\frac{\partial^2 V_\theta}{\partial x^2}-\frac{V_\theta}{r^2}+\frac{2}{r^2}\frac{\partial^2 V_\theta}{\partial \theta^2}\right)$$

$$\rho\left(\frac{\partial V_x}{\partial t}+V_r\frac{\partial V_x}{\partial r}+\frac{V_\theta}{r}\frac{\partial V_x}{\partial \theta}+V_x\frac{\partial V_x}{\partial x}\right)$$

$$=F_x-\frac{\partial p}{\partial x}+\mu\left(\frac{\partial^2 V_x}{\partial r^2}+\frac{\partial^2 V_x}{\partial x^2}+\frac{1}{r}\frac{\partial V_x}{\partial r}+\frac{1}{r^2}\frac{\partial^2 V_x}{\partial \theta^2}\right)$$

. . . (1.2)

Here F_x, F_r and F_θ are the body forces per unit volume, in x, r, and θ directions, respectively.

In addition, the continuity equation must be used for solution of laminar flow problems. For incompressible fluid, this equation in cartesian and cylindrical coordinate systems are

$$\frac{\partial u}{\partial x}+\frac{\partial v}{\partial y}+\frac{\partial w}{\partial z}=0 \qquad \text{. . . (1.3)}$$

and

$$\frac{\partial V_r}{\partial r}+\frac{V_r}{r}+\frac{1}{r}\frac{\partial V_\theta}{\partial \theta}+\frac{\partial V_x}{\partial x}=0 \qquad \text{. . . (1.4)}$$

In the solution of laminar flow problems, the terms, ρ, μ, X, Y and Z or F_x, F_r, and F_θ are known, thus leaving four unknowns u, v, w and p, or V_x, V_r, V_θ and p and four equations. Hence, mathematically speaking, for given boundary conditions, the problems of laminar flow are solvable. Since the real fluid sticks to the wall, $u = 0$, $v = 0$, $w = 0$, at the boundary, represent the boundary conditions if the boundary is stationary. Corresponding boundary conditions in cylindrical coordinate system are $V_x = V_r = V_\theta = 0$ at the stationary boundary.

1.3 CONDITIONS OF SIMILARITY

To examine the conditions that must be satisfied for two laminar flows to be dynamically similar, one can write Navier Stokes equations in dimensionless form. This can be done by choosing the reference velocity U_o, reference length l_o, and reference pressure p_o to nondimensionalise the quantities occurring in Eqs. 1.1. When this is done, the first of the Eqs. 1.1. can be written in the form

$$\left(\frac{\partial u}{\partial t}+u\frac{\partial u}{\partial x}+v\frac{\partial u}{\partial y}+w\frac{\partial u}{\partial z}\right)$$

$$=\frac{Xl_o}{\rho U_o^2}-\frac{p_o}{\rho U_o^2}\frac{\partial p}{\partial x}+\frac{\mu_o}{\rho_o U_o l_o}\left(\frac{\partial^2 u}{\partial x^2}+\frac{\partial^2 u}{\partial y^2}+\frac{\partial^2 u}{\partial z^2}\right) \qquad \text{. . . (1.5)}$$

Here ρ_o and μ_o are mass density and viscosity of the reference fluid, and u, v, w and p are in nondimensional form. According to Eq. 1.5, flows about two geometrically similar bodies of different linear dimensions in streams of different velocities and in fluids of different mass densities and viscosities will be dynamically similar if $U_o/\sqrt{Xl_o/\rho_o}$, $U_o/\sqrt{p_o/\rho_o}$ and $U_o l_o \rho_o/\mu_o$ have the same magnitudes in the two cases.

The first parameter is known as the *Froude number* which can be shown to be equal to the square root of the ratio of inertial force per unit volume ($\sim\rho_o U_o^2/l_o$) to gravity force per unit volume X. It is important in those flows where either a free surface or density stratification exists. Froude number is not important in the case of homogenous fluids if the flow is completely enclosed (e.g. pipe flow), or when body is deeply submerged (e.g. motion of submarines). The second parameter can be either called the *Euler number* (for incompressible fluids), or *Mach number* (for compressible fluids) which is square root of the inertial force per unit volume ($\sim\rho_o U_o^2/l_o$) and pressure force per unit volume, which is $\partial p/\partial x$ ($\sim p_o/l_o$). For incompressible fluids, this is usually taken as a dependent parameter. The third parameter is called the *Reynolds number* which is a ratio of inertial force per unit volume ($\sim\rho_o U_o^2/l_o$) to viscous force per unit volume ($\sim\mu_o U_o/l_o$). The Reynolds number, thus, characterises the relative importance of the viscous force as compared to the inertial force. Since viscosity and mass density appear in Reynolds number as a ratio, this ratio is treated as the property of fluid in itself; the ratio μ/ρ being kinematic in nature and having dimensions of L^2/T is called kinematic viscosity and is usually denoted by ν. The larger the Reynolds number, the less important will be the influence of viscosity on flow pattern. The smaller the Reynolds number, the more important is the role of viscosity on fluid motion. As the Reynolds number increases, inertial force becomes more and more important. Therefore, a Reynolds number approaching infinity will correspond to a flow in which the viscous resistance to deformation plays insignificant role in comparison to inertial resistance to acceleration. Similarly, a value of Reynolds number approaching zero will correspond to a flow in which inertial effects are negligible.

It can, therefore, be seen that, in the case of enclosed flows of incompressible homogenous fluids, equality of Reynolds number is a necessary condition for dynamic similarity of two flows.

1.4 SOLUTIONS OF NAVIER–STOKES EQUATIONS (1)

It can readily be seen that Navier-Stokes equations are second order, nonlinear, partial differential equations of elliptic type. The nonlinearity arises from the convective acceleration terms. As a result, the principle of superposition is not applicable in the solutions of Navier Stokes equations. Indeed, the difficulties in their solution are so formidable that no closed form solution of Navier-Stokes equations exists for the most general case.

In fact only five exact solutions are available for the nonlinear cases with specific boundary conditions. There are some simple problems for which the nonlinear terms disappear on their own, e.g. flow between parallel plates, pipes, rotating cylinder, etc. These are termed as linear exact solutions (see section 1.6).

One can consider several cases where the Navier–Stokes equations can be greatly simplified because of conditions imposed on the flow. In this connection it is worthwhile to mention two simplifications; the first one occurs when the velocities are extremely small and viscosity is very large, i.e. at very small values of Reynolds numbers, or when Reynolds number tends to zero.

In such a case, as a first approximation, the inertial terms on the left hand side of Eqs. 1.1. can be neglected in comparison with viscous terms. This introduces considerable simplification because the equations become linear. In fact, if one considers two dimensional flow in Cartesian Coordinate System, and introduces the stream function ψ, such that

$$\frac{\partial \psi}{\partial y} = u \text{ and } -\frac{\partial \psi}{\partial x} = v$$

Then Eqs. 1.1. and 1.3 along with the above simplification yield

$$\nabla^4 \psi = 0 \qquad \ldots (1.6)$$

This is a linear equation and amenable to mathematical treatment. Such flows are called creeping flows. The omission of inertial terms is permissible from mathematical point of view because the order of the differential equation is not reduced and, hence, all the boundary conditions will be satisfied.

The second limiting case of solution of Navier-Stokes equations arises when inertial terms are much larger than the viscous terms. One is then tempted to put $\mu = 0$ in Eq. 1.1. This, however, reduces the order of differential equations and, hence, it is not possible to satisfy all the boundary conditions for the equations. This led Prandtl to determine the order of magnitude of all the terms in Eq. 1.1. and drop out those terms which have smaller order of magnitude than the others. This gave the well-known boundary layer equations of Prandtl (see Sect. 1.7).

1.5 CREEPING MOTION

For very slow motion, when inertial terms on the left hand side of Eq. 1.1 can be neglected in preference to the viscous terms, one gets

$$\left.\begin{aligned}
\mu\left(\frac{\partial^2 u}{\partial x^2} + \frac{\partial^2 u}{\partial y^2} + \frac{\partial^2 u}{\partial z^2}\right) &= \frac{\partial p}{\partial x} \\
\mu\left(\frac{\partial^2 v}{\partial x^2} + \frac{\partial^2 v}{\partial y^2} + \frac{\partial^2 v}{\partial z^2}\right) &= \frac{\partial p}{\partial y} \\
\mu\left(\frac{\partial^2 w}{\partial x^2} + \frac{\partial^2 w}{\partial y^2} + \frac{\partial^2 w}{\partial z^2}\right) &= \frac{\partial p}{\partial z}
\end{aligned}\right\} \qquad \ldots (1.6)$$

According to Eq. 1.3

$$\frac{\partial u}{\partial x} + \frac{\partial v}{\partial y} + \frac{\partial w}{\partial z} = 0 \qquad \ldots (1.3)$$

which can be written in vector from as

$$\left.\begin{aligned} \mu\, \nabla^2 \bar{V} &= \text{grad } p \\ \text{div } \bar{V} &= 0 \end{aligned}\right\}$$

Where $\bar{V} = i\,u + jv + kw$ is the velocity vector. If the three parts in Eq. 1.6 are differentiated with respect to x, y and z, respectively, and added, one gets

$$\frac{\partial^2 p}{\partial x^2} + \frac{\partial^2 p}{\partial y^2} + \frac{\partial^2 p}{\partial z^2} = \mu \left[\frac{\partial^2}{\partial x^2}\left(\frac{\partial u}{\partial x} + \frac{\partial v}{\partial y} + \frac{\partial w}{\partial z}\right) + \frac{\partial^2}{\partial y^2}\left(\frac{\partial u}{\partial x} + \frac{\partial v}{\partial y} + \frac{\partial w}{\partial z}\right) + \frac{\partial^2}{\partial z^2}\left(\frac{\partial u}{\partial x} + \frac{\partial v}{\partial y} + \frac{\partial w}{\partial z}\right)\right] = 0$$

as a consequence of continuity equation. Hence, one gets

$$\nabla^2 p = 0 \qquad \ldots (1.7)$$

which means that pressure distribution in creeping motion satisfies Laplace equation.

The first example of creeping motion is that of steady flow past a sphere. Stokes in 1851 solved this problem and obtained the following expressions for u, v, w, and p.

$$\left.\begin{aligned} u &= U_o \left[\frac{3}{4}\,\frac{Rx^2}{r^3}\left(\frac{R^2}{r^2} - 1\right) - \frac{1}{4}\,\frac{R}{r}\left(3 + \frac{R^2}{r^2}\right) + 1\right] \\ v &= U_o\, \frac{3}{4}\,\frac{Rxy}{r^3}\left(\frac{R^2}{r^2} - 1\right) \\ w &= U_o\, \frac{3}{4}\,\frac{Rxz}{r^3}\left(\frac{R^2}{r^2} - 1\right) \\ p &= -\frac{3}{2}\,\mu\, U_o\, Rx/r^3 \end{aligned}\right\} \qquad \ldots (1.8)$$

Here U_o is the flow velocity far away from the sphere of radius R, and x, y, z are the co-ordinates of a point which is at a distance r from the centre of the sphere. If one substitutes $r = R$ in the equation for p, pressure distribution on the surface of sphere is obtained. Shear distribution on the surface of the sphere can also be calculated from Eqs. 1.8. Taking the horizontal components of pressure and shear distribution on the surface of the sphere, Stokes showed that the drag force F_D experienced by the sphere is

$$F_D = 3\,\pi\, D\, \mu\, U_o \qquad \ldots (1.9)$$

where D is diameter of the sphere. Out of this, one third of the drag is due to pressure distribution, and two third is due to shear. It can be seen that the drag force is proportional to the first power of velocity, dynamic viscosity, and length dimension. If one defines the drag coefficient C_D as

$$C_D = \frac{F_D/\text{projected area of body normal to flow}}{\rho U_o^2/2}$$

it is apparent that, for the sphere in Stokes's range, C_D is given by

$$C_D = 24/Re \qquad \ldots (1.10)$$

where $Re = U_o\rho D/\mu$. Equations 1.9 and 1.10 are valid for Re less than 0.10. As the Reynolds number exceeds 0.10, the inertia terms cannot be neglected. In fact, at any point in the flow field the ratio of inertial force to viscous force, which in Stokes solution was assumed to be negligibly small, appears as $U_o\rho r/\mu$, where r is the distance from the sphere. Hence, at larger distances from the sphere, this assumption is likely to cause significant error. This was recognised by Oseen who assumed that velocity components can be represented as

$$u = U_o + u', \; v = v' \text{ and } w = w'$$

where u', v' and w' are perturbation terms and are small compared to U_o. Neglecting terms which are proportional to square of perturbation velocities, equations of motion become

$$\left.\begin{aligned} \rho U_o \frac{\partial u'}{\partial x} + \frac{\partial p}{\partial x} &= \mu \nabla^2 u \\ \rho U_o \frac{\partial v'}{\partial y} + \frac{\partial p}{\partial y} &= \mu \nabla^2 v \\ \rho U_o \frac{\partial w'}{\partial z} + \frac{\partial p}{\partial z} &= \mu \nabla^2 w \end{aligned}\right\} \qquad \ldots (1.11)$$

in which $\nabla^2 = \left(\frac{\partial^2}{\partial x^2} + \frac{\partial^2}{\partial y^2} + \frac{\partial^2}{\partial z^2}\right)$. The corresponding expression for drag coefficient becomes

$$C_D = 24/Re\,[1 + 3Re/16] \qquad \ldots (1.12)$$

Experimental results show that Eq. 1.12 is valid approximately upto Reynolds number of five.

One more example of creeping motion that will be discussed is what is known as Helle–Shaw motion. Consider steady 2-dimensional flow between two parallel plates placed distance h apart. Let a cylindrical body be placed between the plates. Assume flow in positive x direction (see Fig. 1.1). Since h is very small, but x and z are large, following approximations can be made:

$$\frac{\partial^2 u}{\partial y^2} \gg \frac{\partial^2 u}{\partial x^2} \text{ or } \frac{\partial^2 u}{\partial z^2}$$

$$\frac{\partial^2 w}{\partial y^2} \gg \frac{\partial^2 w}{\partial x^2} \text{ or } \frac{\partial^2 w}{\partial z^2}$$

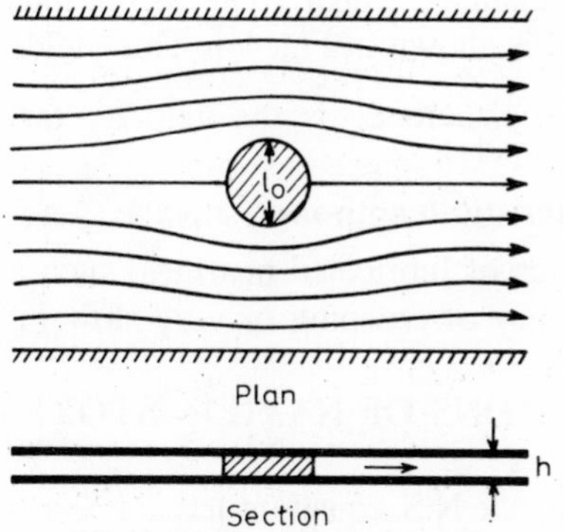

Fig. 1.1 Helle–Shaw motion

These approximations reduce the N.S. equations to the following set of differential equations:

$$\left.\begin{aligned} 0 &= -\frac{\partial p}{\partial x} + \mu\frac{\partial^2 u}{\partial y^2} \\ 0 &= -\frac{\partial p}{\partial y} \\ 0 &= -\frac{\partial p}{\partial z} + \mu\frac{\partial^2 w}{\partial y^2} \end{aligned}\right\} \quad \ldots (1.13)$$

With the boundary conditions $u = 0$ and $w = 0$ at $y = 0$, h and $v = 0$ everywhere, the solution of these equations yields

$$\left.\begin{aligned} u &= \frac{1}{2\mu}\frac{\partial p}{\partial x}(y^2 - hy) \\ w &= \frac{1}{2\mu}\frac{\partial p}{\partial z}(y^2 - hy) \end{aligned}\right\} \quad \ldots (1.14)$$

One can determine the average velocity u_o and w_o at any place be integration of the above equations with respect to y and dividing the result by h. Hence,

$$\left.\begin{aligned} u_o &= -\frac{h^2}{12\mu}\frac{\partial p}{\partial x} \\ \text{and} \qquad w_o &= -\frac{h^2}{12\mu}\frac{\partial p}{\partial z} \end{aligned}\right\} \quad \ldots (1.15)$$

It can be seen that

$$\frac{\partial u_o}{\partial x} + \frac{\partial w_o}{\partial z} = 0$$

and

$$\frac{\partial u_o}{\partial z} - \frac{\partial w_o}{\partial x} = 0$$

which means that the average velocity components u_o and w_o satisfy the continuity equation, and yield irrotational flow. This result is used to obtain irrotational flow around a body by keeping the body between parallel plates kept at a very small distance h apart, and tracing the stream lines. A typical pattern thus obtained is shown in Fig. 1.1. This yields good results provided $\frac{U_o \rho l_o}{\mu} \left(\frac{h}{l_o} \right)^2 << 1.0$, where U_o is the velocity far away from the body, and l_o is the characteristic length in x, z plane.

The hydrodynamics of lubricated bearings, such as slipper bearing, also falls under the category of creeping or very slow motion.

1.6 EXACT SOLUTIONS OF NAVIER–STOKES EQUATIONS

Exact solution of complete N.S. equations has not been possible due, especially, to their nonlinear nature. However, in particular cases, when convective terms vanish, it is possible to get such solutions. A few such solutions are discussed below.

Flow between Parallel Plates

For steady two-dimensional laminar flow between parallel plates distance h apart, N.S. equations reduce to

$$\left.\begin{aligned} 0 &= -\frac{\partial p}{\partial x} + \mu \frac{\partial^2 u}{\partial y^2} \\ 0 &= -\frac{\partial p}{\partial y} \\ 0 &= -\frac{\partial p}{\partial z} \end{aligned}\right\} \quad \ldots (1.16)$$

since v and w are zero and $\partial/\partial z = \partial/\partial x = \partial/\partial t = 0$. Further, continuity equation yields $\partial u/\partial x = 0$, the integration of which yields $u = u\,(y)$. This result, along with first of Eq. 1.16, indicates that $\partial p/\partial x$ must be constant. Then integration of first of Eqs. 1.16 yields

$$u = \frac{y^2}{2\mu} \frac{\partial p}{\partial x} + C_1\, y + C_2 \quad \ldots (1.17)$$

with the boundary conditions $u = 0$ for $y = 0$ and $y = h$.

Hence $C_2 = 0$, and $C_1 = -\frac{h}{2\mu} \frac{\partial p}{\partial x}$

$$\therefore \quad u = -\frac{1}{2\mu} \frac{\partial p}{\partial x} (hy - y^2) \quad \ldots (1.18)$$

This indicates that the velocity distribution is parabolic in nature (see Fig. 1.2). From the above equation, it can be shown that

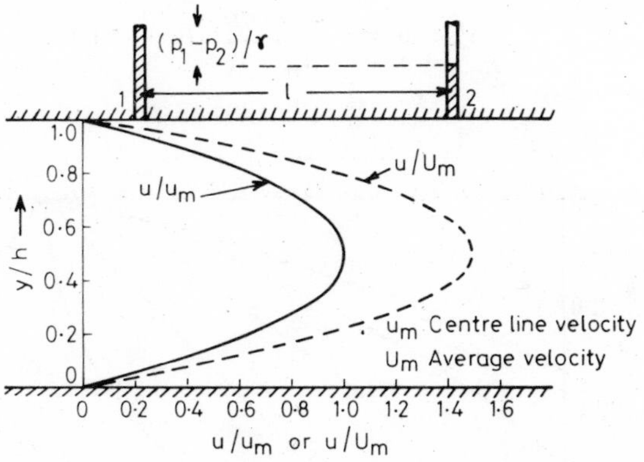

Fig. 1.2 Non-dimensional velocity distribution for laminar flow between parallel plates

$$(p_1 - p_2) = \frac{12\mu\, U_m l}{h^2} \qquad \ldots (1.19)$$

where U_m is the average velocity of flow. In the same manner, one can obtain equations for generalised Couette flow between parallel plates when the top plate is moving at velocity U_o in the forward direction. Substituting the boundary conditions $u = 0$ when $y = 0$ and $u = U_o$ when $y = h$ in Eq. 1.17, one gets

$$u = \frac{yU_o}{h} - \frac{h^2}{2\mu}\frac{\partial p}{\partial x}(1 - y/h)\frac{y}{h} \qquad \ldots (1.20)$$

or

$$\frac{u}{U_o} = \frac{y}{h} - P\,\frac{y}{h}(1 - y/h) \qquad \ldots (1.21)$$

where $P = \dfrac{h^2}{2\mu\, U_o}\dfrac{\partial p}{\partial x}$ is the dimensionless pressure gradient. This velocity distribution is shown in Fig. 1.3.

Hagen–Poiseuille's Equation for Flow in a Pipe

Starting from the Navier Stokes equations in cylindrical coordinate system, steady laminar flow through a straight circular pipe of radius R can be studied. In such a case $\dfrac{\partial}{\partial t} = 0$, $F_x = F_r = F_\theta = 0$, $V_\theta = V_r = 0$ and $\dfrac{\partial}{\partial \theta} = 0$ because of symmetry of flow. As a result, Navier-Stokes equations reduce to the following equations:

$$\left.\begin{aligned} 0 &= -\frac{\partial p}{\partial r} \\ 0 &= -\frac{1}{r}\frac{\partial p}{\partial \theta} \\ 0 &= -\frac{\partial p}{\partial x} + \mu\,(\partial^2 V_x/\partial r^2 + 1/r(\partial V_x/\partial r)) \end{aligned}\right\} \qquad \ldots (1.22)$$

Solutions of these equations show that $\partial p/\partial x$ will be constant and

$$V_x = u = -\frac{1}{4\mu}\frac{\partial p}{\partial x}(R^2 - r^2) \qquad \ldots (1.23)$$

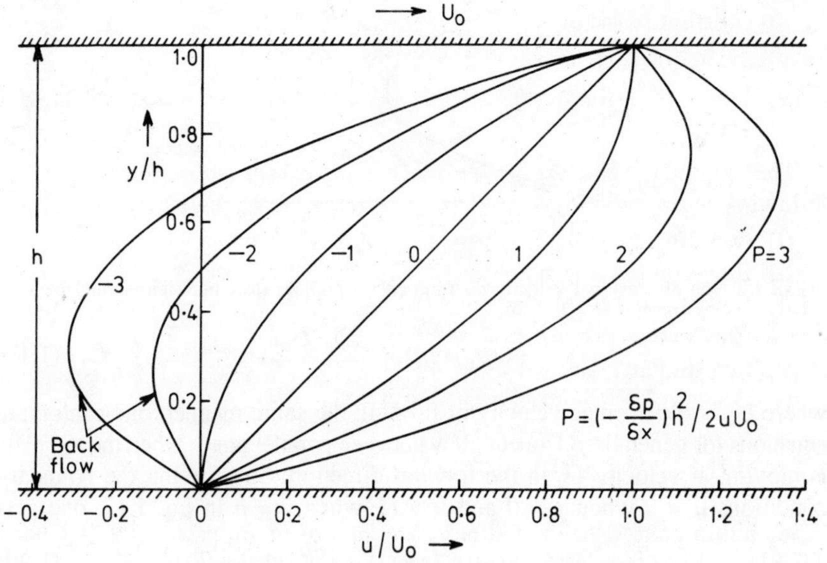

Fig. 1.3 Velocity distribution in generalised Couette flow

Integration of Eq. 1.23 can lead to an expression for average velocity U_m as

$$U_m = -\frac{R^2}{8\mu}\frac{\partial p}{\partial x}$$

and
$$(p_1 - p_2) = \frac{32\mu\, U_m l}{D^2} \qquad \ldots (1.24)$$

which is known as Hagen Poiseuill's equation for laminar flow in a pipe. If one defines friction factor f (known as Darcy-Weisbach friction factor) by the equation

$$\Delta p = f\frac{l}{D}\frac{\rho U_m^2}{2} \qquad \ldots (1.25)$$

it can be shown from Eqs. 1.24 and 1.25 that

$$f = \frac{64}{Re} \qquad \ldots (1.26)$$

where $Re = U_m D\rho/\mu$. Thus, in laminar flow friction factor varies inversely as the Reynolds number.

Other Exact Solutions

Some other exact solutions of N.S. equations include the following cases:

(i) flat plate suddenly set into motion in its own plane with constant velocity U_o;

(ii) flow over an oscillating plate;

(iii) steady flow between parallel plates with injection along the wall at constant velocity;

(iv) axial flow between two concentric tubes; and

(v) flow between two concentric cylinders rotating at different angular velocities.

The nonlinear exact solutions of N.S. equations are available for the following cases:

(i) two dimensional flow between nonparallel plates;

(ii) flow against a normal boundary;

(iii) flow around a rotating disc;

(iv) flow between two rotating discs;

(v) two dimensional and axisymmetric jet issuing in ambient fluid.

These solutions are discussed in many texts on laminar flows (1).

1.7 BOUNDARY LAYER APPROXIMATIONS

When a thin plate is held parallel to the flow of uniform velocity U_o (many times known as *free stream* or *ambient velocity*) of a real fluid, the fluid velocity at the boundary is zero since the fluid sticks to the wall; it gradually increases to U_o some distance away from the wall. This causes retardation of fluid near the wall and a shear stress τ_o on the wall in the direction of flow (and an equal and opposite force is exerted on the fluid by the wall which is known as the shear resistance). Thus, the velocity changes from zero to U_o in the vertical direction but the variation is asymptotic. The vertical distance in which the velocity changes from zero to 0.99 U_o is known as the *nominal boundary layer thickness* δ. As the retarded fluid layer moves in the downstream direction, continued action of shear resistance of the boundary causes retardation at increasing distances across the plate, as a result of which the boundary layer thickness increases in the downstream direction.

Two additional definitions of boundary layer thickness are often used. The *displacement thickness* δ^* indicates the distance by which the external streamlines are shifted owing to the boundary layer formation. It is equal to the distance δ^* such that deficit in discharge due to boundary layer formation is equal to δ^* times the ambient velocity.
In other words

$$\left.\begin{aligned} U_o\,\delta^* &= \int_0^\infty (U_o - u)\,dy \\ \text{or} \quad \delta^* &\sim \int_0^\delta (1 - u/U_o)\,dy \end{aligned}\right\} \quad \ldots (1.27)$$

since beyond y values greater than δ, the local velocity is almost equal to U_o.

The momentum thickness θ is defined as the distance from the boundary such that the momentum flux through the distance θ at ambient velocity is the same as the deficit in the momentum flux due to boundary layer formation.

$$
\left.
\begin{aligned}
&\text{or} \qquad \rho U_o^2 \theta = \int_0^\infty \rho u \, (U_o - u) \, dy \\
&\text{or} \qquad \theta = \int_0^\delta \frac{u}{U_o} \, (1 - u/U_o) \, dy
\end{aligned}
\right\} \quad \ldots (1.28)
$$

It may be mentioned that θ is less than δ* and δ* is less than δ. Figure 1.4 shows the development of boundary on a flat plate.

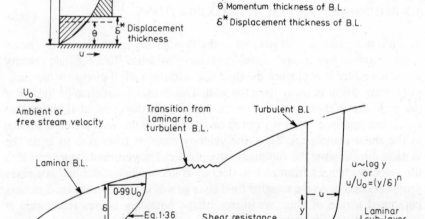

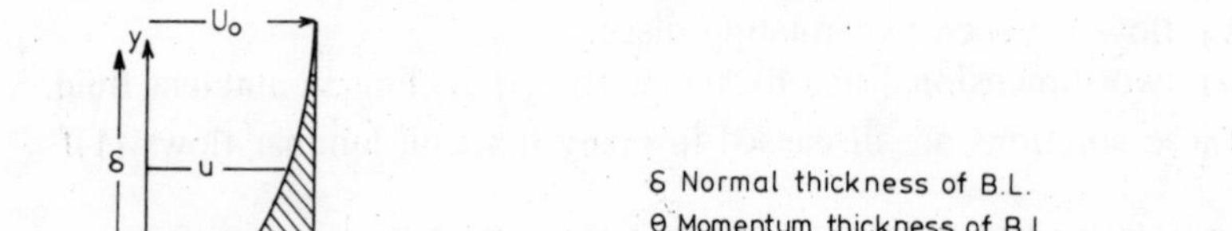

Fig. 1.4 Boundary layer development on flat (a) plate

The concept of boundary layer was first introduced by Prandtl (2) in 1904. He argued that the flow of a real fluid past a solid wall can be divided into two regions; one away from the wall in which the velocity is unaffected by viscosity and can be predicted using potential flow theory, and the other near the wall in which viscous effects cannot be neglected, i.e. boundary layer flow. Assuming that in the boundary layer flow one is dealing with small viscosities and relatively large velocities, the Reynolds numbers will be large and yet the flow in the boundary layer will be laminar. In such a case the inertial terms in Navier–Stokes equations cannot be neglected because they are large. Further, because the viscosity is small, if all terms involving μ on the right hand side of Navier–Stokes equations are omitted, the order of differential equation is reduced from two to one and, hence, some of the specified boundary conditions will not be satisfied. It may be mentioned that for real fluid flow over a stationary boundary $u = v = w = 0$ at $y = 0$ are the boundary conditions.

Hence Prandtl suggested that instead of dropping all the terms involving viscosity, the orders of magnitude of all the terms in Navier–Stokes equations be determined and only those terms in the equation be dropped whose orders of magnitude are much smaller than that of the other terms.

In achieving this simplication, following basic assumptions, in the case of boundary layers without separation formed on a solid wall, are made:

(i) Boundary layer thickness δ is very small compared to length L of the body.

(ii) Outside the boundary layer thickness δ, viscous effects can be neglected and fluid can be considered to be inviscid. Hence, pressure distribution in the flow direction can be determined from potential flow theory.

(iii) U_o is of the order of L. Further, within the boundary layer inertial force and viscous force are of the same order of magnitude. Hence $U_o L/\nu$ is of the order of $(L/\delta)^2$, and is very large.

(iv) Velocity component u in the boundary layer is of the order of U_o and, hence, from continuity equation, v is of the order of $U_o \delta/L$.

(v) $\dfrac{\partial u}{\partial t}$ is of the order of $u \dfrac{\partial u}{\partial x}$ i.e. U_o^2/L.

Considering two dimensional boundary layer flows of an incompressible fluid over a flat plate and having pressure gradient (i.e. free stream velocity U_o changing in the direction of flow), Prandtl showed that the terms $\mu\, \partial^2 u/\partial x^2$, $u \partial v/\partial x$, $\partial v/\partial t$, $\partial v/\partial y$, $\mu \partial^2 v/\partial x^2$ and $\mu \partial^2 v/\partial y^2$ are much smaller than the other terms in Eq. 1.1 and, hence, they can be dropped out. The resulting equations known as boundary layer equations are

$$\left.\begin{aligned} \rho\left(\frac{\partial u}{\partial t} + u\frac{\partial u}{\partial x} + v\frac{\partial u}{\partial y}\right) &= -\frac{\partial p}{\partial x} + \mu\frac{\partial^2 u}{\partial y^2} \\ 0 &= -\frac{\partial p}{\partial y} \end{aligned}\right] \quad \ldots (1.29)$$

Continuity equation is $\dfrac{\partial u}{\partial x} + \dfrac{\partial v}{\partial y} = 0$

With the retention of the term $\mu\, \partial^2 u/\partial y^2$, the order of the differential equation is not changed and, hence, for two dimensional flow the boundary conditions $u = v = 0$ at $y = 0$ will be satisfied. The other boundary condition is that, when $y = \infty$, $u = U_o$.

Solution of Eq. 1.29 for boundary layer formation over a flat plate without pressure gradient was first obtained by Blasius (3) in 1908 by first determining a power series solution for small values of y which satisfies the boundary conditions $u = v = 0$ at $y = 0$, and an asymptotic solution for very large values of y which satisfies the boundary condition $u = U_o$ at $y = \infty$. These solutions are then combined in an appropriate manner (u, $\partial u/\partial y$ and $\partial^2 u/\partial y^2$ are the same at a given point where the two solutions are valid), thereby determining the constants appearing in the two solutions. His analysis showed that δ, δ^* and θ are given by the equations

$$\delta = 5.0/\sqrt{Re_x} \qquad \ldots (1.30)$$

$$\delta^* = 1.73/\sqrt{Re_x} \qquad \ldots (1.31)$$

and

$$\theta = 0.664/\sqrt{Re_x} \qquad \ldots (1.32)$$

where $Re_x = U_o\, x\rho/\mu$. The average shear at distance x from the leading edge is given by

$$c_f = \tau_o/(\rho U_o^2/2) = 0.664/\sqrt{Re_x} \qquad \ldots (1.33)$$

while the average drag froce F over a length L on which laminar boundary layer exists is given by

$$C_f = (F/BL)/(\rho U_o^2/2) = 1.328/\sqrt{Re_L} \qquad \ldots (1.34)$$

where B is width of plate, $Re_L = U_o \rho L/\mu$ and c_f and C_f are known as local and average drag coefficients. The nondimensional form of velocity distribution is given by

$$u/U_o = f'(\eta) \qquad \ldots (1.35)$$

where $\eta = y\sqrt{U_o/\nu x}$. The function f' obtained by Blasius is listed below:

η	0	0.2	0.4	0.6	0.8	1.0	1.6	2.0
$u/U_o = f'$	0	0.0664	0.1128	0.1989	0.2647	0.3298	0.5168	0.6298
η	2.6	3.0	3.6	4.0	4.6	5.0	6.0	7.6
$u/U_o = f'$	0.7725	0.8461	0.9233	0.9555	0.9827	0.9916	0.9990	1.00

This velocity distribution can also be approximated by the equation

$$\frac{u}{U_o} = \frac{3}{2}\left(\frac{y}{\delta}\right) - \frac{1}{2}\left(\frac{y}{\delta}\right)^2 \qquad \ldots (1.36)$$

These results concerning growth of boundary layer thickness and variation of velocity in the vertical have been verified by careful experimental studies by Burgers, Hansen, Nikuradse and others. The shear was directly measured by Lipmann and Dhawan (4) and Dhawan (5). Boundary layer development on flat plate with flow having pressure gradient was studied by Karman and Pohlhausen.

From the basic reasoning, as well as from the above mentioned analysis, one can enumerate the characteristics of boundary layers formed on flat plate without or with pressure gradient. These are: (i) boundary layer thickness increases as x increases: (ii) higher the ambient velocity, smaller will be the boundary layer thickness; (iii) greater the kinematic viscosity, greater will be the thickness of boundary layer; (iv) since Reynolds number characterises the flow, upto a certain value of Re_x the boundary layer will be laminar and then it will change into turbulent; (v) if $\partial p/\partial x$ is negative, as in the case of converging flows, the boundary layer will grow slowly, while when $\partial p/\partial x$ is positive, it will grow faster. In fact, positive pressure gradient can lead to boundary layer separation.

1.8 CHARACTERISTICS OF LAMINAR FLOWS

At this stage it is instructive to review the characteristics of laminar flows. As shown by Reynolds, in the case of laminar flow in a straight pipe, the fluid particles travel in straight lines without losing their identity by mixing between different layers. Thus, fluid particles between two adjacent layers do not mix except by molecular diffusion. If any disturbance in the form of velocity or pressure fluctuations is introduced in the laminar flow, the disturbance eventually damps out because of the action of viscosity; hence the flow again comes back to laminar state. As the Reynolds number exceeds the limit of approximately 2300, the laminar flow in a pipe may become unstable, leading to turbulent flow condition. This aspect is discussed in detail in the next chapter.

It can be seen that the loss of pressure Δp in steady uniform laminar flow in a pipe is directly proportional to the first power of velocity and viscosity, and inversely proportional to D^2. This is true in other laminar flows also, e.g., flow between parallel plates and flow through porous materials. It is also worth noting that, since inertial effects are neglected, Δp is independent of the mass density of the fluid. Another characteristic is the extreme nonuniformity in velocity distribution as indicated by Eqs. 1.18 and 1.23 in the case of flow between parallel plates and in pipes. This gives an energy correction factor of two for pipe flow as against 1.05 to 1.1 for turbulent flow where the velocity distribution follows logarithmic law.

For laminar flow past bodies, the drag force experienced by the body is proportional to first power of velocity and first power of viscosity. This gives drag coefficient to be inversely proportional to Reynolds number.

Lastly, it may be pointed out that as the continuous deformation of fluid is opposed by the viscous stresses that vary directly with the rate of deformation, there is a continuous expenditure of energy. Therefore, in a steady uniform flow, there is a continuous transformation of mechanical energy into heat. This is true for turbulent flow as well, but, because of intense mixing, there is much greater expenditure of energy in turbulent flow than in laminar flow.

REFERENCES

1. Schlichting, H. *Boundary Layer Theory*. McGraw-Hill Series in Mechanical Engineering, New York, Ist Ed., 1955.
2. Prandtl. L. Uber Flussigkeitsbewegung bei sehr Kleiner Reibung. Proc. 3rd Int. Math. Congress. 1904.
3. Blasius, H. *Grenzschichten in Flussigkeiten mit Kleiner Reibung*. Z. Math. u. Phys. Vol. 56, No 1, 1908.
4. Lipmann, H.W. and S. Dhawan. Direct Measurement of Local Skin Friction in Low-Speed and High-Speed Flow. Proc. 1st U.S. National Congress of Applied Mechanics. 1951.
5. Dhawan, S. Direct Measurement of Skin Friction. NACA Rep. No. 1112, 1953.

CHAPTER II

Transition from Laminar to Turbulent Flow

2.1 INTRODUCTION

The characteristics of laminar flow in a pipe, between parallel plates, around a sphere and in a laminar boundary layer have been discussed in Chapter I. It is shown that laminar flow is governed by the characteristic Reynolds number, which is the ratio of inertial force per unit volume to viscous force per unit volume. Hence, as the Reynolds number is increased, it represents relative increase in the inertial force in relation to viscous force. Reynolds number can be increased either by increasing the velocity, and/or length dimension, and/or by decreasing kinematic viscosity. In case of all laminar flows, when the Reynolds number reaches a certain limit, the characteristics of flow begin to change over a small range of Reynolds number and, with further increase in Reynolds number, the flow becomes completely "chaotic" and "random", which is characterised by fluctuations in velocity and pressure. This state of flow is known as *turbulent flow*. This chapter is devoted to the discussion of mechanism of the transition from laminar to turbulent flow, and to the discussion of factors affecting the transition.

The first systematic investigation about the transition were made by Reynolds (1) in 1883. Using glass tubes connected to a reservoir, he observed the flow pattern in the tubes at various velocities of water by injecting a thin stream of dye in the main stream. He found that, at low velocities, the dye filament travelled straight and parallel to the walls of the tube. This is evidently the laminar flow in which the fluid appears to move in distinct parallel layers, one fluid layer sliding over the other. If velocity is increased beyond a certain value, the coloured filament oscillates and, with still further increase in velocity, the filament diffuses over the whole depth and loses its identity. The former is the transitional situation, whereas the latter is that of turbulent condition. By plotting energy gradient versus velocity, one can see that during the laminar stage the energy gradient is proportional to first power of average velocity U_m. The flow starts deviating from such a relation when transition starts. In turbulent flow, it is proportional to U_m^n, where n varies from 1.75 to 2. Reynolds obtained the critical Reynolds number ($U_m D\rho/\mu$) at which flow ceases to be laminar to be 2000 – 2100. This is many times known as the *lower* critical Reynolds number. Barnes and Coker (2) conducted experiments on a 18 mm diameter tube, and were able to maintain the flow laminar upto a Reynolds number value of 5000. Ekman (3) in 1910 was able to maintain laminar flow even upto a Reynolds number value of 24000 by eliminating all the disturbances at the inlet. Recently an approximate value of 50000 has been reached at which the flow remained laminar. If Reynolds number of turbulent flow is reduced, at a certain Reynolds number value, the flow ceases to be turbulent. It is known as *upper critical* Reynolds number. However, once a disturbance was introduced in such a flow, it would turn into turbulent state. At about 2000 and smaller values of Reynolds number, the flow remained laminar even when it was disturbed.

Similar phenomenon has also been observed in the case of laminar boundary layer flow over a flat plate kept at zero incidence when the flow has no streamwise pressure gradient. The transition occurs when $U_o\rho x/\mu$ is 3×10^5 to 6×10^5, where U_o is the free stream velocity and x is the distance from the leading edge. Schubauer and Skramstad (4) found that the critical Reynolds number very much depended on the nature of disturbance imposed on the incoming flow. By eliminating disturbances in the flow, they were able to maintain the boundary layer laminar even upto $U_o\rho x/\mu$ equal to 6×10^6. However, if Re_x is less than 1.12×10^5, any disturbance would be damped and flow remained laminar.

2.2 CONCEPT OF STABILITY

The experimental results discussed above suggest that the transition from laminar to turbulent flow is related to the presence of a disturbance and its amplification. If a small disturbance in the form of velocity fluctuation is superimposed on a laminar flow, its damping or amplification with the passage of time determines the stability of the main flow. If the disturbance is damped, the flow remains stable, and hence laminar; if the disturbance is amplified, the flow is unstable and can lead to turbulence through transition. If the disturbance is neither damped nor amplified, the flow is called neutrally stable. The theory developed to explain the mechanism of transition from laminar to turbulent flow by superimposing a disturbance on the main flow and examining the decay or amplification of the disturbance is known as the theory of stability. The stability of the real flows was first mentioned by Hagen in 1839 and demonstrated experimentally by him in 1854, and independently by Reynolds in 1883. Contributions have also been made by Kelvin, Raleigh, Prandtl, Schlichting, Tollmien, and others.

In order to investigate the stability of the main flow, a disturbance in the form of velocity and pressure fluctuations is superimposed on it. The resulting fluctuations are assumed to be small, so that their squares and higher powers can be neglected. One can then follow either the energy method, or the method of small disturbances for further analysis. Since higher order terms are neglected, such an analysis is called linear analysis.

Energy method is based on the premise that a flow can be unstable only if the energy of the disturbance increases with time. There are two ways in which the energy of the disturbance can change: energy loss due to viscous dissipation, and gain of energy from the main flow. The transfer of energy from main flow is mathematically manifested by nonlinear terms in Navier-

Stokes equations. Thus, the stability of main flow depends on whether the rate of viscous dissipation is greater, or smaller than the rate at which energy is received from the main flow. In the case of method of small disturbances, calculations are made about the development of the disturbance with time by making use of equations of motion. The method of small disturbances and its results are discussed below, with special reference to two dimensional laminar boundary layer flow.

2.3 STABILITY ANALYSIS (5)

Consider the case of two dimensional laminar flow described by the condition

$$\bar{u} = \bar{u}\,(y), \quad \bar{p} = \bar{p}\,(x,y), \quad \text{and } \bar{v} = 0. \qquad \ldots (2.1)$$

These conditions are strictly satisfied by steady laminar flow between parallel plates, and approximately by laminar boundary layer on a flat plate. Upon this main flow is superimposed a two dimensional disturbance given by

$$\left.\begin{aligned} u' &= u'\,(x, y, t) \\ v' &= v'\,(x, y, t) \\ p' &= p'\,(x, y, t) \end{aligned}\right\} \qquad \ldots (2.2)$$

The resulting motion will, therefore, be given by superposition of the disturbance on the main flow, i.e. by the equations

$$\left.\begin{aligned} u &= \bar{u} + u' \\ v &= v' \\ p &= \bar{p} + p' \end{aligned}\right\} \qquad \ldots (2.3)$$

It is assumed that the main motion, as well as the resulting motion described by Eq. 2.3, satisfies Navier-Stokes equations. Further, u', v' and p' are sufficiently small so that the quadratic terms in fluctuating components can be neglected in comparison with lower order terms. Substitution of Eq. 2.3 in Navier-Stokes equations, dropping the higher order terms and eliminating the pressure from the the resulting equation, it can be shown that u' and v' would satisfy the equation

$$\frac{\partial}{\partial t}\left(\frac{\partial u'}{\partial y} - \frac{\partial v'}{\partial x}\right) + \bar{u}\,\frac{\partial}{\partial x}\left(\frac{\partial u'}{\partial y} - \frac{\partial v'}{\partial x}\right) + v'\,\frac{\partial^2 \bar{u}}{\partial y^2}$$

$$= \nu\left(\frac{\partial^3 u'}{\partial x^2\,\partial y} + \frac{\partial^3 u'}{\partial y^3} - \frac{\partial^3 v'}{\partial x\,\partial y^2} - \frac{\partial^3 v'}{\partial x^3}\right) \qquad \ldots (2.4)$$

The disturbance components of velocity can be represented by introduction of a stream function ψ,

$$\psi(x,y,t) = \phi\,(y)\; e^{i(\alpha x - \beta t)} \qquad \ldots (2.5)$$

Here ϕ is an amplitude function which is a complex function, and the physical meaning is attached only to the real part of stream function. The fluctuating components u' and v' can now be expressed in terms of ψ, as $u' = \partial\psi/\partial y$ and $v' = -\partial\psi/\partial x$; and hence Eq. 2.4 reduces to

$$(\bar{u} - c)(\phi'' - \alpha^2 \phi) - \phi\, \bar{u}'' = \frac{-i\nu}{\alpha}(\phi'''' - 2\alpha^2 \phi'' + \alpha^2 \phi) \quad \ldots (2.6)$$

in which the wave number $\alpha = 2\pi/\lambda$, where λ is wave length of the disturbance, $\beta = (\beta_r + i\beta_i)$ in which β_r is the frequency of the disturbance and β_i determines the degree of amplification or damping. The term β_i is known as the *amplification factor.* The term $c = \beta/\alpha = c_r + ic_i$ can be interpreted as follows; c_r is the velocity of propagation of the wave in the x direction, and depending on its sign, c_i determines the degree of damping or amplification of the disturbance. For boundary layer on a flat plate, Eq. 2.6 can be non-dimensionalised by using free stream velocity U_o and nominal boundary layer thickness δ or displacement thickess δ^* as the scaling factors. This yields

$$(\bar{u} - c)(\phi'' - \alpha^2\phi) - \phi\bar{u}'' = -\frac{i}{\alpha Re}(\phi'''' - 2\alpha^2 \phi'' + \alpha^2\phi) \quad \ldots (2.7)$$

where in Eq. 2.7 $\bar{u} = \bar{u}/U_o$, $c = c/U_o$, $\alpha = \alpha\delta$, $Re = (U_o\delta\rho/\mu)$, and all derivatives are with respect to $\eta = y/\delta$. Equation 2.7 is fourth order ordinary differential equation in the amplitude $\phi(\eta)$, and is known as Orr-Sommerfeld equation for stability. The terms on the left hand side are obtained from inertia and pressure, while those on right hand side are obtained from viscous terms. For boundary layer stability the boundary conditions will be

$$\left.\begin{array}{ll} \text{at } y = 0 & u' = v' = 0 \text{ i.e. } \phi = \phi' = 0 \text{ when } \eta = 0 \\ \text{at } y = \infty & u' = v' = 0 \text{ i.e. } \phi = \phi' = 0 \text{ when } \eta = \infty \end{array}\right\} \quad \ldots (2.8)$$

Equation 2.7 contains four parameters α, Re, c_r and c_i, in addition to $\bar{u}(\eta)$. One can assume $\bar{u}(\eta)$, Reynolds number for the flow Re, and wavelength of disturbance λ i.e. $\alpha = 2\pi/\lambda$ to be given. For given $\bar{u}(\eta)$, Re and α, solution of Eq. 2.7 with boundary conditions, Eq. 2.8 will give one eigen-function $\phi(\eta)$ and one complex eigen-value $c = c_r + ic_i$. If c_i is less than zero, the disturbance is damped; if c_i is greater than zero, it is amplified, and when c_i is zero the disturbance is neither amplified nor damped. The last condition is called the condition of *neutral stability.*

Non-Viscous Instability

If one drops the right hand side of Eq. 2.7, physically it would mean that the effect of viscosity on the disturbance is not considered, but the main flow $\bar{u}(\eta)$ is still affected by viscosity. This can be done if Re is large. The resulting equation

$$(\bar{u} - c)(\phi'' - \alpha^2\phi) - \alpha\bar{u}'' = 0 \quad \ldots (2.9)$$

with $\alpha = 0$ when $\eta = 0$ and $\phi = 0$ when $n = \infty$ as the boundary conditions has been studied by Raleigh (5*) and Tollmien (5*). A flow is regarded as possessing *non-viscous instability* when it is unstable with the influence of viscous terms on disturbances omitted; i.e. when the instability is a consequence of frictionless stability (Eq. 2.9). A flow of fluid along a flat wall with density increasing upwards is an example of non-viscous instability. In this connection two theorems of Raleigh are relevant.

According to the first theorem, for a velocity profile for which $\bar{u}''$ is less than zero everywhere, there exists at least one neutral disturbance ($c_i = 0$) for which wave propagation velocity c_r (= c) is equal to local flow velocity, or $\bar{u} - c = 0$. The distance $y = y_k$ where $\bar{u} = c$ is known as the critical layer of the mean flow. According to the analysis, at this distance, ϕ' becomes infinite according to frictionless stability equation. This singularity disappears when friction is taken into consideration.

The second theorem states that existence of a point of inflexion in the mean velocity profile i.e. $\bar{u}'' = 0$ is a necessary and sufficient condition for the amplification (i.e. $c_i > 0$). Thus, stability is very sensitive to the velocity profile of mean flow. Not only $\bar{u}(y)$ but $\bar{u}''(y)$ must also be known accurately. Fjrtoft (6*) has shown that for flow to be unstable, $\bar{u}'' = 0$ must correspond to a point of maximum shear.

Viscous Instability

A flow is regarded as possessing *viscous instability* 1if it is unstable with the influence of viscous terms on the disturbance included, i.e. when it is a result of solution of Eq. 2.7. Boundary layer flow on flat wall with zero or negative pressure gradient can become unstable due to viscous instability.

Computational Results

Tollmien (5*) used the velocity distribution for boundary layer flow on a flat plate with zero pressure gradient as obtained by Blasius, and computed the variation of dimensionless wavelength $\alpha\delta^*$, frequency $\beta_r\delta^*/U_o$ and velocity of propagation of disturbance c_r / U_o for various values of Reynolds number ($U_o\ \delta^*/\nu$) and amplification factor c_i. Neutral stability curves for these parameters as obtained by Tollmien and Schlichting are shown in Figs. 2.1 and 2.2. It can be seen that when Reynolds number for boundary layer based on δ^*, the displacement thickness, is less than 575 disturbances of all wavelengths or α, and frequencies β_r are stable and, hence die out. If $U_o\delta^*/\nu$ is greater than 575, there is range of $\alpha\delta^*$ and $\beta_r\delta^* / U_o$ in which disturbances are unstable. Outside this range they are stable. At $U_o\rho\delta^*/\mu$ equal to 575, there is only one disturbance which produces neutral stability for which

$$c_r / U_o = 0.42,\ \beta_r\delta^* / U_o = 0.118 \text{ and } \alpha\delta^* = 0.28.$$

Since the displacement thickness δ^* for laminar boundary layer is given by $\delta^* = 1.73/\sqrt{\nu x / U_o}$, $U_o\rho\delta^*/\mu = 575$ corresponds to $U_o x/\nu$ equal to 1.12×10^5. Downstream of this section where $U_o\rho x/\mu = 1.12 \times 10^5$ disturbances of

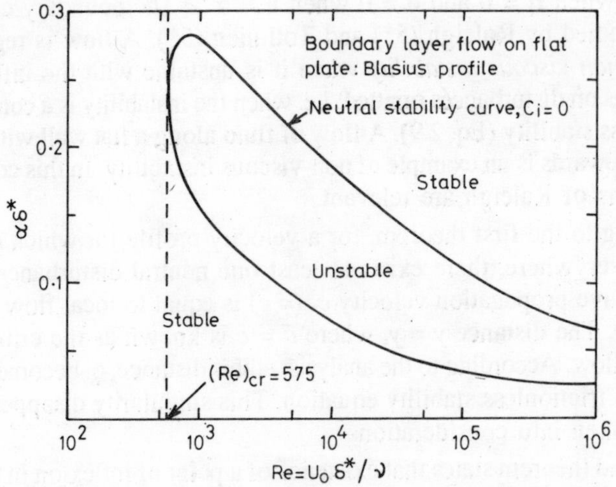

Fig. 2.1 Variation of $\alpha\delta^*$ with $U_o\delta^*/\nu$ for neutral stability curve

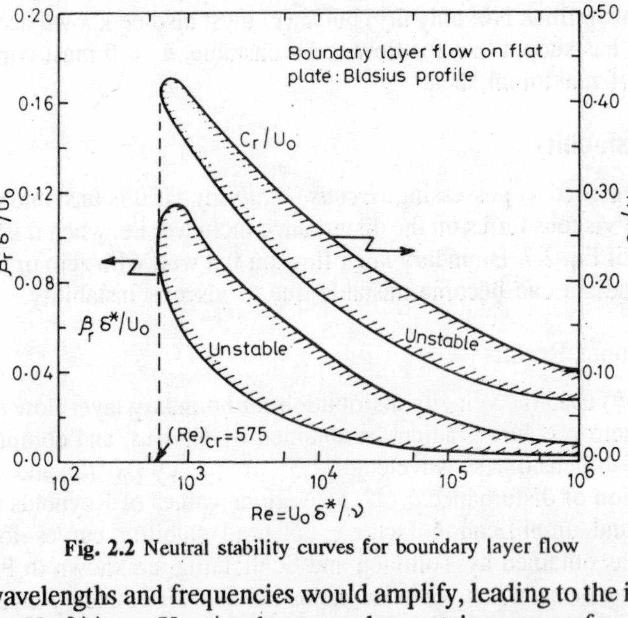

Fig. 2.2 Neutral stability curves for boundary layer flow

certain wavelengths and frequencies would amplify, leading to the instability. For given $U_o\rho\delta^*/\mu$ or $U_o\rho x/\mu$, there are also certain ranges of α and β_r for which the disturbance is stable. If certain disturbances become unstable at a given value of $U_o\rho x/\mu$, some distance would be needed in the downstream direction for these disturbances to grow enough and cause turbulence. Hence, the transition Reynolds number is expected to be greater than 1.12×10^5. Actual transition of laminar boundary layer on a flat plate at zero pressure gradient is found to occur at $U_o\rho x/\mu$ ranging between 3.5×10^5 to 6×10^5. Theoretical results also indicate the degree of amplification of a disturbance of given β_r that is needed before transition sets in. Semiempirical theories proposed by Van Ingen (6*), and Smith and Gamberoni (6*) have stated that when the total amplification exceeds e^9, the transition sets in. However, experimental data indicate that the exponent of e can vary from 6 to 12. Theoretical results of Tollmien and Schlichting have been improved by Shen (7), who has used Lin's (8) procedure to calculate the curve of constant c_i. His neutral stability curve agrees closely with that of Lin in which the critical Reynolds number $U_o\delta^*/\nu$ is found to be closer to 420 than 575. The above numerical results of Tollmien and others are for two-dimensional disturbance. Squire (9) has conclusively shown that by using Blasius profile, two dimensional disturbances become unstable earlier than the three dimensional disturbances. Hence, instability obtained by introduction of two dimensional disturbance gives a conservative value of critical Reynolds number.

2.4 EXPERIMENTAL VERIFICATION

Boundary Layer Flow

The verification of instability of laminar boundary layer flow on a flat plate was carried out by Burgers, Van Zinjnen, Hansen, and Dryden (10), and Schubauer and Skramstad (4). Boundary layer with very low intensity of turbulence of

$$100 \times \frac{\sqrt{(\overline{u'^2} + \overline{v'^2} + \overline{w'^2})/3}}{U_o} = 0.03$$

in the ambient flow of velocity U_o was established in a wind tunnel by Schubauer and Skramstad. Here u', v' and w' are turbulent velocity fluctuations in x, y and z directions, respectively. The disturbances in the laminar boundary layer were caused by oscillating a 300 mm long, 0.05 mm thick and 2.5 mm wide horizontal metallic strip kept at 0.15 mm from the boundary. For imposed disturbances of known frequency and amplitude, the amplification or damping of the disturbance was monitored at various sections in the downstream direction with the help of hot wire anemometer. Figure 2.3 shows variation of c_r/U_o

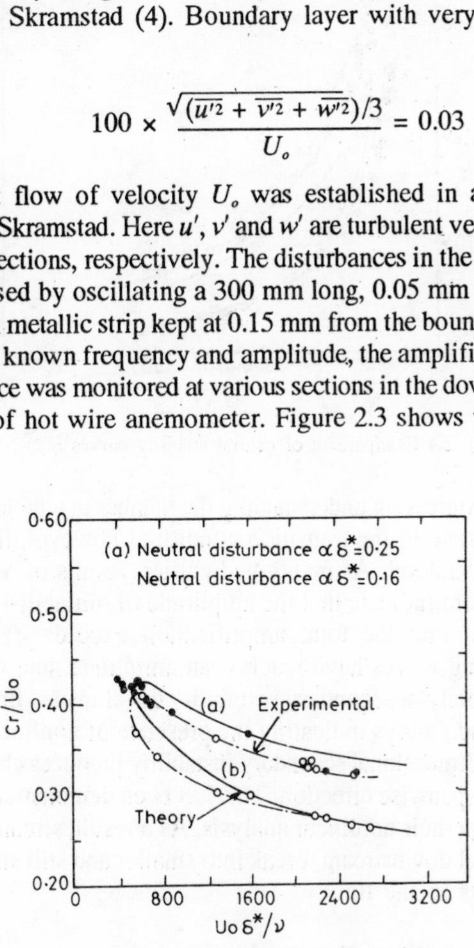

Fig. 2.3 Variation of c_r/U_o with $U_o\delta^*/\nu$

with $U_o\rho\delta^*/\mu$ according to Tollmien's theory and as observed by Schubauer and Skramstad. The instability theory has been further improved by Shen (7), and Barry and Ross (11). Further experimental data were collected by Ross et al. (12). The comparison between theoretical and experimental variation of $\beta_r\nu/U_o^2$ with $U_o\delta^*/\nu$ is shown in Fig. 2.4. These Figures indicate the general validity of instability theory. The difference between the theoretical curves of Tollmien, and Barry and Ross are attributed to the fact that the latter modified Orr-Sommerfeld equation by including more important terms representing the growth of boundary layer.

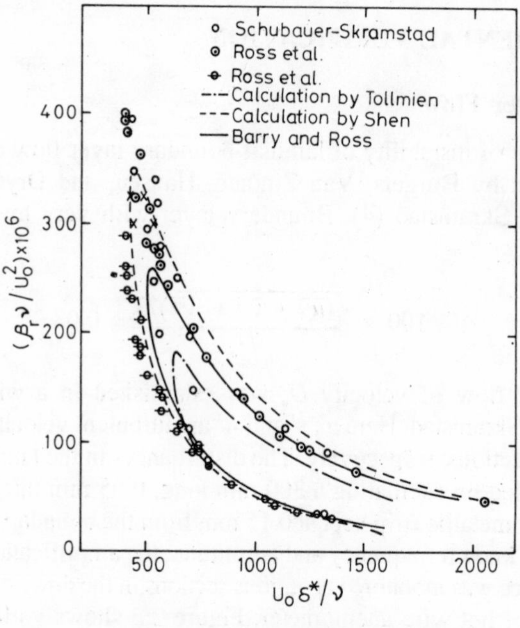

Fig. 2.4 Comparison of neutral stability curves (12)

Research is in progress on understanding the manner in which flow develops from the unstable state to the transition condition; however, the mechanism is not yet fully understood. As mentioned earlier, results of Van Ingen, and Smith and Gamberoni indicate that the amplitude of initial disturbance needs to be amplified so that the total amplification exceeds e^6 to e^{12}. When Tollmien-Schlichting waves have reached an amplitude near one percent of the free stream velocity, a secondary instability develops leading to spanwise regions of peaks and valleys indicating the presence of nonlinear mechanism at work. This three dimensional secondary instability produces changes in mean velocity profiles in spanwise direction. This has been demonstrated by Benney and Lin (13) through their nonlinear analysis. As a result, streamwise vortices develop which travel downstream, break into smaller and still smaller vortices causing randomness in the flow.

Plane Poiseuille and Plane Couette Flows

Orszag and Kells (14) and, Nishioka, Iida and Ichikawa (15) have studied stability and transition in plane Poiseuille and plane Couette flows. Plane Poiseuille flow is flow in rectangular channels with large aspect ratio of width to depth. Orszag and Kells found that plane Poiseuille flow sustained neutrally stable two-dimensional disturbances of finite amplitude at Reynolds number about 2800. Here Reynolds number is defined as $(U_o h/2\nu)$ where h is spacing between plates and U_o is the maximum velocity. No neutrally stable two-dimensional finite amplitude disturbances were found for plane Couette flow. Three dimensional disturbances were shown to have a strong destabilising effect. It was shown that finite amplitude disturbances can cause transition to turbulence in both plane Poiseuille flow and plane Couette flow at Reynolds numbers of the order of 1000. It was also shown that plane Poiseuille flow cannot sustain turbulence at Reynolds number below about 500.

Nishioka et al. (15) performed experiments in a low turbulence wind tunnel in which they were able to maintain plane Poiseuille flow for aspect ratio 27.4, even upto Reynolds number of 8000. This was possible since turbulence intensity in the flow was less than 0.05 percent. Schiller (16*) conducted experiments on plane Poiseuille flow using square and a rectangular duct with aspect ratio of 3.5 and found the critical Reynolds numbers to be 1000 and 1600, respectively, when hydraulic diameter is used as length dimension. On the other hand, Davis and White (17) 1928, Patel and Head (18) 1969, and Kao and Park (16) 1970 have shown that plane Poiseuille flow is unstable to finite amplitude disturbance at Reynolds number as low as 1000. Reichardt (19) 1959 obtained similar result that plane Couette flow becomes turbulent at as low a Reynolds number as 750. This wide variation in critical Reynolds number is believed to be partly due to effect of aspect ratio on the transition.

Kao and Park (16) conducted experimental investigation on a 3.66 m long and 25.4 mm × 203 mm rectangular conduit carrying water. A 0.025 mm thick, 6 mm wide stainless steel ribbon was vibrated across the flow when the ribbon was kept at 1.88 mm and 4.45 mm from the wall. The frequency range used was 1/20 to 50 Hz and mean velocity profiles and disturbance characteristics were measured at various sections using hot film anemometer. Since they used small as well as large amplitude oscillations, both linear as well as nonlinear effects were studied by Kao and Park. The experimental data was plotted as $\beta/2\pi$, c_r/U_o and α_r versus Reynolds number to determine the neutral stability curves. It was found that flow was stable at all disturbances if Reynolds number was less than 2500. At 2500 value there was only one neutral disturbance whose characteristics were $\beta/2\pi = 0.30$, $c_r/U_o = 1.20$ and $\alpha_r = 0.75$ where $\alpha_r = 2\pi$ × amplitude/wavelength. Variation of α_r vs *Re* is shown in Fig. 2.5. Here Reynolds number is defined using $4A/P$ as the length parameter. Figure 2.6 shows a typical hot film output at 0.87 Hz frequency and *Re* equal to 2720 at various distances downstream of the place where oscillations were introduced. Here one can notice a large eddy at 0.33 m. The growth rate is very rapid, and at 0.43 m where the maximum amplitude reached is about 0.8% of U_o, the wave shows incipient breaking with the appearance of dominant higher harmonic. At 0.52 m larger frequencies are now present indicating breaking of waves. These experiments showed also that at Reynolds number greater than critical, the growth rate of the disturbance depended on the frequency.

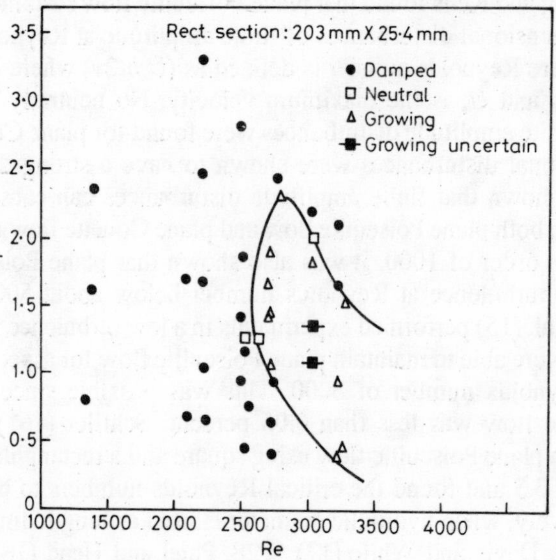

Fig. 2.5 **Neutral stability curve for plane Poiseuille flow as obtained by Kao and Park (16)**

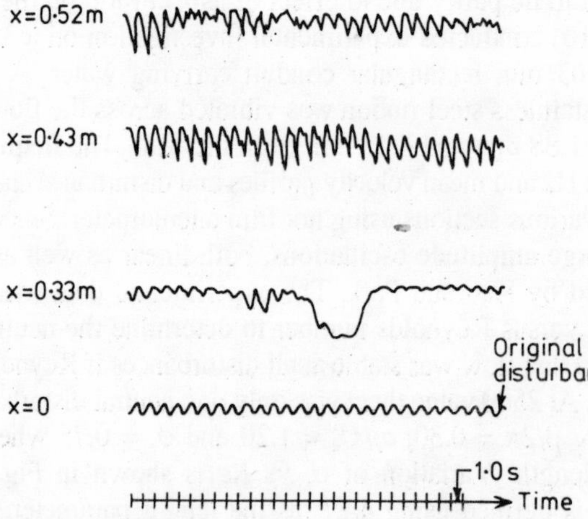

Fig. 2.6 Hot film output for plane Poiseuille flow at *Re* = 2720, and 0.87 Hz frequency (16)

Similar experiments have also been conducted on pipe flow by Fox et al. (20). Typical results of this investigation are shown in Fig. 2.7, where frequency of disturbance is plotted against Reynolds number and neutral stability curve has been drawn. It can be seen that for pipe flow, all the disturbances are damped when Reynolds number is less than 2200.

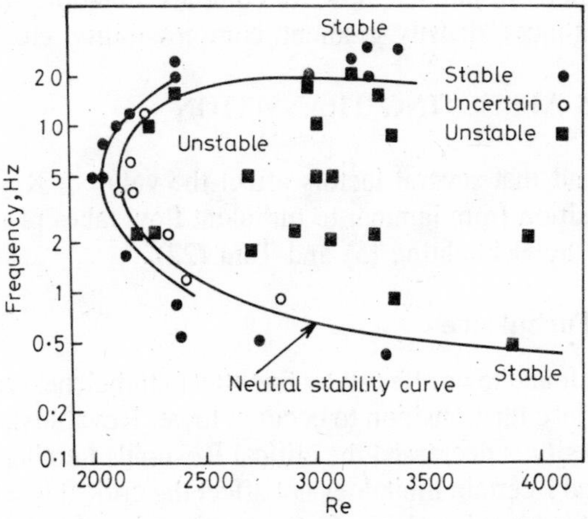

Fig. 2.7 Experimental results on stability of laminar flow in circular pipes (20)

2.5 ROUSE INDEX

Considering the complexities in the analysis of stability of laminar flows, Rouse (21) 1945, attempted to develop a criterion based on mean flow characteristics at a point to determine conditions when a flow at a point would become unstable. From dimensional considerations it was argued that stability of laminar flow at a point would be governed by ρ, $\partial u/\partial y$, the local velocity gradient, y the distance from the boundary and dynamic viscosity of the fluid. Hence dimensional analysis would indicate that

$$\phi\left(\frac{y^2\rho(\partial u/\partial y)}{\mu}\right) = 0$$

This parameter called χ (chi) will be zero at the wall when $y = 0$ and away from the wall where $\partial u/\partial y = 0$; in between, it will assume a maximum value. Under critical condition χ will assume a constant value χ_{cr}. If χ is less than χ_{cr} flow will remain laminar; if χ is greater than χ_{cr} at that point in the flow, instability would develop and flow would become turbulent. Critical values of χ are listed below along with the critical Reynolds number for different flows.

Type of flow	*Critical Re*	χ_{cr}
Laminar boundary layer on a flat plate	3×10^5	560
Circular pipes	2000	590
Flow between parallel plates (Plane Poiseuille flow)	800	380
Flow between cylinders (one rotating)	—	500

From such analysis, Rouse argued that one can take $\chi_{cr} = 500$ at which instability develops in laminar flow. In complex situations χ_{cr} value would depend on roughness, density gradient, compressibility, etc.

2.6 FACTORS AFFECTING TRANSITION

It has been found that several factors affect the value of Reynolds number at which a transition from laminar to turbulent flow takes place. These have been discussed by Schlichting (5) and Tani (22).

Free Stream Turbulence

The transition is found to be affected by free stream turbulene. Larger intensities of turbulence cause the transition to occur at lower Reynolds number. As the turbulence intensity is decreased the critical Reynolds number increases; but reduction beyond a certain limit does not affect the critical Reynolds number. This is shown for boundary layer flow on a flat plate for data collected by Schubauer and Skramstad, Hall and Hislop, and Dryden and Bennet in Fig. 2.8. In these boundary layer flows, turbulence level is controlled by use of proper grid at the upstream section in the wind tunnel. When turbulence level is less than 0.06 to 0.08 percent the critical Reynolds number remains unaffected. Even though certain speculations are made about the reason why turbulence level in main stream affects the transition, no conclusive explanation exists for the same.

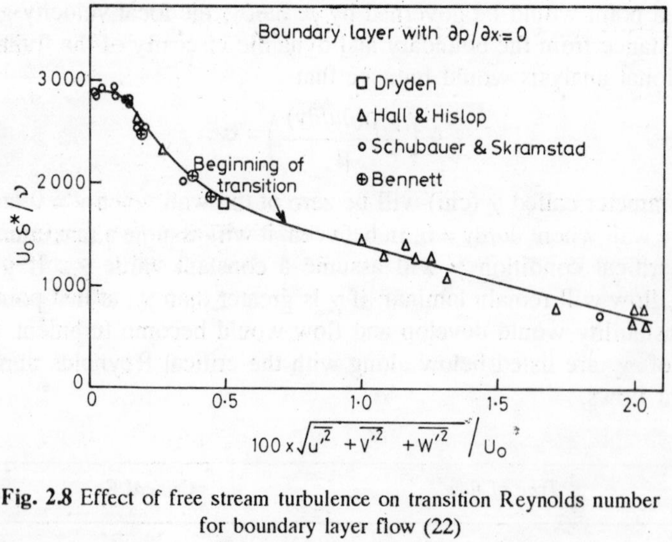

Fig. 2.8 Effect of free stream turbulence on transition Reynolds number for boundary layer flow (22)

Pressure Gradient

The pressure gradient in the streamwise direction exerts an overwhelming influence on the stability of the laminar boundary layer. When the pressure increases in the streamwise direction, the boundary layer exhibits velocity profiles with a point of inflexion suggesting a tendency towards instability according to linearised theory of stability. Results of instability calculations indicate that negative pressure gradient tends to stabilise the flow, while positive pressure gradient has a destabilising effect. Pohlhausen number λ defined as $= (\delta^2/\nu)\partial U_o/\partial x = -(\delta^2/\rho\mu U_o)\,\partial p/\partial x$ can be shown to be related to critical Reynolds number based on displacement thickness for boundary layer as $(U_o\rho\delta^*/\mu)_{cr} \sim e^{0.6\lambda}$ for $-3 < \lambda < 3$. However, Liepmann's experimental results have indicated $(U_o\rho\delta^*/\mu)_{cr}$ varies as $e^{0.08\lambda}$ which indicates that effect of pressure gradient on critical Reynolds number is much less than that predicted by theory. The effect of pressure gradient on critical Reynolds number is also many times expressed by specifying variation of U_o with x as $U_o \sim x^m$ where $m = (x/U_o)(\partial U_o/\partial x)$. The critical Reynolds number $(U_o\rho\delta^*/\mu)$ is then related to m.

When large positive pressure gradient exists along with moderate surface curvature, the laminar boundary layer separates from the surface, but often becomes reattached to the surface as the boundary layer becomes turbulent. Transition to turbulence takes place relatively rapidly in a separated flow.

Surface Curvature

The effect of surface curvature on the stability of two-dimensional disturbances is rather small for surface curvatures—either convex or concave—that are likely to occur in practice. However, the flow on the concave surface exhibits another type of instability due to centrifugal pressure gradient producing a three dimensional system of alternating vortices with axis in streamwise direction; these have been studied by Görtler. The instability is found to occur when $G = (U_o\delta^*/\nu)\sqrt{\delta^*/r} > 1.2$, where r is the radius of curvature of the surface. Therefore, streamwise vortices come into existence at a Reynolds number much below the critical Reynolds number for Tollmien-Schlichting waves, and are considered as mainly responsible for making boundary layer three dimensional. Figure 2.9 shows variation of $(U_o\delta^*/\nu)_{cr}$ with δ^*/r for Liepmann and Aihara's data. Here, negative r corresponds to convex surface. It can be seen that transition Reynolds number is almost unaffected by the convex curvature.

Suction

Suction along the boundary also has a stabilising influence on the boundary layer. Firstly, for a given suction rate, the boundary layer thickness assumes a smaller constant value and a smaller boundary layer is less prone to transition. Secondly, suction modifies the velocity profile in such a manner that it enhances stability.

For continuous suction along the boundary at velocity v_o, after some distance the boundary layer thickness becomes independent of x and its value is given by $\delta^* = \nu/v_o$. The corresponding asymptotic velocity profile is

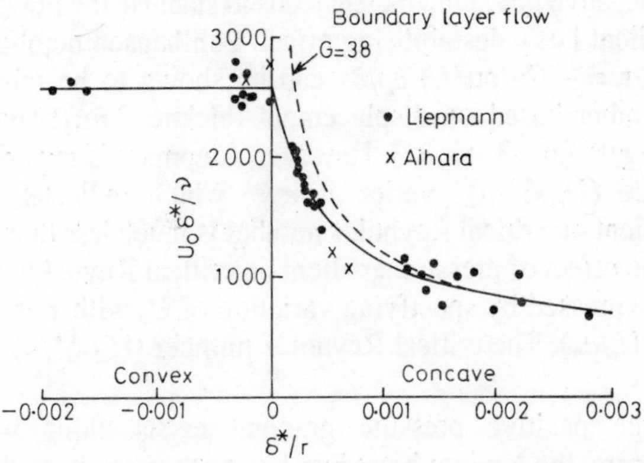

Fig. 2.9 Effect of surface curvature on transition Reynolds number (22)

given by

$$u/U_o = (1 - e^{v_o y/\nu})$$

The analysis by Bussmann and Muenz (23*) showed that $(U_o\rho\delta^*/\mu)_{cr}$ for this case is 70000, which is much higher than that without suction. It can be shown that to achieve this, a very small fraction of U_o is needed as the suction velocity. From the above two expressions it can be shown that $v_o/U_o >$ 1/70000 if $(U_o\rho\delta^*/\mu)_{cr}$ is to be about 70000.

Density Gradient

Density gradient likewise affects the stability of flow. When the density decreases upwards from the boundary, the flow is stable. In such a case the turbulence mixing is impeded because heavier particles must be lifted and lighter particles must be depressed against hydrostatic force. The turbulence can even be suppressed completely if the density gradient is strong enough. In case of increasing density in the upward direction (e.g. when the boundary is heated), there is instability. In the presence of density stratification, critical or transition Reynolds number will depend on the stratification parameter

$$Ri = -\frac{g}{\rho}\frac{\partial\rho}{\partial y} \Bigg/ \left(\frac{\partial u}{\partial y}\right)^2_{y=0}$$

which is commonly known as the Richardson number. Ri equal to zero corresponds to homogeneous fluid; Ri less than zero gives unstable stratification, whereas Ri greater than zero gives a stable condition. Schlichting analysed the stability of flows using Tollmien's theory and showed that the critical Reynolds number increases with increase in Richardson number and $(Re)_{cr}$ = 575 when Ri = 0, and $(Re)_{cr} = \infty$ when Ri = 1/24; see Fig. 2.10 which shows variation of $(U_o\rho\delta^*/\mu)_{cr}$ with Ri for boundary layer flow.

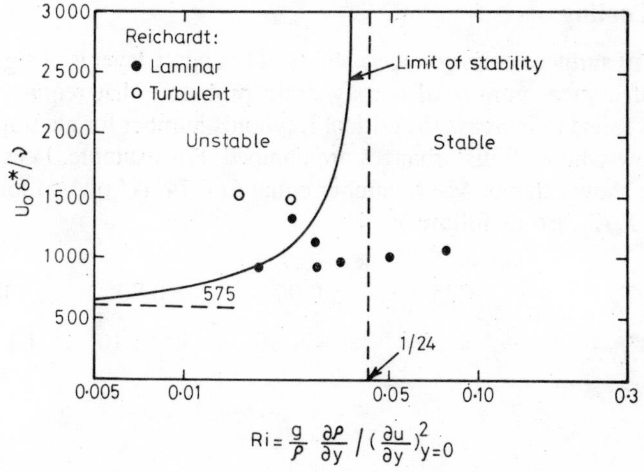

Fig. 2.10 Effect of density gradient on stability of boundary layer flow (5)

Surface Roughness

The presence of surface roughness is found to affect the transition significantly. If a wire of diameter k is placed at a distance x_k from the leading edge of a smooth plate on which fluid is flowing parallel to it at a velocity U_o, the transition occurs earlier at a distance x_t, where x_t is greater than x_k. As U_o or k is increased, x_t decreases and approaches x_k. When k/δ^*_k is less than 0.30, the transition Reynolds number $U_o\delta^*_t / \nu$ is related to k/δ^*_k by the equation

$$U_o\delta^*_t / \nu = 826\ (\delta^*_k / k)$$

where δ^*_k and δ^*_t are values of displacement thickness at the position of roughness and transition. For $U_o\rho\delta^*_k / \mu$ less than 0.30 the transition occurs at $x_t = x_k$. This corresponds to $U_o\rho k / \mu = 826$, or $(U_o\rho\delta^*_t / \mu) = 2.96\ (x_k / k)\ (k/\delta^*_k)$.

Analysis of experimental data of Tani, Iuichi and Yamamoto by Gibbings has shown that this critical value of $U_o k/\nu$ decreases in boundary layer flow with negative pressure gradient as

$$U_o k\rho/\mu = 826\ e^{-0.90\lambda}$$

for $0 < \lambda < 0.30$ where λ is Pohlhausen number $= -\ (\delta^2 / \rho\mu U_o)\ \dfrac{\partial p}{\partial x}$.

There is little quantitative information available on the effect of presence of distributed roughness elements on the location of transition. Evidently the shape and concentration of roughness elements play an important role. Feindt has used sand grains as roughness elements and shown that if $U_o k\rho/\mu$ is less than 120, the transition Reynolds number is unaffected. However, for larger

$U_o k/\nu$ values the critical Reynolds number drops appreciably with increase in $U_o k/\nu$ for given pressure gradient.

Surface Cooling

The effect of surface cooling on the stability of boundary layer is of significance because of a great variety of aerodynamic problems that require cooling. Cooling is found to increase the critical Reynolds number for neutral stability curve below which all disturbances are damped. For example, Lees and Lin (23), have shown that at Mach number equal to 0.70, $(U_o\rho\delta^*/\mu)$ for various values of T_W/T_F are as follows:

T_W/T_F	1.25	0.90	0.80	0.70
$(U_o\rho\delta^*/\mu)_{cr}$	2×10^2	1.4×10^3	3.5×10^3	1.1×10^4

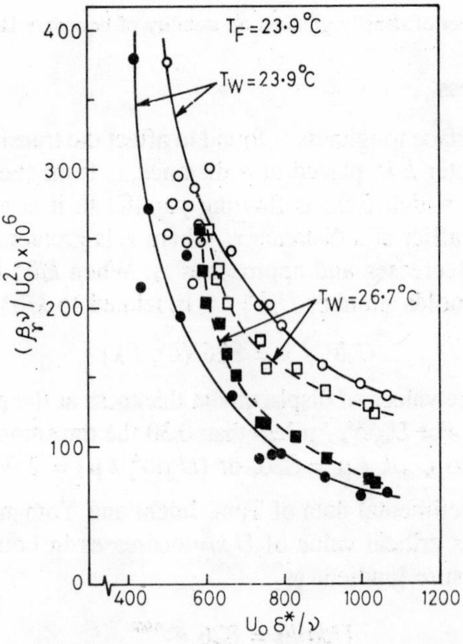

Fig. 2.11 Experimental neutral-stability results for unheated and heated flat-plate boundary layers in water (Strazisar's data)

Here T_W and T_F are the wall and fluid temperatures respectively. It is curious that, for boundary layers in water, the effect of cooling is destabilising whereas the effect of heating is stabilising. This has been found by Wazzan, Okamura, and Smith as reported by Reshotko (24). Because the viscosity of water decreases sharply with increase in temperature, heating yields a fuller velocity profile whereas cooling tends to give an inflected velocity profile. Figure 2.11 shows variation of $U_o\rho\delta^*/\mu$ with $\beta_r\nu/U_o^2$ when wall and water temperature are the same and different. It can be seen that heated wall yields a greater critical Reynolds number.

REFERENCES

1. Reynolds, O. An Experimental Investigation of the Circumstances which Determine Whether the Motion of Water Shall be Direct or Sinuous and of the Law of Resistance in Parallel Channel. Phil., Trans. RSL, London, Ser. A 174, 1883.
2. Barnes, H.T. and E.G. Coker. The Flow of Water Through Pipes. Proc. RSL, Vol. 74, 1905.
3. Ekman, V.W. On the Change from Steady to Turbulent Motion of Liquids. Archiv. Für Maths , Astr. Och. Fysik, Vol. 6, No. 12, 1910.
4. Schubauer, G.B. and H.K. Skramstad. Laminar Boundary Layer Oscillations and Stability of Laminar Flow. NACA Wartime Report, W-8, 1943.
5. Schlichting, H. *Boundary Layer Theory*. Mc-Graw Hill Series in Mechanical Engineering, New York, 1955, Chapter 16.
6. Landhal, M.T. and E.M. Christensen. *Turbulence and Random Processes in Fluid Mechanics*. Cambridge University Press, London, 1986.
7. Shen, S.F. Calculated Amplified Oscillations in Plane Poiseuille and Blasius Flows. Jour. Aero. Sci. Vol. 21, 1954.
8. Lin, C.C., *The Theory of Hydrodynamic Stability*. Cambridge University Press, London 1955.
9. Squire, H.B. On the Stability of the Three Dimensional Disturbances of Viscous Flow between Parallel Walls. Proc. RSL, Series A 142, 1933.
10. Dryden, H.L. Boundary Layer Flow Near Flat Plates. Proc. 4th Int. Conf. for Applied Mechanics, Cambridge, 1934.
11. Barry, M.D.S. and M.A.S. Ross. The Flat Plate Boundary Layer. Pt. 2. The Effect of Increasing Thickness on Stability JFM, Vol. 43, Pt. 4, 1970.
12. Ross. J.A., F.H. Barnes, J.G. Burns and M.A.S. Ross. The Flat Plate Boundary Layer. Pt. 3 Comparison of Theory with Experiment. JFM, Vol. 43, Pt. 4, 1970.
13. Benney, D.J. and C.C. Lin. On the Secondary Motion Induced by Oscillations in a Shear Flow. Physics of Fluids, Vol. 3, 1960.
14. Orszag. S.A. and L.C. Kells. Transition to Turbulence in Plane Poiseuille and Plane Couette Flows. JFM Vol. 96, Pt. 1, 1980.
15. Nishioka, M., S. Iida and Y. Ichikawa. An Experimental Investigation of Stability of Plane Poiseuille Flow. JFM, Vol. 72, 1975.
16. Kao, T.W. and C. Park. Experimental Investigations of the Stability of Channel Flows. Pt. 1. Flow of a Single Liquid in a Rectangular Channel. JFM, Vol. 43, Pt. 1, 1970.
17. Davis, S.S. and C.M. White. An Experimental Study of Flow of Water in Pipes of Rectangular Section. Proc. RSL, Ser. A. 119, 1928.
18. Patel, V., and M.R. Head. Some Observations on Skin Friction and Velocity Profiles in Fully Developed Pipe and Channel Flows. JFM, Vol. 38, 1969.
19. Reichardt, H. Gesetzmässigkeiten der Gerandlingen Turbulenten Couette Strömung. Mitt. Max-Planck Inst. Ström. Forsch. 22, Göttingen 1955.
20. Fox, J.A., M. Lessen and W.V. Bhat. Experimental Investigation of the Stability of Hagen Poiseuille Flow. The Physics of Fluids, Vol. 11, No. 1, 1968.

21. Rouse, H. A General Stability Index for Flow Near Plane Boundaries. Jour. Aero. Sci. Vol. 12, No. 4, 1945.

22. Tani, I. Boundary Layer Transition. Annual Review of Fluid Mechanics. Annual Reviews Inc., U.S.A. Vol. 1, 1969.

23. Lees, L. and C.C. Lin. Investigation of the Stability of the Laminar Boundary Layer in a Compressible Fluid. NACA Tech. Rep. No. 1115, 1946.

24. Reshotko, E. Boundary Layer Stability and Analysis. Annual Review of Fluid Mechanics. Annual Review Inc. U.S.A., Vol. 8, 1976.

CHAPTER III

Nature of Turbulence

3.1 INTRODUCTION

It has been pointed out in Chapter II that when the Reynolds number exceeds a certain limit, disturbances of certain wavelengths and frequencies are amplified. When amplification is large, the disturbance breaks down into chaotic motion. Repeated breakdown of disturbances leads to completely chaotic sustained motion, commonly known as turbulence. Of all the flows encountered in hydraulic, environmental, chemical and aeronautical engineering as well as in meteorology and oceanography, most of the flows are turbulent in character. Flows in rivers, canals and pipes, air flow over land surfaces, ocean currents, flow past large bodies such as aeroplanes, torpedoes, ships and bridge piers, flows in expansions and contractions in channels and conduits, flows in free and wall jets and in the wakes, and flows in many unit processes in chemical and environmental engineering involving mixing or diffusion are some of the examples of turbulent flows. Because of such common occurrence, there is a great need of understanding the intricacies of turbulent motion.

Gross motion in turbulent flow is far from orderly as can be realised by observing the smoke patterns from chimneys, wakes behind bridge piers and dust patterns behind cars, buildings, trees and in corners. It is difficult to give a precise definition of turbulence; in 1937 Taylor and Von Karman (1*) defined turbulence as follows:

"Turbulence is an irregular motion which, in general, makes its appearance in fluids, gases or liquids, when they flow past solid surface or even when neighbouring streams of the same liquid flow past, over one another."

However, the definition given by Hinze seems more satisfactory. Hinze (1) has defined turbulence as follows:

"Turbulent fluid motion is an irregular condition of flow in which various quantities show a random variation with time and space co-ordinates so that statistically distinct averages can be discerned."

Thus, even in an otherwise steady turbulent flow the velocity at a given point will randomly fluctuate with time, whereas even in uniform turbulent flow instantaneous velocity at all points along a streamline would vary randomly. The same statement is true about other flow parameters such as pressure, temperature, sediment concentration at a point, or force on structures such as smoke stacks and suspension bridges. Turbulent flow can be considered to consist of a mean motion superimposed on which are the randomly varying components of velocity in all the three directions.

It may be recalled that the lateral movement of fluid particles in the case of laminar flow takes place due to molecular diffusion, and it is very small indeed. In turbulent flow, lumps of fluid particles move laterally and longitudinally giving rise to the concept of eddy motion. *Eddy* can be defined as a large group of fluid particles which move laterally or longitudinally in the flow field; while undergoing this motion an eddy can change its shape or stretch, and rotate or break into two or more eddies. The motion of a smoke ring released from a cigarette and moving in the air gives a realistic picture of an eddy. Figure 3.1 shows an instantaneous flow pattern in turbulent flow in a pipe. Eddies are generated in the region of high shear in the mean flow field i.e. near the boundary in a pipe or channel flow, or in the vicinity of interface between two streams flowing at different velocities and parallel to one another. At high shear, which is analogous to saying that at high Reynolds numbers, the size of larger eddies is governed by the size of flow and its geometry.

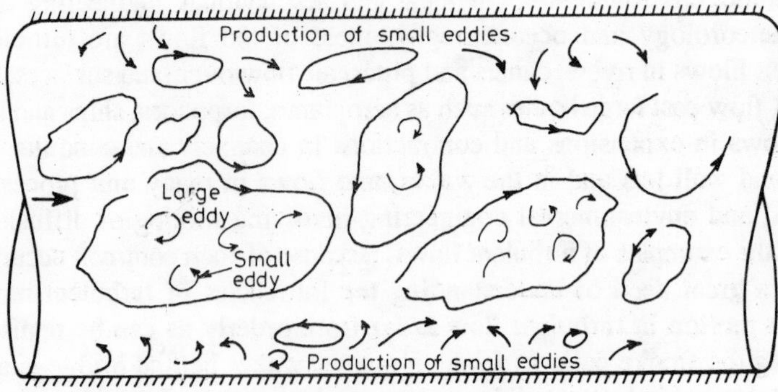

Fig. 3.1 Instantaneous eddy representation in a pipe

At any given time the flow contains eddies of various sizes. The largest eddy will be of the size of the flow, e.g. pipe diameter or flow depth in a channel, whereas the smallest eddy will be very small, say of the order of a millimeter. However, it may be emphasised that the size of the smallest eddy is still sufficiently large compared to the size of molecules of the fluid. These eddies are three dimensional in nature, even though flow may be one dimensional. Eddies of various sizes are embedded in each other and are impermanent in nature. That is larger eddies, which are continually formed, break into smaller and still smaller eddies until they are finally dissipated through viscous shear.

If one were to focus attention at a point in flow field, passage of smallest, small and large eddies through this point would induce velocity fluctuations of small magnitude and large frequency, and large magnitude and small frequency. This would yield a fluctuating velocity field. Figure 3.2 shows instantaneous velocity variation with time for the River Teshio in Japan (2). Turbulent frequencies ranging from 1 Hz to 10000 Hz have been observed in the flow of air and water. With this type of flow it is desirable to express instantaneous velocity, pressure or any other flow parameter as

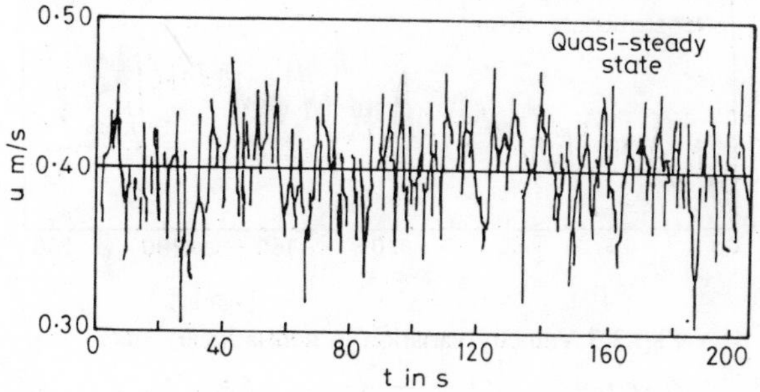

Fig. 3.2 Velocity variation in stationary state

$$\left.\begin{aligned} u &= \bar{u} + u' \\ v &= \bar{v} + v' \\ w &= \bar{w} + w' \\ p &= \bar{p} + p' \end{aligned}\right\} \qquad \ldots (3.1)$$

in which u, v, w and p are the instantaneous values, $\bar{u}$, $\bar{v}$, $\bar{w}$ and $\bar{p}$ are the average values, and u', v', w' and p' are the fluctuation components.

3.2 AVERAGING PROCEDURES

Turbulent flow field is called *stationary* when statistical parameters related to the flow do not change with respect to time; it can alternately be called quasi-steady; such a flow is shown in Fig. 3.2. Further comment on stationary random process is made in Chapter V.

In such a case, even though instantaneous value of velocity is changing, an average constant value with respect to time is apparent. Here one defines the time average $\bar{u}$ as

$$\bar{u} = \frac{1}{2\,T}\int_{-T}^{T} u \; dt \qquad \ldots (3.2)$$

at a given point (x, y, z) in space. The sampling time $2\,T$ should be sufficiently large compared to time scale of small eddies, but small compared to time scale of large eddies. The time scale for a eddy can be considered as the time required for eddy to pass past a given point. It can be easily seen that

$$\frac{1}{2\,T}\int_{-T}^{T} u' \; dt = 0 \qquad \ldots (3.3)$$

i.e. the time average of turbulent fluctuations u', namely $\bar{u}' = 0$. In the case of quasi-unsteady, or nonstationary flow, such an average $\bar{u}$ taken over times will be different, and hence less useful (see Fig. 3.3).

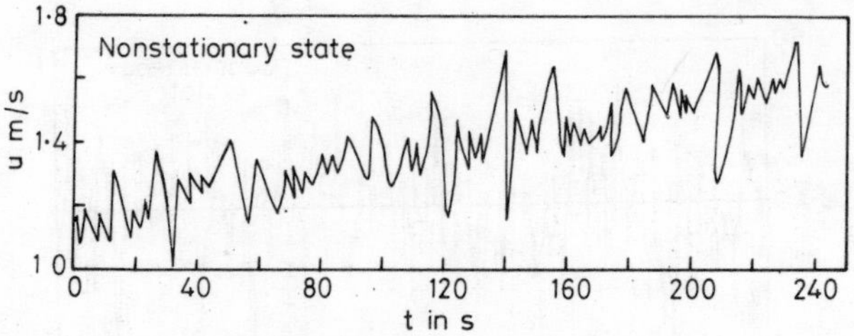

Fig. 3.3 Velocity variation in nonstationary state

If flow is quasi-uniform, the instantaneous values of u at different points along a streamline would give a constant average value. In such a case, one defines

$$\bar{u} = \frac{1}{2x}\int_{-x}^{x} u \, dx \qquad \ldots (3.4)$$

Such averaging is essential for homogeneous turbulence. Extending this definition further, one can define a space average at any time as follows:

$$\bar{u} = \frac{1}{\forall}\int_{\forall} u d\forall \qquad \ldots (3.5)$$

where $\forall$ is the volume over which averaging is carried out.

In statistics one also defines an ensemble average. Consider an infinite number of identical macroscopic flows in identical flumes. If u is measured at N corresponding points in these flows, the ensemble average can be defined as

$$< u(t) > = \lim_{N \to \infty} \frac{1}{N}\int_{i=1}^{N} u_i(t) \qquad \ldots (3.6)$$

As can be realised, space average or ensemble average is very difficult to determine. Therefore, it is of interest to know the relationship between time average and ensemble average. If the process is *ergodic*, the two averages are the same. Ergodic theorem gives the sufficient conditions for ensemble average to be equal to time average. Turbulence phenomenon is assumed to be ergodic and, therefore, one can deal with time averages.

The following rules, commonly known as Reynolds rules of averaging, govern the mathematical operations about varying quantities. If $f = \bar{f} + f'$ and $g = \bar{g} + g'$, Reynolds rules state

$$\overline{f+g} = \bar{f} + \bar{g}$$

$$\overline{cf} = c\bar{f} \text{ where } c \text{ is a constant} \qquad \ldots (3.7)$$

$$\overline{fg} = \bar{f}\,\bar{g}$$

$$\overline{\text{Lim} f} = \text{Lim}\,\bar{f} \text{ where } f \text{ is a sequence of a function}$$

$$\overline{\partial f/\partial x} = \partial \bar{f}/\partial x$$

As a consequence $\overline{fg} = \bar{f}\,\bar{g} + \overline{f'g'}$

These will be used in Chapter IV while deriving equations of motion.

3.3 CHARACTERISTICS OF TURBULENT FLOWS

As mentioned earlier, turbulent flow possesses entirely different characteristics than those typical of laminar flow. The first characteristic of turbulent flow, which results from movement of large lumps of fluid (i.e. eddies) in longitudinal and lateral directions, is the greater ability for mixing or difffusion. Transportable quantities related to flow, such as momentum, heat, sediment or other pollutants spread much more rapidly in turbulent flow than in laminar flow. This ability is several hundred times greater in turbulent flow than in laminar flow. Greater ability for diffusion is beneficial in many cases, and harmful in others.

A detailed analysis of such a flow is extremely complex; however, in a simplified analysis, one can ignore directly the turbulent motion and take its effect into account by introducing a property such as eddy dynamic viscosity η, eddy heat conductivity or eddy sediment transfer coefficient. Such a flow would be governed by the relationship for turbulent shear: $\bar{\tau} = \eta\, d\bar{u}/dy$. This concept was first introduced by Boussinesq. However, unlike the laminar flow, in which dynamic viscosity is only fluid property, the above mentioned coefficients depend on the flow characteristics as well. Figure 3.4 shows variation of η in the pipe flow which is turbulent; η is made dimensionless by dividing it by Ru_* where R is radius of pipe and u_* is shear velocity i.e. $\sqrt{\tau_o/\rho}$, τ_o being the bed shear. It can be seen that η is zero at the walls, reaches a maximum value slightly away from wall and then reduces towards the centre. The term $\varepsilon = \eta/\rho$ is known as eddy kinematic viscosity having dimensions L^2/T, the same as that of kinematic viscosity. Since ε can be interpreted as $L \times L/T$, for pipe flow one can say that it will be proportional to the size of eddy and turbulent fluctuations $\sqrt{\overline{v'^2}}$ at any point. In the case of jets and wakes, it is found to be proportional to the flow width and maximum difference in mean velocity of flow.

As a direct consequence of greater diffusion due to movement of eddies, the second characteristic of turbulent flow is that the velocity distribution in turbulent flow is much more uniform than in laminar flow. This is shown in Fig. 3.4 for flow in a circular pipe where velocity distribution in nondimensional form as u/U_m vs r/R is shown for laminar flow, and turbulent flow in hydrodynamically smooth pipe of Reynolds numbers 10^4 and 10^6. It can be seen that in turbulent flow, velocity distribution has very little effect of Reynolds number, except close to the wall. Thus, in turbulent flow, velocity distribution changes very little with Reynolds number, and flow field shows tendency for similarity as it grows in size.

Figure 3.5 shows variation of turbulent fluctuations $\sqrt{\overline{u'^2}}/u_*$, $\sqrt{\overline{v'^2}}/u_*$ and

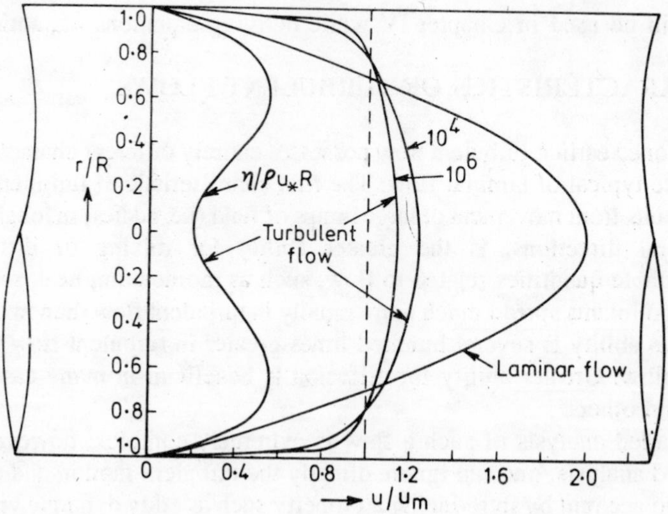

Fig. 3.4 Velocity and eddy dynamic viscosity variation in a pipe

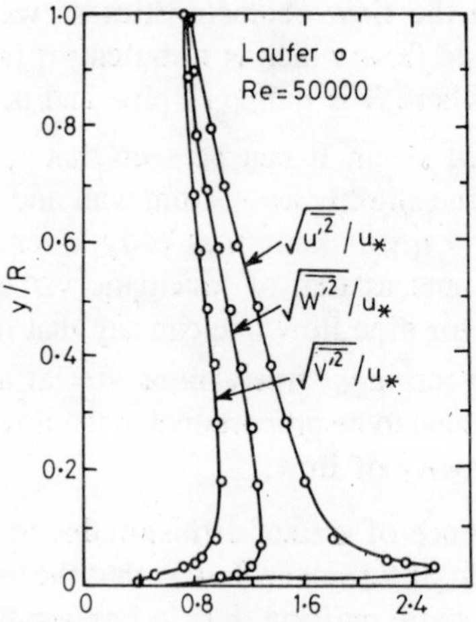

Fig. 3.5 Relative fluctuations in velocity

$\sqrt{\overline{v'^2}}/u_*$ plotted against y/R for turbulent flow in pipe. It may be noted that turbulent fluctuations reach a maximum value very close to the wall and they are about 1 and 3 times u_*. Away from the wall, they reduce and reach low value at the centre. Further very close to the wall $\sqrt{\overline{u'^2}}$ is much greater that $\sqrt{\overline{v'^2}}$ or $\sqrt{\overline{w'^2}}$, however, near the centre they tend to be equal, approaching the condition of isotropy. Since average size of a turbulent eddy increases as y

increases, and since $\sqrt{\overline{v'^2}}$ varies in a manner depicted in Fig. 3.5, similarity between shapes of curves showing variation of $\eta/(\rho u_* R)$ and $\sqrt{\overline{v'^2}}$ seems logical.

Thirdly, as a consequence of greater mixing due to lateral movement of eddies, large shear stresses are developed between different fluid layers. As is shown in Chapter IV, the additional stresses developed in turbulent flow, known as Reynolds stresses, have nine components, three being the normal stresses and six shear stresses. The Reynolds stress tensor is

$$\begin{bmatrix} -\rho\overline{u'^2} & -\rho\overline{u'v'} & -\rho\overline{u'w'} \\ -\rho\overline{v'u'} & -\rho\overline{v'^2} & -\rho\overline{v'w'} \\ -\rho\overline{w'u'} & -\rho\overline{w'v'} & -\rho\overline{w'^2} \end{bmatrix} \quad \ldots (3.8)$$

Evidently, $\overline{u'v'} = \overline{v'u'}$, $\overline{u'w'} = \overline{w'u'}$ and $\overline{v'w'} = \overline{w'u'}$. Hence there are three normal stresses $-\rho\overline{u'^2}$, $-\rho\overline{v'^2}$ and $-\rho\overline{w'^2}$, and three shear stresses $-\rho\overline{u'v'}$, $-\rho\overline{v'w'}$ and $-\rho\overline{u'w'}$. Because of large magnitudes of Reynolds stresses, there is much greater energy loss in turbulent flow than in laminar flow. As shown by Hagen–Poseuille's equation, energy loss in laminar flow is proportional to U_m, whereas in turbulent flow in pipes it is proportional to U_m^n where n varies between 1.75 and 2.0. Thus, existence of Reynolds stresses is responsible for higher energy loss.

Since the turbulent eddies break into smaller and still smaller eddies which ultimately die out due to viscous dissipation, the kinetic energy associated with the eddies reduces. If this secondary motion of eddies is to be maintained at a quasi-steady state, then there must be some mechanism which supplies energy to the secondary motion at the same rate at which its energy is dissipated. This energy supply is obtained from the mean flow through a mechanism which requires existence of measurable mean shear. If mean shear is not present, turbulence cannot sustain itself and will decay. Turbulence generated by grids in the entrance to the wind tunnel is of decaying type, since over the central portion of the cross-section the mean velocity is constant and hence, there is no mean shear. As a result, if one were to measure $\overline{u'^2}$ along the wind tunnel length, it would decrease.

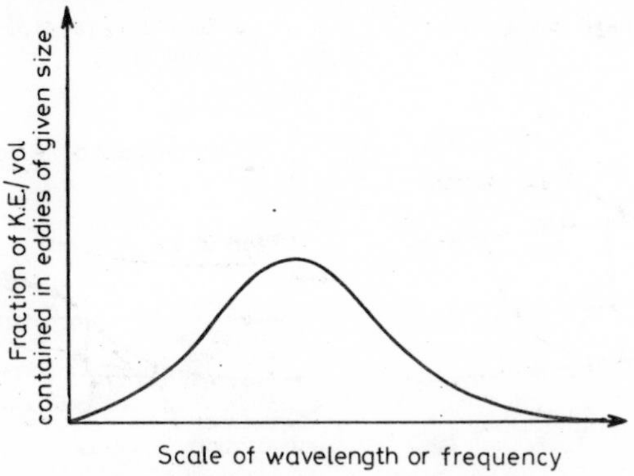

Fig. 3.6 Concept of turbulence spectrum

Lastly, it is apparent that each eddy is associated with kinetic energy. Since eddies show a large variation in size, one can prepare a graph showing fraction of total kinetic energy of turbulence contained in eddies of different sizes. The graph would show the variation of energy as depicted in Fig. 3.6. This graph shows resemblance to the distribution of light energy as a function of wave number; hence Taylor called this a spectrum of turbulence. For spectrum of turbulence, the abscissa can be size or scale, wave length or frequency associated with the eddies.

3.4 TYPES OF TURBULENT FLOWS

Turbulent flows can be classified in two ways, either according to the manner in which they are produced, or according to the variation of statistical parameters of the turbulence. As first suggested by Karman, turbulence can be generated either when the fluid flows past solid surface or when the neighbouring parallel streams of the same fluid flow one over the other. In both these cases mean shear is present, and turbulence will be sustained. Because of the presence of mean shear, both the flows are called *turbulent shear flows*. Turbulent shear flows are further classified as *wall turbulent shear flows* and *free turbulent shear flows*, depending on whether turbulence is generated at the boundary or at the interface of two streams where shear is high.

Typical wall turbulent shear flows are shown in Fig. 3.7. Figure 3.7 (a) shows airflow over earth's surface. The turbulence is generated at the boundary which diffuses within the atmospheric boundary layer. Atmospheric boundary

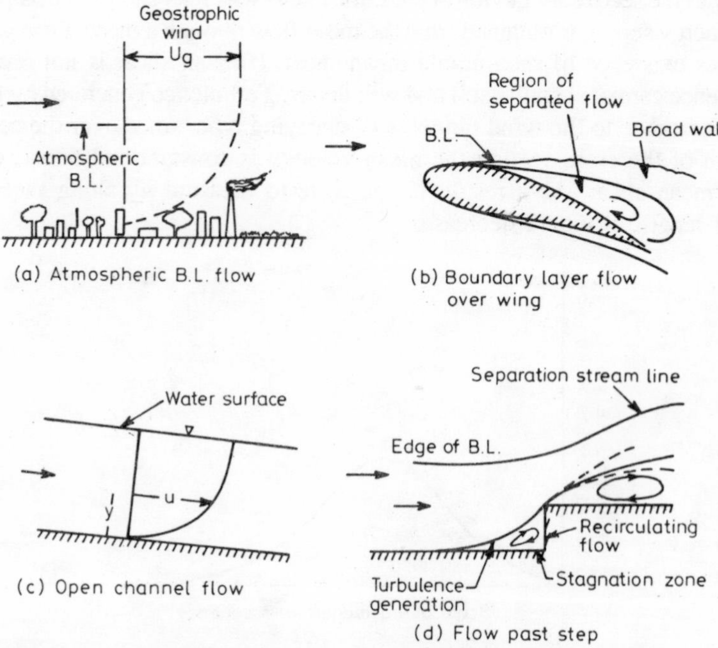

Fig. 3.7 Wall turbulent shear flows (External)

layer can be of the order of 300 m thick in open areas, whereas in urban areas it can be 500–600 m thick. Because of large roughnesses in the form of trees and buildings, the velocity shows greater reduction near the earth's surface in the case of boundary layer in urban areas than in the case of boundary layer over open areas. The velocity within atmospheric boundary layer can be expressed as

$$\frac{u}{U_o} = \left(\frac{y}{\delta}\right)^n \qquad \ldots (3.9)$$

where n changes from terrain to terrain. For urban areas n can be 0.30 to 0.45, whereas in open areas it can be of the order of 0.2. Largest eddies in atmospheric boundary layer flow can be very large in size since the flow is not bounded on the other side. Outside the boundary layer, the flow structure is governed by diurnal variation of large scale pressure changes which affect the flow. Within the atmospheric boundary layer which is turbulent, eddies of various sizes are present. Figure 3.7 (b) shows turbulent flow over the wing of an aeroplane. Similar situations arise in case of blades of fan, wind mill or propeller. In all these cases, turbulence is limited to boundary layer flow, and outside the flow is undisturbed. Further, the domain of wall turbulence increases in the downstream direction with the growth of the boundary layer. Separation of boundary layer flow causes a large region of separation at the edge of which turbulence is generated due to high shear. Diffusion of this turbulence produces a broad wake. Figure 3.7 (c) shows flow in an open channel. Turbulence is generated at the wall, but its growth is restricted by the water surface. For adequate channel length and steep slope, water waves known as roll waves can form and disturb the turbulence field completely at high Froude number values. The artificial roughnesses on the bed and sides, e.g. rock outcrops, trees, natural undulations on the botton known as dunes and bars, and irregularities in plan form e.g. meanderings, make the flow more complex. All these flows can be called *external flows* since they are restrained by the boundary only on one side.

One can get complicated external wall turbulent shear flows when a regular or irregular, two or three dimensional body is fixed on the boundary past which turbulent flow occurs, or when there is an abrupt change in boundary shape, e.g. a positive or negative step [see Fig. 3.7 (d)]. In all such cases, regions of stagnation and separation develop; the latter creates zone of high shear and, hence, intense turbulence is developed.

Wall turbulent shear flows which are enclosed on all sides are known as *internal flows*. Developing and developed flows in straight pipes, flows in conical contractions and diffusers, flows in sudden expansion and contraction, and flow in a single or series of bends fall under this category. These are discussed in detail by Sorran (3). In all such cases, when Reynolds number is sufficiently high, boundary geometry and curvature affect the flow characteristics in the vicinity of wall. However, because of diffusion, away from the boundary these effects decrease with increasing distance and turbulence would tend to forget its historical background. Figure 3.8 shows these examples of internal wall turbulent shear flows. Considerable work is done in understanding internal turbulent flows.

Turbulent flows are termed *free turbulent flows* when they are not confined by the solid walls. A few examples of free turbulent flows can be given. When two streams moving at different velocities in the same direction meet, a surface of discontinuity develops at which there is high shear which leads to turbulence generation and diffusion, see Fig. 3.9 (a). A free jet occurs when fluid is discharged from an orifice or nozzle or two dimensional opening. Consider a jet of fluid being discharged in an infinite fluid as shown in Fig. 3.9 (b). A small distance from its origin the jet becomes turbulent. Fluid from surrounding area is entrained by the jet, and the jet decelerates. In the process,

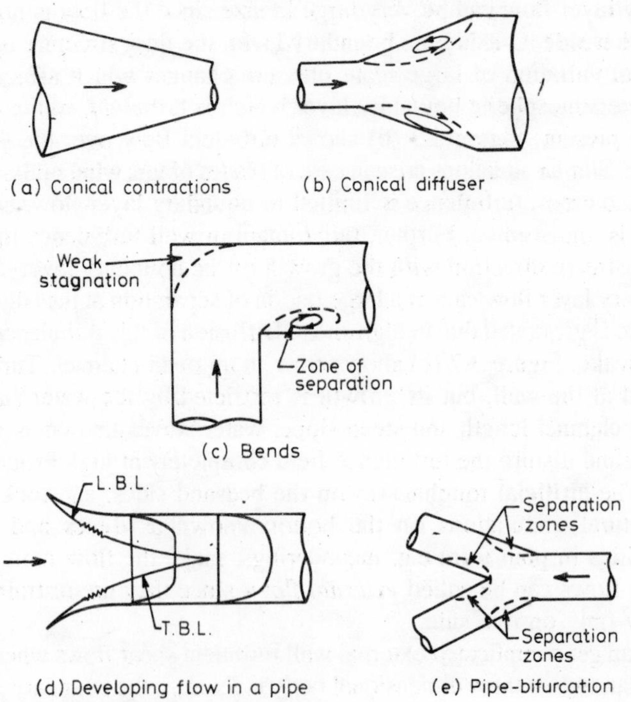

Fig. 3.8 Wall turbulent shear flows (Internal)

the jet dimensions increase continually in the downstream direction as shown in Fig. 3.9 (b). A wake is formed behind a solid body past which fluid flows, or when the solid body is dragged through a stationary fluid. Typical flow pattern in the wake behind a solid body kept in a flowing fluid is shown in Fig. 3.9 (c). It can be seen that the velocities in the wake are much smaller than those outside, and that wake dimensions go on increasing in the downstream direction.

Turbulence can also be classified according to the structure of turbulence and its spatial variation. *Pseudo turbulence* is a hypothetical or idealised turbulence created artificially, in which the flow shows a regular repeatable pattern; i.e. it has a distinct constant periodicity in time and space (1). Turbulence produced by cyclic oscillations of a grid in a liquid, as was done by Rouse (4) in 1938 in this experiments on sediment suspension, or by Bouvard and Petkovic (5) in 1973 when they studied the turbulence characteristics in the presence of material in suspension, would fall under this category. Pseudo turbulent flow fields are extremely useful for simulating turbulent flows, but they have their own limitations and need to be interpreted with caution.

Turbulence is said to be *isotropic* if the statistical average parameters have no preferred direction, which means that they remain invariant with respect to reflection or rotation of the axes; in other words, at a point $\overline{u'^2} = \overline{v'^2} = \overline{w'^2}$. Later it will be shown that, in isotropic turbulence, turbulent shear stresses are zero, i.e. there is no mean velocity gradient. True isotropy is hypothetical since no truly isotropic flow condition can be obtained. Turbulence behind grid in the wind tunnel is nearly isotropic. Because of its relative simplicity, isotropic turbulence is more amenable to theoretical treatment and, hence, has been extensively studied. Many results of such studies are applicable to actual turbulent flows. In all other situations, when mean velocity shows a gradient, turbulence will be *nonisotropic* or *anisotropic*. For example, boundary layer flow or flow in a pipe near the boundary is anisotoropic since $\overline{u'^2} \neq \overline{v'^2} \neq \overline{w'^2}$ at a point.

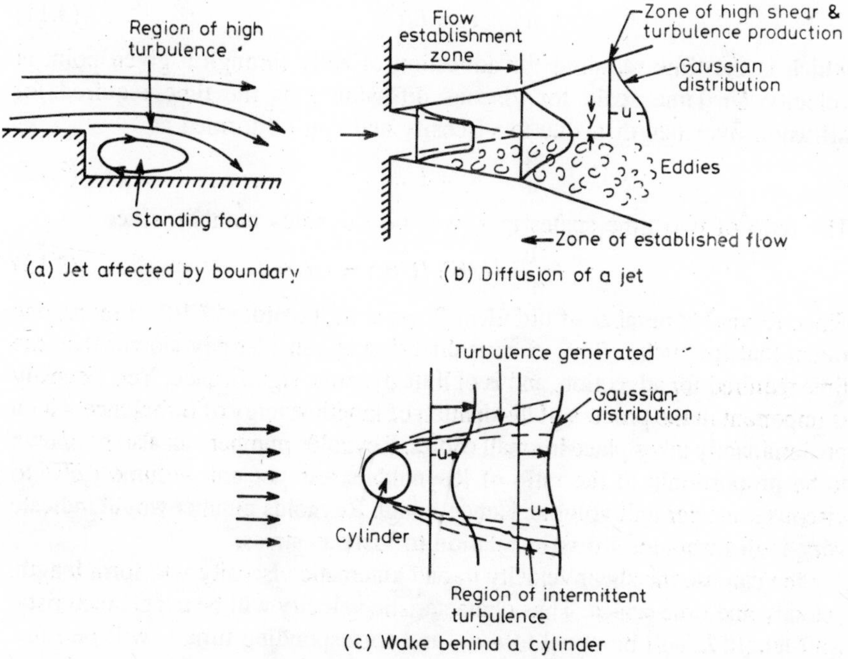

Fig. 3.9 Free turbulent shear flows

If the turbulence has the same qualitative structure in all parts of the flow, the turbulence is said to be *homogeneous*. Thus, if one considers two points a and b in the flow field, and if

$$(\overline{u'^2})_a = (\overline{u'^2})_b, \; (\overline{v'^2})_b = (\overline{v'^2})_b \text{ and } (\overline{w'^2})_a = (\overline{w'^2})_b \quad \ldots (3.10)$$

the turbulence will be homogeneous. Flow downstream of grid in a wind tunnel is isotropic homogeneous, since turbulence dies out rather slowly so that homogeneity can be assumed. It can be shown that homogeneous shear turbulence exists only if main motion has a constant velocity in a given direction, say along x axis, and a constant lateral velocity gradient throughout the flow field. Plane Couette turbulent flow falls under this category (6). In all other situations where the mean velocity shows a varying gradient, the turbulence is anisotropic and nonhomogeneous.

3.5 CHARACTERISTIC SCALES OF TURBULENCE AND ORDERS OF MAGNITUDES

Landahl and Mollo-Christensen (7) have discussed the concepts of various scales of turbulence that are commonly used along with physical significance of dimensionless parameters related to turbulence.

As discussed earlier, the turbulent flow consists of eddies of various sizes. The largest eddy can be taken to be of the size of flow itself, e.g. diameter D of pipe or depth of flow. Let it be denoted by L. If U is the characteristic velocity, one can determine the characteristic overall time scale t_a as

$$t_a = L/U \qquad \ldots (3.11)$$

which is the time required for advection of eddy through a given point at velocity U. Time scale for viscous diffusion t_v is the time required for diffusion over length L due to viscosity and can be formed as

$$t_v = L^2/\nu \qquad \ldots (3.12)$$

The ratio of two time scales t_v / t_a will be Reynolds number, since

$$t_v / t_a = (L^2/\nu)\ (U/L) = UL/\nu \qquad \ldots (3.13)$$

Since Reynolds number of turbulent flows is of the order of 10^5 or more, one infers that for such a flow viscous diffusion is considerably slower than the time required for advection, and is of little dynamic significance. Yet, viscosity is important in the process of dissipation of kinetic energy of turbulence which predominantly takes place in small eddies. Reynolds number can also be shown to be proportional to the ratio of Reynolds stress per unit volume $\rho\overline{u'v'}$ to viscous force per unit volume. Hence, a high Reynolds number would indicate very high turbulent stress in relation to viscous stress.

One can use the shear velocity u_* and kinematic viscosity ν to form length, velocity and time scales. Thus characteristic velocity will be u_*; characteristic wall length l_* will be equal to ν/u_*, and corresponding time t_* will be l_*/u_*, i.e. ν/u_*^2. These wall related characteristic scales are useful in description of mean flow quantities in the wall region of turbulent boundary layer. Thus, turbulent fluctuations near the boundary, as well as the length scale of energy dissipating eddies, are multiples of u_*. For example it, can be seen that in a pipe, the maximum turbulent fluctuations are about 1 to 3 times u_*.

In the case of stratified flows, variation of $\bar{u}$ and ρ with y can be used to obtain the characteristic scales. Thus, mean flow period

$$t_r = (d\bar{u}/dy)^{-1} \qquad \ldots (3.14)$$

can be compared with intrinsic oscillation period of the fluid

$$t_s = (-\,g/\rho.\ d\rho/dy)^{-1/2} \qquad \ldots (3.15)$$

and N, which is equal to t_s^{-1}, is known as the Brunt-Vaisala frequency. Richardson number R_i is equal to the square of the two time scales

$$Ri = (t_r / t_s)^2 = -\frac{-(g/\rho)\ (d\rho/dy)}{(du/dy)^2} \qquad \ldots (3.16)$$

Richardson number is very often used in the study of turbulence effects in stratified flows. If Ri is greater than 1.0, the flow has strong static stability. If Ri is less than 0.25, shear induced instability and turbulence may set in. It may be noted that bar over u and ρ is omitted in the above equations for convenience.

To study the small size eddies which are responsible for energy dissipation in a turbulent flow, one can use viscous dissipation rate per unit mass of fluid, ε and kinematic viscosity ν. Combining these two, Kolmogorov obtained the length scale l_k and velocity scale v_k as

$$l_k = (\nu^3/\varepsilon)^{1/4} \qquad \ldots (3.17)$$

and

$$v_k = (\nu\varepsilon)^{1/4} \qquad \ldots (3.18)$$

These can be interpreted as the size and velocity of small energy dissipating eddies. For these eddies the characteristic Reynolds number will be

$$\frac{v_k\, l_k}{\nu} = \left(\frac{\nu\varepsilon}{1}\ \frac{\nu^3}{\varepsilon}\right)^{1/4} \frac{1}{\nu} = 1$$

To get an idea about the order of magnitude of Kolmogorov's length and velocity scales, consider a 75 W household mixer used for mixing one littre of liquid.

$$\varepsilon = 75\ \text{W/1kg} = 75\ \text{W/kg}$$

$$\nu = 10^{-6}\ m^2/s$$

$$l_k = (10^{-18}/75)^{1/4} = 0.107 \times 10^{-4}\ \text{m or } 0.0107\ \text{mm}$$

$$v_k = (10^{-6} \times 75)^{1/4} = 0.093\ \text{m/s or } 9.3\ \text{cm/s}$$

In atmospheric boundary layers one uses a length scale L_{Mo} called Monin-Obukov length scale defined as

$$L_{Mo} = (-\,\rho_0\, u_*^3\, C_p/K\alpha q g)$$

Here ρ_o is the mean surface density, K is Karman constant, α is thermal expansion coefficient, C_p is the specific heat at constant pressure, and q is the vertical heat flux which is taken as positive in the upward direction. Monin-Obukov length gives a measure of the relative contribution of energy supplied to turbulence by buoyancy forces to that supplied by heat generated from friction. For large values of L_{Mo}, the vertical heat flux has a small influence on the structure of atmospheric boundary layer near the surface.

REFERENCES

1. Hinze, J.O. *Turbulence: An Introduction to Its Mechanism and Theory.* McGraw Hill Series in Mechanical Engineering, New York, 1959 Chapt. 1.
2. Yoshida, S. and S. Yagi. Development of Laser Dopler Anemometry (LDA) for River Flow Measurement. Jour. of Hydroscience and Hyd. Engg. (Japan), Vol. 5, No. 1, July 1987.
3. Sorran, G. *Fluid Mechanics of Internal Flows.* Elsevier Publishing Company, Inc. 1967.
4. Rouse, H. Experiments on Mechanics of Sediment Suspensions. Proc. 5th Congress of App. Mechanics, Cambridge, 1938.
5. Bouvard, M. and S. Petkovic. Modification des Characteristiques d'une Turbulence Sons l'influence de Particules Solides en Suspensions. La Houille Blanche, No. 1, 1973.
6. Bradshaw, P. *An Introduction to Turbulence and its Measurement.* Pergamon Press, Oxford (U.K.) First Ed. 1971.
7. Landahl, M.J. and E. Mollo-Christensen. *Turbulence and Random Processes in Fluid Mechanics.* Cambridge University Press, Cambridge, First Ed. 1976.

CHAPTER IV

Equations of Motion

4.1 INTRODUCTION

Flow problems involving incompressible fluid can be analysed using three basic principles, namely, conservation of mass principle, which is known as the continuity equation, dynamic equation, or its other form conservation of momentum principle leading to momentum equation, and the conservation of energy principle which leads to the energy equation. These equations are developed and discussed in this chapter for the case of turbulent flow.

4.2 CONTINUITY EQUATION

Principle of conservation of mass states that the net rate of mass inflow in a given volume in space must be equal to the rate of increase of mass within the volume. In cartesian co-ordinate system, it can be written as

$$\frac{\partial \rho}{\partial t} + \frac{\partial(\rho u)}{\partial x} + \frac{\partial(\rho v)}{\partial y} + \frac{\partial(\rho w)}{\partial z} = 0 \qquad \ldots (4.1)$$

If fluid is incompressible (homogeneous or nonhomogeneous), ρ is constant with respect to time, and mass of a unit volume does not change as it is convected. Hence Equation 4.1 reduces to

$$\frac{\partial u}{\partial x} + \frac{\partial v}{\partial y} + \frac{\partial w}{\partial z} = 0 \qquad \ldots (4.2)$$

In the analysis of turbulent flow, Eq. 4.2 is considered to be valid for instantaneous velocity components u,v, and w. Substituting for u,v,w

$$u = \bar{u} + u'$$

$$v = \bar{v} + v'$$

$$w = \bar{w} + w'$$

one gets

$$\frac{\partial(\bar{u} + u')}{\partial x} + \frac{\partial(\bar{v} + v')}{\partial y} + \frac{\partial(\bar{w} + w')}{\partial z} = 0 \qquad \ldots (4.3)$$

If Reynolds rules of averages are applied and time average of the above equation is taken, one has

$$\frac{\partial \bar{u}}{\partial x} + \frac{\partial \bar{v}}{\partial y} + \frac{\partial \bar{w}}{\partial z} = 0 \qquad \ldots (4.4)$$

indicating that mean velocity components satisfy continuity equation. Subtraction of Eq. 4.4 from Eq. 4.3 shows that

$$\frac{\partial u'}{\partial x} + \frac{\partial v'}{\partial y} + \frac{\partial w'}{\partial z} = 0 \quad \ldots (4.5)$$

indicating that fluctuating velocity components u', v', w' also satisfy the continuity equation. In tensor notation the x, y, z coordinates are designated as x_1, x_2, x_3 and the corresponding velocity components as u_1, u_2, u_3, etc. In tensor notation Eqs. 4.3, 4.4 and 4.5 are written as

$$\frac{\partial(\bar{u}_i + u'_i)}{\partial x_i} = 0 \quad \ldots (4.3)$$

$$\frac{\partial \bar{u}_i}{\partial x_i} = 0 \quad \ldots (4.4)$$

$$\frac{\partial u'_i}{\partial x_i} = 0 \quad \ldots (4.5)$$

If in a term the subscript is repeated as in Eqs. 4.3, 4.4 and 4.5, the subscript is to be given all possible values, i.e. 1, 2, 3 and terms added.

For incompressible fluid, the continuity equation in cylindrical polar co-ordinate system is

$$\frac{\partial V_x}{\partial x} + \frac{\partial V_r}{\partial r} + \frac{V_r}{r} + \frac{1}{r}\frac{\partial V_\theta}{\partial \theta} = 0 \quad \ldots (4.6)$$

Substituting $V_x = \bar{V}_x + V'_x$, $V_r = \bar{V}_r + V'_r$ and $V_\theta = \bar{V}_\theta + V'_\theta$ in Eq. 4.6, and taking time-average, one gets

$$\frac{\partial \bar{V}_x}{\partial x} + \frac{\partial \bar{V}_r}{\partial r} + \frac{\bar{V}_r}{r} + \frac{1}{r}\frac{\partial \bar{V}_\theta}{\partial \theta} = 0 \quad \ldots (4.7)$$

Subtraction of Eq. 4.7 from the equation involving instantaneous velocity components yields

$$\frac{\partial V'_x}{\partial x} + \frac{\partial V'_r}{\partial r} + \frac{V'_r}{r} + \frac{1}{r}\frac{\partial V'_\theta}{\partial \theta} = 0 \quad \ldots (4.8)$$

4.3 REYNOLDS EQUATIONS OF MOTION (1)

Dynamic equation of motion is obtained by applying Newton's Second Law of Motion to a unit volume of fluid. For laminar flow of an incompressible fluid, Navier-Stokes equations are the dynamic equations of motion. In cartesian coordinate system, they are given in Chapter 1; in tensor notation, Navier-Stokes equations can be written as

$$\frac{\partial u_i}{\partial t} + u_j \frac{\partial u_i}{\partial x_j} = X_i - \frac{1}{\rho}\frac{\partial p}{\partial x_i} + \nu \frac{\partial^2 u_i}{\partial x_j \partial x_j} \quad \ldots (4.9)$$

where X_i is now component of gravity force per unit mass in *i*-direction. It is assumed that Eq. 4.9 is applicable to turbulent motion of incompressible fluid of constant viscosity. Hence, one can substitute $u_i = \bar{u}_i + u'_i$ and $u_j = \bar{u}_j + u'_i$ in Eq. 4.9 to obtain,

$$\frac{\partial(\bar{u}_i + u'_i)}{\partial t} + (\bar{u}_j + u'_j)\frac{\partial(\bar{u}_i + u'_i)}{\partial x_j} = X_i - \frac{1}{\rho}\frac{\partial(\bar{p} + p')}{\partial x_i} + \nu\frac{\partial^2(\bar{u}_i + u'_i)}{\partial x_j \partial x_j}$$

When Reynolds rules of averages are applied, the above equation reduces to

$$\frac{\partial \bar{u}_i}{\partial t} + \frac{\partial \overline{u'_i}}{\partial t} + \bar{u}_j\frac{\partial \bar{u}_i}{\partial x_j} + \overline{u'_j}\frac{\partial \bar{u}_i}{\partial x_j} + \bar{u}_j\frac{\partial \overline{u'_i}}{\partial x_j} + \overline{u'_j\frac{\partial u'_i}{\partial x_j}}$$

$$= \bar{X}_i - \frac{1}{\rho}\frac{\partial \bar{p}}{\partial x_i} - \frac{1}{\rho}\frac{\partial \overline{p'}}{\partial x_i} + \nu\frac{\partial^2 \bar{u}_i}{\partial x_j \partial x_j} + \nu\frac{\partial^2 \overline{u'_i}}{\partial x_j \partial x_j} \quad \ldots (4.10)$$

The second, fourth and fifth terms on the left hand side, and third and fifth terms on the right hand side are zero. However, the sixth term on left hand side of the equation can be written as

$$u'_j\frac{\partial u'_i}{\partial x_j} = \frac{\partial(u'_i u'_j)}{\partial x_j} - u'_i\frac{\partial u'_j}{\partial x_j} \quad \ldots (4.11)$$

and the second term on right hand side of Eq. 4.11 is zero because of continuity. Hence,

$$\overline{u'_j\frac{\partial u'}{\partial x_j}} = \frac{\partial(\overline{u'_i u'_j})}{\partial x_j}$$

Hence, Eq. 4.10 reduces to

$$\frac{\partial \bar{u}_i}{\partial t} + \bar{u}_j\frac{\partial \bar{u}_i}{\partial x_j} = X_i - \frac{1}{\rho}\frac{\partial \bar{p}}{\partial x_i} + \left(\nu\frac{\partial^2 \bar{u}_i}{\partial x_j \partial x_j} - \frac{\partial(\overline{u'_i u'_j})}{\partial x_j}\right) \quad \text{(a)} \ldots (4.12)$$

$$\text{or,}\quad \rho\left(\frac{\partial \bar{u}_i}{\partial t} + \bar{u}_j\frac{\partial \bar{u}_i}{\partial x_j}\right) = X_i - \frac{\partial \bar{p}}{\partial x_i} + \left(\mu\frac{\partial^2 \bar{u}_i}{\partial x_j \partial x_j} - \frac{\partial(\rho\overline{u'_i u'_j})}{\partial x_j}\right) \quad \text{(b)}$$

It may be noted that in Eq. 4.12 (b), X_i is the body force per unit volume. Equation 4.12 is a short hand description of three equations commonly known as *Reynolds equations of motion.* In extended form these are

$$\rho\left(\frac{\partial \bar{u}}{\partial t} + \bar{u}\frac{\partial \bar{u}}{\partial x} + \bar{v}\frac{\partial \bar{u}}{\partial y} + \bar{w}\frac{\partial \bar{u}}{\partial z}\right)$$

$$= X - \frac{\partial \bar{p}}{\partial x} + \mu\left(\frac{\partial^2 \bar{u}}{\partial x^2} + \frac{\partial^2 \bar{u}}{\partial y^2} + \frac{\partial^2 \bar{u}}{\partial z^2}\right) - \left(\frac{\partial \rho\overline{u'^2}}{\partial x} + \frac{\partial \rho\overline{u'v'}}{\partial y} + \frac{\partial \rho\overline{u'w'}}{\partial z}\right)$$

$$\rho\left(\frac{\partial \bar{v}}{\partial t} + \bar{u}\frac{\partial \bar{v}}{\partial x} + \bar{v}\frac{\partial \bar{v}}{\partial y} + \bar{w}\frac{\partial \bar{v}}{\partial z}\right)$$

$$= Y - \frac{\partial \bar{p}}{\partial y} + \mu\left(\frac{\partial^2 \bar{v}}{\partial x^2} + \frac{\partial^2 \bar{v}}{\partial y^2} + \frac{\partial^2 \bar{v}}{\partial z^2}\right) - \left(\frac{\partial \rho\overline{u'v'}}{\partial x} + \frac{\partial \rho\overline{v'^2}}{\partial y} + \frac{\partial \rho\overline{v'w'}}{\partial z}\right)$$

$$\rho\left(\frac{\partial \bar{w}}{\partial t}+\bar{u}\,\frac{\partial \bar{w}}{\partial x}+\bar{v}\,\frac{\partial \bar{w}}{\partial y}+\bar{w}\,\frac{\partial \bar{w}}{\partial z}\right)$$

$$= Z-\frac{\partial \bar{p}}{\partial z}+\mu\left(\frac{\partial^2 \bar{w}}{\partial x^2}+\frac{\partial^2 \bar{w}}{\partial y^2}+\frac{\partial^2 \bar{w}}{\partial z^2}\right)-\left(\frac{\partial \rho\overline{u'w'}}{\partial x}+\frac{\partial \rho\overline{v'w'}}{\partial y}+\frac{\partial \rho\overline{w'^2}}{\partial z}\right) \quad (4.13)$$

It should be noted that X_i in Eq. 4.12 (a) denotes components of body force per unit mass, whereas in Eqs. 4.12 (b) and 4.13, they represents body force per unit volume. Reynolds equations of motion bear considerable similarity with Navier-Stokes equations for laminar flow. In fact, the two terms on the left hand side and the first three terms on the right hand side in Reynolds Eqs. 4.12 and 4.13 and in N.S. equations are identical, except that the velocity and pressure terms used in Reynolds equations are now time averaged. The fourth term on the right hand side of Reynolds equations comprising three terms each, and a total of nine terms in the three equations is new, in that it involves velocity fluctuations only. In fact these fluctuations have caused nine stresses constituting what is commonly known as the Reynolds stress tensor

$$\begin{bmatrix} -\rho\overline{u'^2} & -\rho\overline{v'u'} & -\rho\overline{w'u'} \\ -\rho\overline{u'v'} & -\rho\overline{v'^2} & -\overline{w'v'} \\ -\rho\overline{u'w'} & -\rho\overline{v'w'} & -\rho\overline{w'^2} \end{bmatrix} \quad \ldots (3.8)$$

Out of these, three are normal stresses, viz. $-\rho\overline{u'^2}$, $-\rho\overline{v'^2}$, and $-\rho\overline{w'^2}$, while the other six are shear stresses. Further, it is apparent that $-\rho\overline{u'v'} = -\rho\overline{v'u'}$, $-\rho\overline{v'w'} = -\rho\overline{w'v'}$ and $-\rho\overline{u'w'} = -\rho\overline{w'u'}$. Thus, there are three normal and three shear stresses. Reynolds stresses are often called apparent stresses. It may be pointed out that Reynolds stresses arise from inertia terms, but have been grouped together, for convenience, with viscous stresses in Eq. 4.12, and their roles are different. Further, Reynolds stresses can be much larger than viscous stresses. Viscous terms are responsible for the eventual dissipation of energy of mean as well as turbulent motion.

One can get a better insight into occurrence of Reynolds stresses through the following explanation given by Schilichting (1). Consider an elementary area dA perpendicular to x axis. If u, v, w are the velocity components at the centre of area dA,

$\rho u dA \cdot dt$ = Mass of fluid passing through dA in x direction in time dt

Momentum flux in x direction, $dM_x = \rho u^2\, dA \cdot dt$

Momentum flux in y direction, $dM_y = \rho uv\, dA \cdot dt$

Momentum flux in z direction, $dM_z = \rho\, uw\, dA \cdot dt$

Assume ρ to be constant and substitute $u = \bar{u} + u'$, $v = \bar{v} + v'$, and $w = \bar{w} + w'$ in the above equations for dM_x, dM_y and dM_z. When time average is taken, one gets

$$\overline{dM_x} = \rho\overline{(\bar{u} + u'^2)\, dA\, dt} = \rho(\bar{u}^2 + \overline{u'^2})\, dA \cdot dt$$

$$\overline{dM_y} = \rho\overline{(\bar{u} + u')\,(\bar{v} + v')\, dA\, dt} = \rho(\overline{uv} + \overline{u'v'})\, dA \cdot dt$$

$$\overline{dM_z} = \rho\overline{(\bar{u} + u')\,(\bar{w} + w')\, dA\, dt} = \rho(\overline{uw} + \overline{u'w'})\, dA \cdot dt$$

Since the momentum flux per unit area per unit time is equal to the stress acting on the fluids in the opposite direction, one gets the following stresses

$-\rho(\overline{u}^2 + \overline{u'^2}),\ -\rho(\overline{uv} + \overline{u'v'}),\ -\rho(\overline{uw} + \overline{u'w'})$

It can be seen that the second part in each of the above stresses is Reynolds stress component.

It can be seen that equations of continuity along with Reynolds equations of motion contain ten unknowns, namely $\overline{u}, \overline{v}, \overline{w}, \overline{u'^2}, \overline{v'^2}, \overline{w'^2}, \overline{u'v'}, \overline{u'w'}, \overline{v'w'}$, and $\overline{p}$ and four equations. Therefore, there are more unknowns than the number of equations, and unless some assumptions are made about these variables, these equations cannot be solved even in simplified form. Further, they are highly nonlinear. The phenomenological theories discussed in Chapter VI attempt to express some of these variables in terms of known quantities.

Reynolds equations in cylindrical polar co-ordinate system can be obtained in a manner similar to that used in obtaining corresponding equations in cartesian co-ordinate system. These are listed below (2):

$$\rho\left(\frac{\partial \overline{V}_x}{\partial t} + \overline{V}_x\frac{\partial \overline{V}_x}{\partial x} + \overline{V}_r\frac{\partial \overline{V}_x}{\partial r} + \frac{\overline{V}_\theta}{r}\frac{\partial \overline{V}_x}{\partial \theta}\right)$$

$$= F_x - \frac{\partial \overline{p}}{\partial x} + \mu\nabla^2\overline{V}_x - \left(\frac{\partial \rho \overline{V_x'^2}}{\partial x} + \frac{1}{r}\frac{\partial(\rho\overline{V_x'V_r'})}{\partial r} + \frac{1}{r}\frac{\partial(\rho\overline{V_\theta'V_x'})}{\partial \theta}\right)$$

$$\rho\left(\frac{\partial \overline{V}_r}{\partial t} + \overline{V}_x\frac{\partial \overline{V}_r}{\partial x} + \overline{V}_r\frac{\partial \overline{V}_r}{\partial r} + \frac{\overline{V}_\theta}{r}\frac{\partial \overline{V}_r}{\partial \theta} - \frac{\overline{V}_\theta^2}{r}\right) = F_r - \frac{\partial \overline{p}}{\partial r} + \mu\left(\nabla^2\overline{V}_r - \frac{\overline{V}_r}{r^2}\right.$$

$$\left. - \frac{2}{r^2}\frac{\partial \overline{V}_\theta}{\partial \theta}\right) - \left(\frac{\partial(\rho\overline{V_x'V_r'})}{\partial x} + \frac{1}{r}\frac{\partial(r\rho\overline{V_r'^2})}{\partial r} + \frac{1}{r}\frac{\partial(\rho\overline{V_r'V_\theta'})}{\partial \theta} - \frac{\rho\overline{V_\theta'^2}}{r}\right)$$

$$\rho\left(\frac{\partial \overline{V}_\theta}{\partial t} + \overline{V}_x\frac{\partial \overline{V}_\theta}{\partial x} + \overline{V}_r\frac{\partial \overline{V}_\theta}{\partial r} + \frac{\overline{V}_\theta}{r}\frac{\partial \overline{V}_\theta}{\partial \theta} - \frac{\overline{V}_r\overline{V}_\theta}{r}\right)$$

$$= F_\theta - \frac{1}{r}\frac{\partial \overline{p}}{\partial \theta} + \mu\left(\nabla^2\overline{V}_\theta + \frac{2}{r^2}\frac{\partial \overline{V}_\theta}{\partial \theta} - \frac{\overline{V}_\theta}{r^2}\right)$$

$$-\left(\frac{\partial(\rho\overline{V_x'V_\theta'})}{\partial x} + \frac{\partial \rho\overline{V_r'V_\theta'}}{\partial r} + \frac{1}{r}\frac{\partial \overline{V_\theta'^2}}{\partial \theta} - \frac{2\rho\overline{V_r'V_\theta'}}{r}\right)$$

Where $$\nabla^2 = \left(\frac{\partial^2}{\partial x^2} + \frac{\partial^2}{\partial r^2} + \frac{1}{r}\frac{\partial}{\partial r} + \frac{1}{r^2}\frac{\partial^2}{\partial \theta^2}\right)$$

Since all the turbulent fluctuations must die out at the boundary, the boundary conditions to be satisfied at the wall (for cartesian co-ordinate system) are $\overline{u} = \overline{v} = \overline{w} = 0$, $\overline{u'^2} = \overline{v'^2} = \overline{w'^2} = 0$, $\overline{u'v'} = \overline{u'w'} = \overline{v'w'} = 0$ at $y = 0$. Similar conditions can be written in cylindrical polar co-ordinate system.

4.4 FLOW BETWEEN PARALLEL PLATES AND IN A PIPE

To illustrate the complexities involved in the solution of Reynolds equations of motion, two cases will be considered, namely steady uniform turbulent flow between parallel plates and in a circular pipe.

Consider two dimensional steady uniform turbulent flow in x direction between parallel plates kept at a distance h apart. Assume the fluid to be incompressible. Hence,

(i) ρ = constant;

(ii) $\frac{\partial}{\partial t} = 0$ because the flow is steady;

(iii) $\frac{\partial}{\partial z} = 0$ because the flow is two-dimensional;

(iv) $\bar{v} = \bar{w} = 0$ since mean flow is in x direction and plates are parallel.

Therefore, continuity equation gives $\frac{\partial \bar{u}}{\partial x} = 0$, which on integration yields

$$\bar{u} = \bar{u}\,(y) \qquad \ldots (4.15)$$

The three Reynolds equations (Eq. 4.13) reduce to the following differential equations.

$$\left.\begin{aligned} \frac{\partial \bar{p}}{\partial x} &= \mu\left(\frac{\partial^2 \bar{u}}{\partial y^2}\right) - \frac{\partial(\rho\overline{u'v'})}{\partial y} \quad &(a)\\ \frac{\partial \bar{p}}{\partial y} &= -\frac{\partial(\rho\overline{v'^2})}{\partial y} \quad &(b)\\ 0 &= -\frac{\partial(\rho\overline{v'w'})}{\partial y} \quad &(c) \end{aligned}\right] \qquad \ldots (4.16)$$

Integration of Eq. 4.16 (c) with respect to y yields

$\overline{u'w'}$ = constant over the depth h.

However, $\overline{v'w'} = 0$ at $y = 0$ and h. Hence, $\overline{v'w'} = 0$ over the whole depth, which means that there is no correlation between v' and w'. Integration of Eq. 4.16 (b) yields

$$\bar{p} + \rho\overline{v'^2} = f\,(x) \qquad \ldots (4.17)$$

Since $\overline{v'^2} = 0$ at the walls, and since $\bar{p} + \rho\overline{v'^2}$ must remain constant for given x, it follows that $\bar{p}$ must be maximum at $y = 0$ and h. Further $\bar{p}$ must vary with y because $\overline{v'^2}$ will vary with y. This result is unlike in the case of laminar flow where p is constant across h. However, variation of $\bar{p}$ with y in turbulent flow is very small, and is usually neglected. Equation 4.15 indicates $\bar{u} = \bar{u}\,(y)$ whereas Eq. 4.16 (a) indicates $\bar{u}$ to be function of x and y if $\frac{\partial \bar{p}}{\partial x}$ depends on x. This shows that $\frac{\partial \bar{p}}{\partial x}$ must be independent of x and must be constant. Integration of Eq. 4.16 (a) with respect to y, taking $\partial\bar{p}/\partial x$ as constant, yields,

$$\left(\frac{\partial \bar{p}}{\partial x}\right) y = \mu\left(\frac{\partial \bar{u}}{\partial y}\right) - \rho\overline{u'v'} + c_1 \qquad \ldots (4.18)$$

When $y = 0$, $\overline{u'v'} = 0$ and $\mu \left(\frac{\partial \bar{u}}{\partial y} \right)_{y=0} = \tau_o$ the boundary shear.

Hence $c_1 = -\tau_o$. Further, at $y = h/2$, $\left(\frac{\partial \bar{u}}{\partial y} \right) = 0$ and $\rho \overline{u'v'} = 0$

$$\therefore \quad \left(\frac{\partial \bar{p}}{\partial x} \right) \frac{h}{2} = -\tau_o \quad \text{or} \quad \frac{\partial \bar{p}}{\partial x} = -\frac{2\tau_o}{h}$$

Equation 4.18 then yields

$$-2\tau_o \frac{y}{h} = \mu \left(\frac{\partial \bar{u}}{\partial y} \right) - \rho \overline{u'v'} - \tau_o$$

or

$$\tau_o \left(1 - \frac{2y}{h} \right) = \mu \left(\frac{\partial \bar{u}}{\partial y} \right) - \rho \overline{u'v'} \qquad \ldots (4.19)$$

In order to obtain variation of $\bar{u}$ with y, one must have knowledge of variation of $\overline{u'v'}$ with respect to y and, of course, τ_o. This, therefore, needs further assumptions.

In a similar manner one can simplify Reynolds equations of motion in cylindrical polar co-ordinate system, and study steady turbulent flow of an incompressible fuild in a pipe (2). For this case

ρ = constant, $\frac{\partial}{\partial t} = 0$, $\frac{\partial}{\partial x} = 0$, since flow is uniform, and $\frac{\partial}{\partial \theta} = 0$ because of symmetry. It can also be shown that $\partial \bar{p}/\partial x$ is constant. Also for fully developed turbulent flow $\bar{V}_r = \bar{V}_\theta = 0$. Hence, continuity equation yields

$$\frac{\partial \bar{V}_x}{\partial x} = 0 \text{ or } \bar{V}_x = f(r)$$

Reynolds equations then reduce to

$$\left.\begin{aligned}
0 &= -\frac{\partial \bar{p}}{\partial x} + \mu \left(\frac{\partial^2 \bar{V}_x}{\partial r^2} + \frac{1}{r} \frac{\partial \bar{V}_x}{\partial r} \right) - \frac{1}{r} \frac{\partial (r\rho \overline{V'_x V'_r})}{\partial r} && \text{(a)} \\
0 &= -\frac{\partial \bar{p}}{\partial r} - \left(\frac{1}{r} \frac{\partial (r\rho \overline{V_r'^2})}{\partial r} - \frac{\rho \overline{V_\theta'^2}}{r} \right) && \text{(b)} \\
0 &= -\frac{\partial (\rho \overline{V'_r V'_\theta})}{\partial r} + 2 \frac{\rho \overline{V'_r V'_\theta}}{r} && \text{(c)}
\end{aligned}\right\} \cdots (4.20)$$

Integration of Eq. 4.20 (c) indicates that $\overline{V'_r V'_\theta} = 0$ everywhere since $\overline{V'_r V'_\theta} = 0$ at the wall for all x values. Equation 4.20 (a) can be written as

$$\frac{\partial \bar{p}}{\partial x} = -\frac{\rho}{r} \frac{\partial (r\overline{V'_x V'_r})}{\partial r} + \frac{\mu}{r} \frac{\partial}{\partial r} \left(r \frac{\partial \bar{V}_x}{\partial r} \right)$$

or

$$\frac{\partial \bar{p}}{\partial x} = -\frac{\rho}{r} \frac{\partial}{\partial r} \left(r(\overline{V'_x V'_r}) - \nu \frac{\partial \bar{V}_x}{\partial r} \right) \qquad \ldots (4.21)$$

$$\text{Also } \frac{\partial \bar{p}}{\partial r} = \rho\left(-\frac{1}{r}\frac{\partial}{\partial r}(r\overline{V_r'^2}) + \frac{\overline{V_\theta'^2}}{r}\right)$$

Differentiating the above equation with respect to x gives $\dfrac{\partial^2 \bar{p}}{\partial r \partial x} = 0$

which indicates $\dfrac{\partial \bar{p}}{\partial x}$ to be independent of r. Integration of Eq. 4.21 yields

$$\frac{r^2}{2}\frac{\partial \bar{p}}{\partial x} = \rho\left[-r\left(\overline{V_x'V_r'} - \nu\frac{\partial \bar{V}_x}{\partial r}\right) + A(x)\right] \quad \ldots (4.22)$$

and integration of equation for $\partial \bar{p}/\partial r$ yields

$$\frac{\bar{p}}{\rho} + \overline{V_r'^2} - \int_r^R \frac{\overline{V_\theta'^2} - \overline{V_r'^2}}{r}\, dr = B(x) \quad \ldots (4.23)$$

where R is radius of the pipe. The boundary conditions at $r = 0$ are $\dfrac{\partial \bar{V}_x}{\partial r} = 0$,

$\overline{V_x' V_r'} = 0$; therefore, $A(x) = 0$.

Further, for steady uniform flow, the pressure force will be balanced by the frictional force at the boundary. Hence,

$$-\frac{\partial \bar{p}}{\partial x} \times \pi R^2 \times \delta x = 2\pi \times R \times \delta x \times \tau_o$$

$$\frac{\partial p}{\partial x} = \frac{-2\tau_o}{R} = \frac{-2\rho u_*^2}{R}$$

where u_* is the shear velocity and τ_o is the boundary shear;

or
$$\frac{1}{\rho}\frac{\partial \bar{p}}{\partial x} = -\frac{2u_*^2}{R} \quad \ldots (4.24)$$

Integration of Eq. 4.24 yields

$$\frac{\bar{p}}{\rho} = -\frac{-2\,u_*^2}{R}x + C(r) + C_1$$

where $C(r)$ is a function of r. It can be seen that at $r = 0$, $C(r) = 0$. If $\bar{p} = \bar{p}_r$ at $x = 0$ and $r = R$

$$\frac{\bar{p}_r}{\rho} = C_1$$

and
$$\frac{\bar{p}}{\rho} = -\frac{2\,u_*^2}{R}x + C(r) + \frac{\bar{p}_r}{\rho} \quad \ldots (4.25)$$

Comparison of Eqs. 4.23 and 4.25 allows evaluation of $B(x)$, and then Eqs. 4.22 and 4.23 can be written in the final form as

$$\overline{V_x'V_r'} = \nu \frac{\partial \overline{V}_x}{\partial r} + \frac{ru_*^2}{R} \qquad \ldots (4.26)$$

$$\overline{V_r'^2} + \int_r^R \frac{(\overline{V_r'^2} - \overline{V_\theta'^2})}{r} dr = \frac{\overline{p}_r}{\rho} - \frac{\overline{p}}{\rho} - \frac{2u_*^2}{R} x \qquad \ldots (4.27)$$

This is all the information that one can get from Reynolds equations as far as the variation of $\overline{V}_x$ with r is concerned, as well as the variation of $\overline{p}$ with r. Again, as in the case of flow between parallel plates, Eq. 4.26 indicates strong dependence of $\overline{V}_x$ on variation of $\overline{V_x'V_r'}$ with r. On the other hand, turbulence terms cause only a very small departure from constant pressure over the cross-section.

4.5 LINEAR MOMENTUM EQUATION (3, 4)

Consider Reynolds equations of motion in the cartesian co-ordinate system written in tensor notation

$$\frac{\partial \overline{u}_i}{\partial t} + \overline{u}_j \frac{\partial \overline{u}_i}{\partial x_j} + \frac{\partial(\overline{u_i'u_j'})}{\partial x_j} = X_i - \frac{1}{\rho}\frac{\partial \overline{p}}{\partial x_i} + \frac{\mu}{\rho}\frac{\partial^2 \overline{u}_i}{\partial x_j \partial x_j}$$

The second term on the left hand side can be written as $\frac{\partial(\overline{u}_i\overline{u}_j)}{\partial x_j}$ since $\overline{u}_i \frac{\partial \overline{u}_j}{\partial x_j} = 0$ because of continuity equation. Multiply each term of the above equation by $d\forall$ and integrate each term over a volume $\forall$ through which the fluid moves. Hence, one gets

$$\int_\forall \frac{\partial \overline{u}_i}{\partial t} d\forall + \int_\forall \frac{\partial(\overline{u}_i\overline{u}_j)}{\partial x_j} d\forall + \int_\forall \frac{\partial(\overline{u_x'u_r'})}{\partial x_j} d\forall$$

$$= \int_\forall X_i \, d\forall - \frac{1}{\rho}\int_\forall \frac{\partial \overline{p}}{\partial x_i} d\forall + \frac{\mu}{\rho}\int_\forall \frac{\partial^2 \overline{u}_i}{\partial x_j \partial x_j} d\forall \qquad \ldots (4.28)$$

Volume integral can be converted into surface integral by Gaussian rule which states that

$$\int_\forall \frac{\partial(\ )}{\partial x_j} d\forall = \int_S (\ ) \frac{\partial x_j}{\partial n} dS \qquad \ldots (4.29)$$

where S is the surface area over which integration is performed, and which encloses the volume $\forall$, n is the outward normal to the surface, and ( ) is the quantity of which the volume intregral is to be evaluated. Hence Eq. 4.28, after multiplying each term by $d\forall$ and applying Gaussian rule, becomes

$$\frac{\partial}{\partial t}\int_{\forall}(\rho\bar{u}_i)\,d\forall+\int_S(\rho\bar{u}_i\bar{u}_j)\frac{\partial x_j}{\partial n}\,dS+\int_S(\rho\overline{u'_iu'_j})\frac{\partial x_j}{\partial n}\,dS$$

$$=\int_{\forall}(\rho X_i)\,d\forall-\int_S\bar{p}\frac{\partial x_i}{\partial n}\,dS+\int_S\mu\frac{\partial\bar{u}_i}{\partial x_j}\frac{\partial x_j}{\partial n}\,dS \qquad \ldots(4.30)$$

Momentum equation is a vector equation, and, hence, Eq. 4.30 represents three equations which are obtained when $i = 1, 2, 3$ are successively taken in Eq. 4.30, and the equation expanded. The first term on the left represents the rate of change of momentum within volume $\forall$ with time. This term will be zero for steady flow.

The second term represents the net rate of momentum flux of the mean flow going out of the region, while the third term represents net rate of momentum flux of turbulence going out of the region. The first term on the right hand side gives the components of the body force in the given direction, while the second term gives components of the mean pressure force in the same direction. The third term represents the three components of the mean tangential force exerted on the surface. To get an idea about the long form of Eq. 4.30, the momentum equation in x direction is written below. Thus, when $i = 1$,

$$\frac{\partial}{\partial t}\int_{\forall}(\rho\bar{u}_1)\,d\forall+\int_S(\rho\bar{u}_1\bar{u}_1)\frac{\partial x_1}{\partial n}\,dS+\int_S(\rho\bar{u}_1\bar{u}_2)\frac{\partial x_2}{\partial n}\,dS+\int_S(\rho\bar{u}_1\bar{u}_3)\frac{\partial x_3}{\partial n}\,dS$$

$$+\int_S(\rho\overline{u'^2_1})\frac{\partial x_1}{\partial n}\,dS+\int_S(\rho\overline{u'_1u'_2})\frac{\partial x_2}{\partial n}\,dS+\int_S(\rho\overline{u'_1u'_3})\frac{\partial x_3}{\partial n}\,dS=$$

$$\int_{\forall}(\rho X_1)\,d\forall-\int_S\bar{p}\frac{\partial x_1}{\partial n}\,dS+\int_S\mu\frac{\partial\bar{u}_1}{\partial x_1}\frac{\partial x_1}{\partial n}\,dS+\int_S\mu\frac{\partial\bar{u}_1}{\partial x_2}\frac{\partial x_2}{\partial n}\,dS$$

$$+\int_S\mu\frac{\partial\bar{u}_1}{\partial x_3}\frac{\partial x_3}{\partial n}\,dS \qquad \ldots(4.31)$$

As an illustration, one can reduce the above equation for the case of stationary hydraulic jump on the horizontal floor of a wide rectangular channel as shown in Fig. 4.1 (a) and (b). Figure 4.1 (a) shows the hydraulic jump in a wide channel, roller formation and forces acting on the control volume $\forall$ between sections 1 and 2 in the figure. The control volume $\forall$ enclosed between sections 1 and 2, water surface and channel floor, includes the nonuniformity of flow. Figure 4.1 (b) shows idealised water surface along with values of $\frac{\partial x_i}{\partial n}$ on the surface of $\forall$. At sections 1 and 2, pressure distribution can be assumed to be hydrostatic. Since the flow is steady, $\frac{\partial}{\partial t} = 0$. It is also assumed that $\frac{\partial \bar{u}}{\partial y} = 0$ at the water surface. Hence, integral momentum equation Eq. 4.31 reduces to

$$\int_0^{y_2} \rho \bar{u}^2 dy - \int_0^{y_1} \rho \bar{u}^2 dy + \int_0^{y_2} \rho \overline{u'^2} dy - \int_0^{y_1} \rho \overline{u'^2} dy$$

$$= \frac{\gamma y_1^2}{2} - \frac{\gamma y_2^2}{2} - \int_0^{L_j} \tau_o(x) dx$$

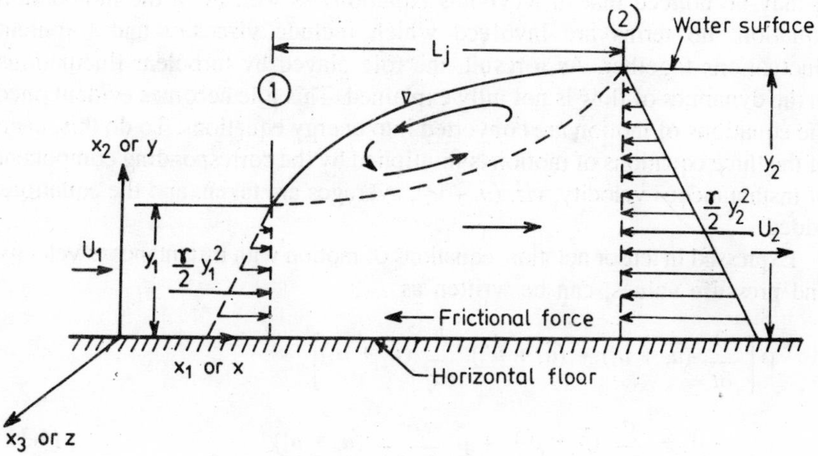

Fig. 4.1 (a) Stationary hydraulic jump

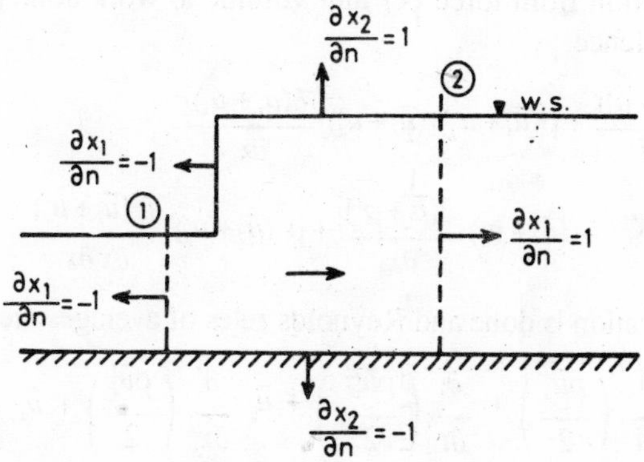

Fig. 4.1 (b) Values of $\partial x_i / \partial n$ for idealised jump geometry

where L_j is the length of hydraulic jump and $\gamma = \rho g$ is the unit weight of fluid. The last term in the above equation represents the frictional force exerted by the floor on the fluid. If one assumes that mean velocity is constant at sections 1 and 2, turbulence intensities at sections 1 and 2 are negligibly small (which is not true, especially at Section 2) and boundary friction is negligible, the above equation reduces to conventional momentum equation

$$(\rho U_2^2 y_2 - \rho U_1^2 y_1) = \frac{1}{2} \gamma (y_1^2 - y_2^2)$$

where U_1 and U_2 are the average velocities before and after the hydraulic jump, and y_1 and y_2 are the depths of flow before and after the jump, respectively.

4.6 ENERGY EQUATION (4, 5)

It may be noticed that in Reynolds equations as well as in the momentum equation, no terms are involved which include viscosity and turbulent fluctuations together. As a result, the role played by turbulent fluctuations in the dynamics of flow is not fully explained. This role becomes evident once the equations of motion are converted into energy equations. To do this, each of the three equations of motion is multiplied by the corresponding component of instantaneous velocity, *viz.* $(\bar{u}_i + u_i')$, averages are taken, and the equations added.

Expressed in tensor notation, equations of motion with instantaneous velocity and pressure values, can be written as

$$\rho\left[\frac{\partial}{\partial t}(\bar{u}_i + u_i') + (\bar{u}_j + u'_j)\frac{\partial}{\partial x_j}(\bar{u}_i + u_i')\right]$$

$$= X_i - \frac{\partial}{\partial x_i}(\bar{p} + p') + \mu\frac{\partial^2}{\partial x_j \partial x_j}(\bar{u}_i + u_i') \qquad \ldots (4.31)$$

Multiply each term of Eq. 4.31 by $(\bar{u}_i + u_i')$, which will change each term in the above equation from force per unit volume to work done per second per unit volume. Hence,

$$\rho(\bar{u}_i + u_i')\frac{\partial(\bar{u}_i + u_i')}{\partial t} + \rho(\bar{u}_j + u'_j)(\bar{u}_i + u_i')\frac{\partial(\bar{u}_i + u_i')}{\partial x_j}$$

$$= (\bar{u}_i + u_i')X_i - (\bar{u}_i + u_i')\frac{\partial(\bar{p} + p')}{\partial x_i} + \mu(\bar{u}_i + u_i')\frac{\partial^2(\bar{u}_i + u_i')}{\partial x_j \partial x_j} \qquad \ldots (4.32)$$

When multiplication is done and Reynolds rules of averages are used, one gets

$$\text{LHS} = \frac{\partial}{\partial t}\left(\frac{\rho\bar{u}_i^2}{2}\right) + \frac{\partial}{\partial t}\left(\frac{\rho\overline{u_i'^2}}{2}\right) + \bar{u}_j\frac{\partial}{\partial x_j}\left(\frac{\rho\bar{u}_i^2}{2}\right) + \bar{u}_j\frac{\partial}{\partial x_j}\left(\frac{\rho\overline{u_i'^2}}{2}\right)$$

$$+ \frac{\partial}{\partial x_j}(\bar{u}_i\,\rho\overline{u_i' u'_j}) + \overline{u_j'\frac{\partial}{\partial x_j}\left(\frac{\rho u_i'^2}{2}\right)}$$

$$\text{RHS} = X\bar{u}_i - \bar{u}_i\frac{\partial\bar{p}}{\partial x_i} - \overline{u_i'\frac{\partial p'}{\partial x_i}} + \mu\,\bar{u}_i\frac{\partial^2\bar{u}_i}{\partial x_j \partial x_j} + \mu\,\overline{u_i'\frac{\partial^2 u_i'}{\partial x_j \partial x_j}} \qquad \ldots (4.33)$$

Substituting i = 1, 2, 3, one gets three equations from Eq. 4.33 and, energy being a scalar quantity, these three equations can be added. Calling $\bar{u}_1^2 + \bar{u}_2^2 + \bar{u}_3^2 = \overline{V}^2$, and $\overline{u_1'^2} + \overline{u_2'^2} + \overline{u_3'^2} = \overline{V'^2}$, one gets from Eq. 4.33 the final form of energy equation in cartesian co-ordinate system as

$$\left.\begin{aligned}&\frac{\partial}{\partial t}\left(\frac{\rho \bar{V}^2}{2}\right)+\frac{\partial}{\partial t}\left(\frac{\rho \overline{V'^2}}{2}\right)+\bar{u}_j\frac{\partial}{\partial x_j}\left(\frac{\rho \bar{V}^2}{2}\right)+\bar{u}_j\frac{\partial}{\partial x_j}\left(\frac{\rho \overline{V'^2}}{2}\right)\\&+\overline{u'_j\frac{\partial}{\partial x_j}\left(\frac{\rho V'^2}{2}\right)}+\frac{\partial}{\partial x_j}(\bar{u}_i\,\rho\overline{u'_iu'_j})=\bar{u}_i X_i-\bar{u}_i\frac{\partial \bar{p}}{\partial x_i}-\overline{u'_i\frac{\partial p'}{\partial x_i}}\\&+\mu\,\bar{u}_i\frac{\partial^2\bar{u}_i}{\partial x_j\partial x_j}+\overline{\mu\,u'_i\frac{\partial^2 u'_i}{\partial x_j\partial x_j}}\end{aligned}\right\}\quad(4.34)$$

Equations 4.34 shows that many of the fluctuating stresses which disappeared from the momentum equation during the averaging process are now retained. It can also be seen that there is a consistent parallel between the terms pertaining to mean flow and those pertaining to turbulence. In fact, for the convenience of discussion. Eq. 4.34 can be written in two parts—one pertaining to mean flow and the other pertaining to turbulent motion, as shown below.

Mean Flow Equation

$$\underset{\text{I}}{\frac{\partial}{\partial t}\left(\frac{\rho \bar{V}^2}{2}\right)}+\underset{\text{II}}{\bar{u}_j\frac{\partial}{\partial x_j}\left(\frac{\rho \bar{V}^2}{2}\right)}+\underset{\text{III}}{\bar{u}_i\frac{\partial}{\partial x_j}(\rho\overline{u'_iu'_j})}$$

$$=\underset{\text{IV}}{X_i\,\bar{u}_i}-\underset{\text{V}}{\bar{u}_i\frac{\partial\bar{p}}{\partial x_i}}+\underset{\text{VI}}{\mu\,\bar{u}_i\frac{\partial^2\bar{u}_i}{\partial x_j\,\partial x_j}}\quad\ldots(4.35)$$

Turbulent Motion Equation

$$\underset{\text{I}}{\frac{\partial}{\partial t}\left(\frac{\rho \overline{V'^2}}{2}\right)}+\underset{\text{II}}{\bar{u}_j\frac{\partial}{\partial x_j}\left(\frac{\rho \overline{V'^2}}{2}\right)}\pm\underset{\text{III}}{\rho\overline{u'_iu'_j}\frac{\partial\bar{u}_i}{\partial x_j}}+\underset{\text{IV}}{\overline{u'_j\frac{\partial}{\partial x_j}\left(\frac{\rho V'^2}{2}\right)}}$$

$$=-\underset{\text{V}}{\overline{u'_i\frac{\partial p'}{\partial x_i}}}+\underset{\text{VI}}{\overline{\mu\,u'_i\frac{\partial^2u'_i}{\partial x_j\partial x_j}}}\quad\ldots(4.36)$$

The following interpretation of terms occurring in the two equations can be given. The first term in each equation represents the rate of change of kinetic energy per unit volume with respect to time, for mean flow and turbulent motion, respectively. If the flow is steady these terms will be zero. Since turbulence is dissipative in nature, energy of turbulence will decrease unless there is a continuous supply of energy from mean flow. At the same time, energy of turbulence per unit volume will change due to convection, and diffusion. These phenomena are mathematically described by the energy equation as under.

Term	Mean Flow	Term	Turbulent Motion
I	Rate of change of kinetic energy of mean flow per unit volume.	I	Rate of change of kinetic energy of turbulent motion per unit volume.
II	Rate of change of kinetic energy of mean flow per unit volume due to convection	II	Rate of change of kinetic energy of turbulent motion per unit volume due to convection
III	Rate at which Reynolds stresses perform work on mean flow. This is transferred from mean flow to turbulent motion, and hence, is loss to mean flow	III	Rate of production of turbulent energy; it is negative.
		IV	Rate of change of kinetic energy of turbulence due to diffusion
IV & V	Rate at which work is done by body force and mean pressure	V	Rate at which work is done by fluctuating pressure force
VI	Rate at which kinetic energy of mean flow is dissipated by viscous stresses	VI	Rate at which energy of turbulent motion is dissipated by viscous action

As in the case of momentum equation, the energy equation can also be integrated over a given volume in space. Some of the volume integrals can be converted into surface integrals using Gauss rule. Then the energy equations for mean flow and turbulent motion take the form (4)

Mean Flow

$$\int_{\forall} \frac{\partial}{\partial t}\left(\frac{\rho \bar{V}^2}{2}\right) d\forall + \int_S \frac{\rho \bar{V}^2}{2}\, \bar{u}_j \frac{\partial x_j}{\partial n}\, dS + \int_S \rho \overline{u'_i u'_j}\, \bar{u}_i \frac{\partial x_j}{\partial n}\, dS$$

$$- \int_{\forall} \rho \overline{u'_i u'_j} \frac{\partial \bar{u}_i}{\partial x_j}\, d\forall = \int_{\forall} X_i \bar{u}_i\, d\forall - \int_S \bar{p}\, \bar{u}_i \frac{\partial x_i}{\partial n}\, dS$$

$$+ \int_S \mu\left(\frac{\partial \bar{u}_i}{\partial x_j} + \frac{\partial \bar{u}_j}{\partial x_i}\right) \bar{u}_i \frac{\partial x_j}{\partial n}\, dS - \int_{\forall} \mu\left(\frac{\partial \bar{u}_i}{\partial x_j} + \frac{\partial \bar{u}_j}{\partial x_i}\right) \frac{\partial \bar{u}_i}{\partial x_i}\, d\forall \quad \ldots (4.37)$$

Turbulent Motion

$$\int_{\forall} \frac{\partial}{\partial t}\left(\frac{\rho \overline{V'^2}}{2}\right) d\forall + \int_S \frac{\partial \overline{V'^2}}{2}\, \bar{u}_j \frac{\partial x_j}{\partial n}\, dS + \int_{\forall} \rho \overline{u'_i u'_j} \frac{\partial \bar{u}_i}{\partial x_j}\, d\forall$$

$$+ \int_S \frac{\rho \overline{V'^2}}{2}\, u'_j \frac{\partial x_j}{\partial n}\, dS = - \int_S \overline{p' u'_i} \frac{\partial x_i}{\partial n}\, dS + \int_S \mu\left(\frac{\partial u'_i}{\partial x_j} + \frac{\partial u'_j}{\partial x_i}\right) u'_i \frac{\partial x_j}{\partial n}\, dS$$

$$- \int_{\forall} \mu\left(\frac{\partial u'_i}{\partial x_j} + \frac{\partial u'_j}{\partial x_i}\right) \frac{\partial u'_i}{\partial x_j}\, d\forall \quad \ldots (4.38)$$

Note that the last two terms on the right hand side of Eqs. 4.37 and 4.38 represent (a) the rates at which work is done by the viscous stresses of the mean flow over the surface and throughout the interior of the region, and (b) the rates at which work is done by the viscous stresses of turbulence over the surface of the region and throughout the interior, respectively. The energy equations have been reduced to a simpler form by Rouse, et al. (3) for the case of hydraulic jump in a horizontal wide rectangular channel. Energy equation has been discussed in detail by Rouse (5), and given for axisymmetric turbulent flow by Chaturvedi (6), and Narasimhan (7).

4.7 BERNOULLI'S EQUATION FOR MEAN FLOW (8, 9)

It is instructive at this point to make a few comments on Bernoulli's equation for mean turbulent flow as given by Rouse (8, 9). For steady flow of an inviscid, incompressible fluid along a streamline, Euler's equation states that

$$\frac{\partial}{\partial S}\left(\frac{\rho V^2}{2}\right) = -\frac{\partial}{\partial S}(p + \gamma Z) \qquad \ldots (4.39)$$

the integration of which yields

$$\frac{\rho V_1^2}{2} + p_1 + \gamma Z_1 = \frac{\rho V_2^2}{2} + p_2 + \gamma Z_2 \qquad \ldots (4.40)$$

Here Z_1 and Z_2 are elevations of points 1 and 2 along the streamline, and V_1 and V_2 are the velocities. If Eq. 4.39 is multiplied by V, as in energy equation, and by $dA\ dS$, one gets

$$V\frac{\partial}{\partial S}\left(\frac{\rho V^2}{2}\right) dA\ dS = -V\frac{\partial}{\partial S}(p + \gamma Z)\ dA\ dS \qquad \ldots (4.41)$$

However, $VdA = dQ$ is the flow through a stream tube of area dA and is independent of S. Hence, one can integrate Eq. 4.41. A second integration with respect to A is thereafter performed to obtain power relationship for gross flow section. However, if Eq. 4.41 is divided by VdA first, and then integrated, Eq. 4.40 would result again. This process looks trivial, but it is only through such process that a true dissipation term is obtained in Bernoulli's equation for viscid fluids. This is achieved by starting from Reynolds equations along tangential, normal and binormal directions for the mean motion, and each term is multiplied by corresponding velocity component. Or

$$\bar{V}\frac{\partial}{\partial S}\left(\frac{\rho \bar{V}^2}{2}\right) + \bar{u}_i\frac{\partial}{\partial S}(\overline{\rho u_i' u_j'}) = -\bar{V}\frac{\partial}{\partial S}(\bar{p} + \gamma Z) + \bar{u}_i\frac{\partial \bar{\tau}_{ij}}{\partial S} \qquad \ldots (4.42)$$

Since the velocity $\bar{V}$ is tangential, $\bar{u}_1 = \bar{V}$ and $\bar{u}_2 = \bar{u}_3 = 0$.

The intensity of shear τ_{ij} can be expressed as the sum of dynamic viscosity multiplied by the rate of deformation and Reynolds stress $-\rho\overline{u_i' u_j'}$. Hence

$$\bar{\tau}_{ij} = \mu\left(\frac{\partial \bar{u}_i}{\partial x_j} + \frac{\partial \bar{u}_j}{\partial x_i}\right) - \rho\overline{u_i' u_j'} \qquad \ldots (4.43)$$

If Eq. 4.43 is substituted in Eq. 4.42 and it is multiplied by $dS\ dA$, it will have a form corresponding to Eq. 4.41, namely

$$\bar{V}\frac{\partial}{\partial S}\left(\frac{\rho\bar{V}^2}{2}\right) dA\ dS = -\bar{V}\frac{\partial}{\partial S}(\bar{p}+\gamma Z)\ dA\ dS + \bar{u}_i\frac{\partial\bar{\tau}_{ij}}{\partial x_j}\ dA\ dS. \quad \ldots (4.44)$$

This equation can be integrated along a stream tube after dividing each term by $dQ = \bar{V}\ dA$, which gives

$$\frac{\rho\bar{V}_1^2}{2} + \bar{p}_1 + \gamma Z_1 = \frac{\rho\bar{V}_2^2}{2} + \bar{p}_2 + \gamma Z_2 - \int_{S_1}^{S_2}\frac{\bar{u}_i}{\bar{V}}\frac{\partial\bar{\tau}_{ij}}{\partial x_j}\ dS \quad \ldots (4.45)$$

It may be noted that no explicit terms for kinetic energy of turbulent motion are present in Eq. 4.45. Moreover, the fact that the term at the end of right hand side of Eq. 4.45 is not restricted as to the sign, indicates that the viscous and turbulent stresses can cause unit energy along a particular stream tube to either increase or decrease. The reduction in the Bernoulli sum $\left(\frac{\rho\bar{V}^2}{2} + \bar{p} + \gamma Z\right)$ along a stream tube would mean a negative value of the integral; this can be produced not only by energy dissipation but also by energy transfer to the neighbouring stream tube. A positive value of integral will represent a local gain in energy in excess of local dissipation at the expense of surrounding flow. Thus, it is erroneous to assume that the customary additive term in Bernoulli's equation represents an irreversible loss.

Since
$$\bar{u}_i\frac{\partial\bar{\tau}_{ij}}{\partial x_j} = \frac{\partial(\bar{u}_i\,\bar{\tau}_{ij})}{\partial x_j} - \bar{\tau}_{ij}\frac{\partial\bar{u}_i}{\partial x_j} \quad \ldots (4.46)$$

Eq. 4.45 with the substitution of Eq. 4.46 yields

$$\frac{\rho\bar{V}_1^2}{2} + \bar{p}_1 + \gamma Z_1 = \frac{\rho\bar{V}_2^2}{2} + \bar{p}_2 + \gamma Z_2 - \int_{S_1}^{S_2}\frac{1}{\bar{V}}\frac{\partial(\bar{u}_i\,\bar{\tau}_{ij})}{\partial x_j}\ dS + \int_{S_1}^{S_2}\frac{\bar{\tau}_{ij}}{\bar{V}}\frac{\partial\bar{u}_i}{\partial x_j}\ dS \quad \ldots (4.47)$$

The first integral on the right hand side represents the transfer of energy through shear to (i.e. work done upon) the surrounding fluid. The second integral represents the loss suffered by the mean flow through either viscous shear or generation of turbulence. The shear $\bar{\tau}$ in Eq. 4.47 can be replaced by Eq. 4.43. However, in many cases the viscous stresses (i.e. first term on right hand side of Eq. 4.43) are small in comparison with the Reynolds stresses of turbulence, and, hence, can be neglected without appreciable error. In such a case Eq. 4.47 reduces to

$$\frac{\rho\bar{V}_1^2}{2} + \bar{p}_1 + \gamma Z = \frac{\rho\bar{V}_2^2}{2} + \bar{p}_2 + \gamma Z + \rho\int_{S_1}^{S_2}\frac{1}{\bar{V}}\frac{\partial(\bar{u}_i\,\overline{u_i'u_j'})}{\partial x_j}\ dS - \rho\int_{S_1}^{S_2}\frac{\overline{u_i'u_j'}}{\bar{V}}\frac{\partial\bar{u}_i}{\partial x_j}\ dS \quad \ldots (4.48)$$

The last term on the r.h.s. of the above equation represents loss to the mean flow, but it is not a dissipation because it does not contain viscosity. Thus energy is transferred from mean flow to turbulent motion. Thus, if Bernoulli's equation along a stream tube is written as

$$\left(\frac{\rho\overline{V_1^2}}{2} + \bar{p}_1 + \gamma Z_1\right) = \left(\frac{\rho\overline{V_2^2}}{2} + \bar{p}_2 + \gamma Z_2\right) + B$$

B is made up of transfer of energy through shear to the surrounding fluid, and loss suffered by mean flow through either viscous dissipation or turbulence generation. Hence, depending on their relative magnitudes, the energy along a stream tube may increase or decrease. The former happens in the case of a wake.

If Eq. 4.47 is used over the entire region of flow, the following points need attention. Firstly, over the area A, the velocity may not be constant; hence, the kinetic energy term must be replaced by $\left(\alpha \frac{\rho Q^2}{2A^2}\right)$, where α is the energy correction factor. Secondly, since $\bar{p}$ will vary over the section, one can introduce a pressure correction factor η, replacing the term $(\bar{p} + \gamma Z)$ by $\eta\ (\bar{p} + \gamma Z)_c$. $(\bar{P} + \gamma Z)_c$ is the value of $(\bar{p} + \gamma Z)$ at the centre of the cross section. Thirdly, if the boundary is nonmovable, as is usually the case, the integral of the transfer term $\int_A\int_S \frac{1}{\overline{V}} \frac{\partial(u_i\ \bar{\tau}_{ij})}{\partial x_j}\, dA\ dS$ is almost zero. Hence, the weighted mean loss will be $-\int_A\int_S \frac{\bar{\tau}_{ij}}{\overline{V}} \frac{\partial \bar{u}_i}{\partial x_j}\, dA\ dS = L_m$

Hence, $\alpha \dfrac{\rho Q^2}{2A^2} + (\bar{p} + \gamma Z)_c\,\eta + L_m = \text{constant.}$

4.8 EXPERIMENTAL RESULTS

In recent times valuable experimental data for mean and turbulent motions have been collected for flows in two dimensional and circular conduits, axisymmetric expansions, hydraulic jump, and separation zones, in order to evaluate the various terms occurring in momentum and energy equations. The objective in such studies has been to investigate the complete mechanism of these flows.

Laufer (10) made measurements in steady two dimensional turbulent flow of air between parallel plates and then in a steady turbulent flow of air in a circular pipe (11). Figure 4.2 shows the results of energy balance in turbulent motion in the central part of the pipe at Reynolds number of the order of 2.5×10^5. In this figure the ordinate represent nondimensional energy obtained by multiplying energy per unit volume by $R\,\bar{\tau}_o^{-3/2}$, where R is pipe radius and $\bar{\tau}_o$ is boundary shear; the abscissa is $(1 - r/R)$, where r is the radial distance. It can be seen from this figure that production and dissipation of energy of turbulent motion are nearly equal for $(1 - r/R)$ values less than 0.50. Further, nearer the centre dissipation exceeds production and turbulence level is maintained by the transport

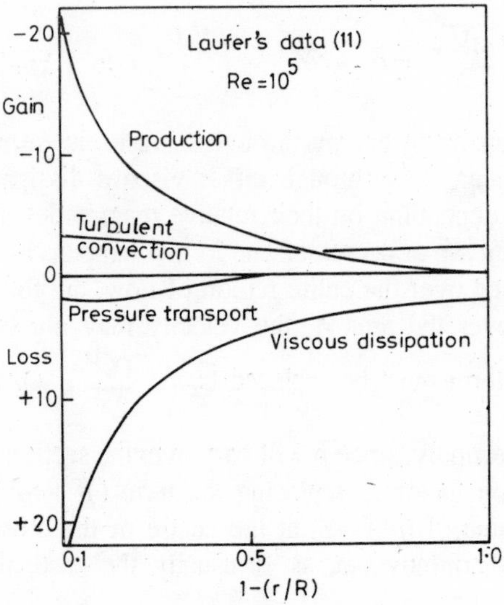

Fig. 4.2 Energy balance outside constant shear layer in a pipe

of energy through turbulent diffusion. Transfer of energy due to fluctuating pressure forces is appreciable only when $(1 - r/R)$ is less than 0.50.

Chaturvedi (6) has presented results of momentum analysis and energy analysis for mean as well as turbulent motions for axisymmetric expansion with half expansion angles of 15°, 30° 45° and 90°, and diameter ratio of two. Similar analysis for aerodynamic model of hydraulic jump was earlier carried out by Rouse et al. (3). Energy and momentum balance has been studied for separated flows at conical afterbodies by Narasimhan (7). However, in such studies, tripple velocity correlation terms and pressure velocity correlation terms were not measured. Based on the results of Townsend (12), it was assumed that these two terms, which have opposite signs, tend to balance over any section.

REFERENCES

1. Schlichting, *H. Boundary Layer Theory,* McGraw Hill Book Co. Inc., New York, 1955.
2. Ward-Smith, A.J. *Internal Fluid Flow—The Fluid Dynamics of Flow in Pipes and Ducts.* Clarendon Press. Oxford, 1980 (Chapter C).
3. Rouse, H., T.T. Sio and T. Nagratnam. Turbulence Characteristics of Hydraulic Jump. Trans. ASCE, Vol. 124, 1959.
4. Rouse, H. (Ed.) *Advanced Mechanics of Fluids.* John Wiley and Sons, Inc. New York, 1959.
5. Rouse, H. Repartition de l'energie dans des Zones de d'ecoulement. La Houille Blanche, July 4, 1959.
6. Chaturvedi, M.C. Flow Characteristics of Axisymmetric Expansions. JHD. Proc. ASCE, Vol. 89, No HY-3, May 1963.

7. Narasimhan, S. Characteristics of Separation at Conical Afterbodies. JEM, Proc. ASCE, Vol. 93, No. EM5, Oct. 1967.

8. Rouse, H. On the Bernoulli's Theorem for Turbulent Flow. Tollmien Festschrift Miszellaneen der Angewandten Mechanik, Academie-Verlag, Berlin, 1962.

9. Rouse, H. Work Energy Equation for the Streamline. JHD, Proc. ASCE, Vol. 96, NO. HY-5, May 1970.

10. Laufer, J. Investigation of Turbulent Flow in a Two-Dimensional Channel. NACA Rep. No. 1033, 1951.

11. Laufer, J. The Structure of Turbulence in Fully Developed Pipe Flow, NACA Rep. No. 1174, 1955.

12. Townsend, A.A. *The Structure of Turbulent Shear Flow.* Cambridge Monographs on Mechanics and Applied Mathematics. Cambridge University Press, Cambridge, 1956.

CHAPTER V

Statistical Theory of Turbulence

5.1 INTRODUCTION

In Chapter IV it has been pointed out that Reynolds equations together with continuity equation cannot be solved, because in four equations there are ten unknowns. In order to obtain mean flow characteristics, even for simple flows, assumptions have to be made relating Reynolds stresses to the mean flow, or one must have additional equations. This approach to the solution of turbulent flow problems is known as mathematical models of turbulence and has been discussed in Chapter VI.

In the early stages of development of the theory of turbulence, it was realised that, since the turbulent fluctuations are caused by the generation, convection, diffusion and dissipation of eddies, they vary from time to time, and from place to place in a very complex and random manner. Therefore, a deterministic approach is unlikely to yield satisfactory results. Hence, Taylor initiated the application of statistical techniques to turbulent flows. Since 1938, several mathematicians, physicists, engineers and meteorologists have developed this approach; among them, mention may be made of Karman, Batchelor, Kolmogorov, Heisenberg, Kempe'de Feriet, Burgers, Hopf, Lin, Chandrasekhar, Panchev and Obukhov. Only some of the basic concepts of the statistical approach will be presented in this chapter without going through the mathematical details.

5.2 SOME DEFINITIONS

A variable which, as a result of experimentation under defined conditions, can assume any one of a number of possible numerical values which would be impossible to predict in advance is called a *random variable*. Thus, if instantaneous value of velocity in x direction at a given point in turbulent flow is measured at successive times, its variation would be random. Such random data can be continuous or discrete (i.e. taken at regular time interval). In order to comprehend the statistical laws of random variables or quantities, it is necessary to know the probability of occurrence of given values. To do this, the concept of a function of distribution of frequency, known as probability distribution function, must be introduced.

Probability Distribution Function

In the case of a continous random variable A, which is a function of time t, let T be the total time for which record of variation of A with t is available, as shown in Figure 5.1.

If $T_1 = t_1 + t_2 + t_3 + t_4 + \ldots.$ is the total time during which the magnitude of A lies between A_1 and $(A_1 + \Delta A)$, the ratio T_1/T gives the probability, or frequency ΔP of occurrence of magnitude of A between A_1 and $(A_1 + \Delta A)$. As ΔA becomes small, T_1/T, which is denoted by ΔP, also

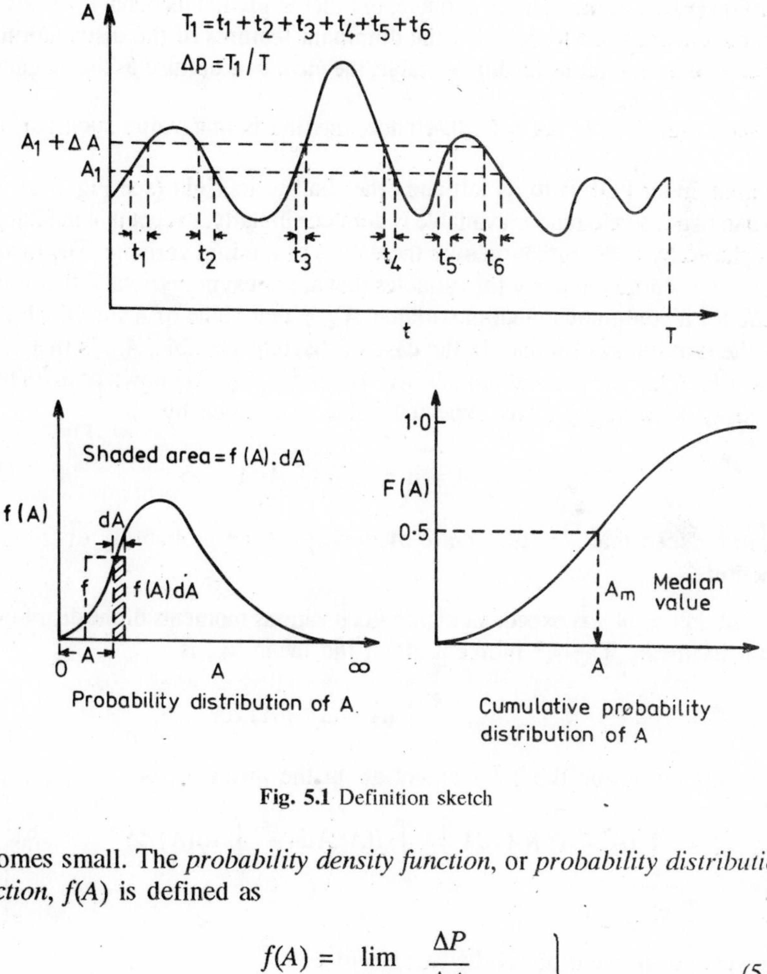

Fig. 5.1 Definition sketch

becomes small. The *probability density function*, or *probability distribution function*, $f(A)$ is defined as

$$\left.\begin{aligned} f(A) &= \lim_{\Delta A \to 0} \frac{\Delta P}{\Delta A} \\ \text{or} \qquad \Delta P &= f(A)\, \Delta A \end{aligned}\right\} \qquad \ldots (5.1)$$

Thus, knowing ΔP and ΔA, one can determine $f(A)$ for various values of A. If the magnitude of A varies from 0 to ∞, the cumulative probability that A will lie between 0 and A is

$$F(A) = \int_0^A f(A)\, dA$$

(see Fig. 5.1). Also the probability that A will lie between 0 and ∞ is evidently $\int_0^\infty f(A)\, dA = 1.0$. Further, the probability that A will lie between A_1 and A_2 is

$$\int_{A_1}^{A_2} f(A)\, dA = \int_0^{A_2} f(A)\, dA - \int_0^{A_1} f(A)\, dA = F(A_2) - F(A_1)$$

Strictly speaking, one must have complete knowlege of probability distribution function (PDF) of the random variable, in order to predict its behaviour. Several parameters are used to describe the dominant features of the distribution. In the case of a continuous random variable, the *median* is defined as the magnitude A_m such that $\int_0^{A_m} f(A)\, dA = 0.50$. Hence, median is that value such that half the area under PDF is to its left and other half to its right (see Fig. 5.1). The median of a discrete random variable is defined similarly, except that the integral is replaced by the summation over the values of random variable. The median shows the central tendency for variables that are not symmetrically distributed. *Mode* for a continuous random variable, A_{mo} is that value of A for which PDF has the maximum ordinate. In the case of discrete variable, A_{mo} is that value of A which has the highest probability. The *mean* $\bar{A}$ (also known as *arithmetic average*, or average) or its expected value μ is given by

$$\bar{A} = \mu = \int_0^\infty f(A).\, A dA \qquad \ldots (5.2)$$

It can be seen that $\bar{A}$ is the centre of gravity of the probability distribution function.

At this point it is expedient to introduce various moments of the distribution about its mean. The Kth moment about the mean, μ_{ck} is

$$\mu_{ck} = \int_{-\infty}^{\infty} (A - \mu)^k f(A)\, dA \qquad \ldots (5.3)$$

With this definition the 1st moment about the mean μ_{c1} is

$$\mu_{c1} = \int_{-\infty}^{\infty} (A - \mu)\, f(A)\, dA = \int_{-\infty}^{\infty} f(A).A dA - \int_{-\infty}^{\infty} \mu f(A)\, dA$$

$$= \bar{A} - \mu = 0$$

The second moment about the mean will be

$$\mu_{c2} = \int_{-\infty}^{\infty} (A - \mu)^2 f(A)\, dA \qquad \ldots (5.4)$$

If $(A - \mu)$ is denoted by A', then μ_{c2} will be

$\mu_{c2} = \int_{\infty}^{-\infty} A'^2 f(A)\, dA = \overline{A'^2}$, which is known as the variance of the distribution. The positive square root of variance $\sqrt{\overline{A'^2}}$ is known as the *standard deviation* σ of the distribution which has the same units as A. The dimensionless parameter σ/μ is known as the *coefficient of variation*. It may be mentioned that standard deviation or coefficient of variation gives information about the spread of the distribution. The third moment about the mean, μ_{c3} is related to the asymmetry, or skewness of the distribution, and is given by

$$\mu_{c3} = \int_{-\infty}^{\infty} (A - \mu)^3 f(A)\, dA$$

When it is made dimensionless as μ_{c3}/σ^3, it is known as the *skewness coefficient* γ_1. If $\gamma_1 > 0$, the PDF is skewed to the left and has long tail to the right. If $\gamma_1 < 0$, the PDF is skewed to the right and has long tail to the left. When $\gamma_1 = 0$, the curve is symmetrical, e.g. normal or Gaussian distribution. In the same manner, the fourth moment about the mean μ_{c4} gives an idea about the peakedness of the distribution;

$$\mu_{c4} = \int_{-\infty}^{\infty} (A - \mu)^4 f(A)\, dA$$

The *kurtosis* is defined as $\gamma_2 = \mu_{c4}/\sigma^4$. For normal or Gaussian distribution, $\gamma_1 = 0$ and $\gamma_2 = 3$. If γ_2 is less than 3, the distribution has a sharper peak than the normal distribution; if it is greater than 3, the peak is flatter than that of normal distribution.

One of the most commonly used probability distribution functions for random variables is the *normal* or *Gaussian* which is a symmetrical bell-shaped curve. This distribution is given by

$$f(A) = \frac{1}{\sigma\sqrt{2\pi}} = e^{-(A-\mu)^2/2\sigma^2} \quad \ldots (5.5)$$

This is shown in Fig. 5.2(a). At $A = \bar{A}$ or μ, $f(A)$ has a maximum value of $1/\sigma\sqrt{2\pi}$.

If $\dfrac{(A-\mu)}{\sigma}$ is taken as the normalised variable Z and if Z varies from $-\infty$ to $+\infty$, the cumulative probability that Z lies between $-\infty$ and Z is

$$F(Z) = \frac{1}{\sqrt{2\pi}} \int_{-\infty}^{Z} e^{-Z^2/2}\, dZ \quad \text{if } Z < 0$$

and

$$F(Z) = \frac{1}{\sqrt{2\pi}} \int_{0}^{Z} e^{-Z^2/2}\, dZ + 0.50 \quad \text{if } Z > 0 \quad \ldots (5.6)$$

Further, it can be shown from Eq. 5.6 that for A between $\mu \pm \sigma$, $\mu \pm 2\sigma$ and $\mu \pm 3\sigma$, the cumulative probability for Gaussian distribution is 68.26%, 95.45% and 99.73%, which means that values of A will be less than $(\mu - 3\sigma)$ and greater than $(\mu + 3\sigma)$ for only 0.27 percent of the time. The turbulent fluctuations in velocity and pressure in the case of flow behind grids, pipe

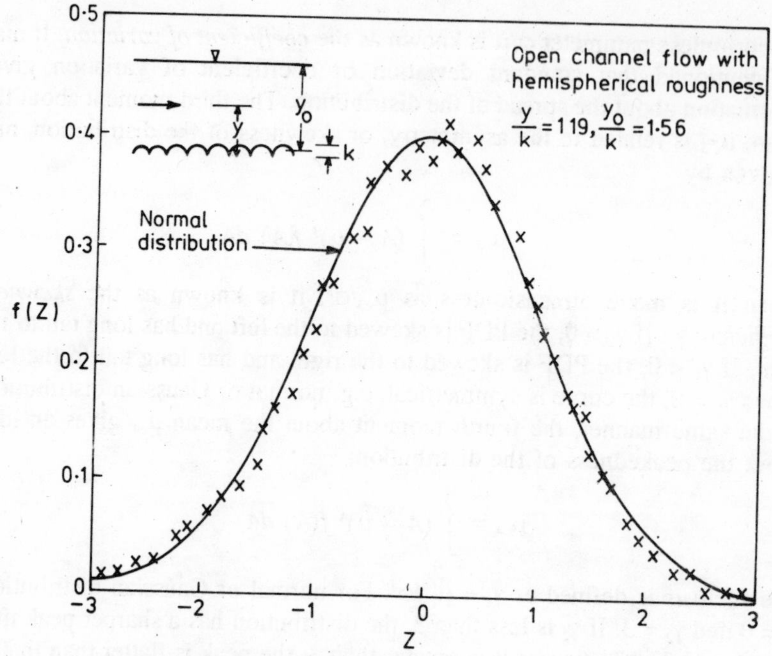

Fig. 5.2 (a) Velocity fluctutations in open channel flow

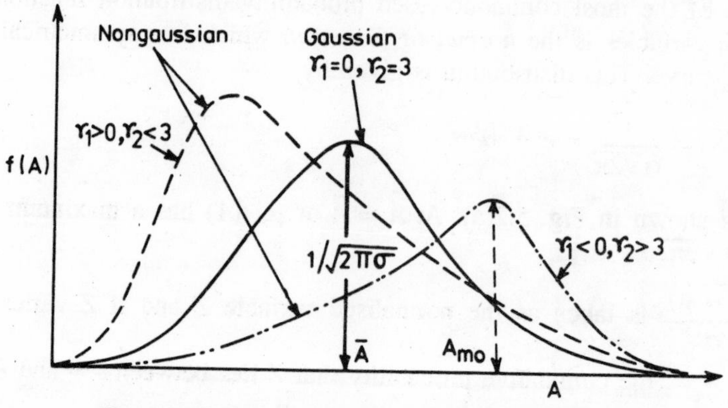

Fig. 5.2 (b) Gaussian and non Gaussian distributions

flow, jets, and boundary layers are found to follow nearly Gaussian distribution, see Fig. 5.2. These data were collected by Bayazit (1) in an open channel with hemispherical roughnesses on the bed. Since the distribution is nearly Gaussian, skewness is very small. In such a case the average value obtained using Eq. 5.2 and the formula $\bar{A} = \frac{1}{T}\int_0^T A\,dt$ will be identical.

It has been found that for very rough boundaries, the turbulent fluctuations close to the boundary deviate from Gaussian distribution; however, as the distance from the boundary increases, the distribution approaches the Gaussian distribution. If one wants to analyse the dynamics of turbulence in detail, the non Gaussian properties become important, and need to be included in the analysis. However, if one is concerned with the effects of turbulence, one may, after experimental verification, find that the Gaussian distribution may be adequate (2).

The probability distribution of the derivatives of fluctuating velocity components with respect to spatial co-ordinates are found to be not normally distributed.

Joint Probability Distribution

Consider two varying quantities A and B; then following the definition of probability distribution function, one can define the probability ΔP that the magnitude of A will lie between A and $(A + dA)$, and B between B and $(B + dB)$ simultaneously, as

$$\Delta P = f(A, B)\, \Delta A \cdot \Delta B$$

The function $f(A, B)$ is known as the *joint probability distribution function* of A and B. The joint probability distribution of the velocities at two different points has been measured, and it has been found that the joint probabilities of the velocity components are not normally distributed in turbulent shear flows, even though in isotropic turbulence it is nearly Gaussian (3). Since the quantities such as $\overline{u'v'}$, $\overline{v'w'}$ and $\overline{u'w'}$ appear in Reynolds equations of motion, these quantities may not be normally distributed.

Stationary Process

A random process is said to be *stationary* if the mean values and the variances are constant, while the correlation function (see below) only depends upon the time difference $(t_2 - t_1)$ and not on t_2 or t_1 (4). In practice, conditions of stationarity can be directly verified, if the mean value and the variance for different instants of time are constant.

Ensemble Average

Stationary random processes have a remarkable characteristic which is called *ergodicity*, and which consists of the fact that the statistical characteristic of the random process attained by averaging a set of realisations of this random process at a given time is, with a probability arbitrarily near unity, equal to the characteristic obtained by averaging a single realisation for a sufficiently long interval of time (4). In other words, a stationary random variable is ergodic if its statistical properties calculated from one sample record are equal to those calculated from other samples of the variable taken during the same time interval which is sufficiently long.

Consider steady uniform turbulent flow in a channel, and assume that instantaneous velocity u in x direction is measured at a given point for sufficiently long time. These data can be used to obtain the time averaged velocity. If now we have an infinite number of identical channels having identical steady uniform turbulent flow, instantaneous velocity in x direction can be measured at the homologous point. The average of these values will be the *ensemble average* and, according to the property of ergodicity or according to ergodic theorem, these two averages will be identical.

The proof of ergodic theorem for mean values is given by Panchev (4). Stationary turbulence is assumed to follow ergodic theorem; hence, the statistical average and the time average are equal. This facilitates the determination of ensemble average, since time average is relatively easy to determine.

Correlation Tensor

If the turbulent fluctuations u', v' and w' are completely unrelated, or not correlated, the turbulent stresses such as $-\rho\overline{u'v'}$, $-\rho\overline{u'w'}$ and $-\rho\overline{v'w'}$ would be zero. The existence of Reynolds stresses depends on the existence of correlation between various components of velocity fluctuations at a point. In a more general analysis, it depends on the double velocity correlation between two points a and b (see Fig. 5.3). Q_{ij} is defined as

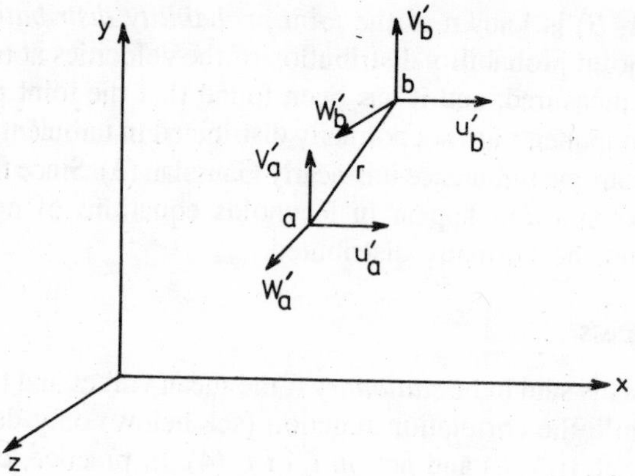

Fig. 5.3 Definition sketch

$$Q_{ij} = \begin{bmatrix} \overline{u'_a u'_b} & \overline{u'_a v'_b} & \overline{u'_a w'_b} \\ \overline{v'_a u'_b} & \overline{v'_a v'_b} & \overline{v'_a w'_b} \\ \overline{w'_a u'_b} & \overline{w'_a v'_b} & \overline{w'_a w'_b} \end{bmatrix} \quad \ldots (5.7)$$

Here the subscript i corresponds to velocity component at a and j corresponds to velocity component at b. Thus, when $i = 2, j = 3$ it is $\overline{v'_a w'_b}$. The correlation Q_{ij} is, thus, a 2nd order tensor having nine components. As the distance r between a and b tends to zero, Q_{ij} gives Reynolds stress tensor if it is multiplied by $(-\rho)$. The *correlation coefficient* is obtained from correlation tensor by nondimensionaling it. Thus,

$$R_{ij} = \frac{\overline{(u'_i)_b\,(u'_j)_b}}{\sqrt{\overline{(u'_i)^2_a}}\,\sqrt{\overline{(u'_j)^2_b}}} \quad \text{or} \quad R_{11} = \frac{\overline{u'_a u'_b}}{\sqrt{\overline{u'^2_a}}\,\sqrt{\overline{u'^2_b}}} \quad \ldots (5.8)$$

When the distance $ab = 0$, the correlation is called *autocovariance*, and R_{ij} is called *autocorrelation coefficient*. Further, Q_{ij} and R_{ij} are called *Eulerian space correlation* and correlation coefficients, respectively. It may be mentioned that for a quasi-steady flow field, that is for a flow field where average flow pattern does not change with time, the double correlation, or double correlation coefficient, is a function of locations of a and b alone, and, hence, has a fixed value.

The magnitude of correlation coefficient will be maximum when $r = 0$, will decrease as r increases and will be zero beyond a certain magnitude of r. The maximum value of correlation coefficient can be $\pm$ 1.0. The shape of correlation coefficient vs. r curve, in general, depends on the sizes of eddies present in the flow.

Figure 5.4 illustrates the physical meaning of correlation. A perfect correlation occurs when one variable is a linear function of the other. In such case, simultaneously measured values of two variables, say u' and v', would lie on a line. Figure 5.4 (a) shows a situation when $R = -1.0$. No correlation occurs if u' and v' are completely independent, in which case $R = 0$. Simultaneously measured values of two variables A and B will scatter randomly around origin as shown in Fig. 5.4 (b). This happens in case of u' and v' in isotropic turbulence. In turbulent shear flows, such as in pipes and in open channels, the condition is between these two extremes as shown in Fig. 5.4 (c). Here R is about –0.40. to –0.70.

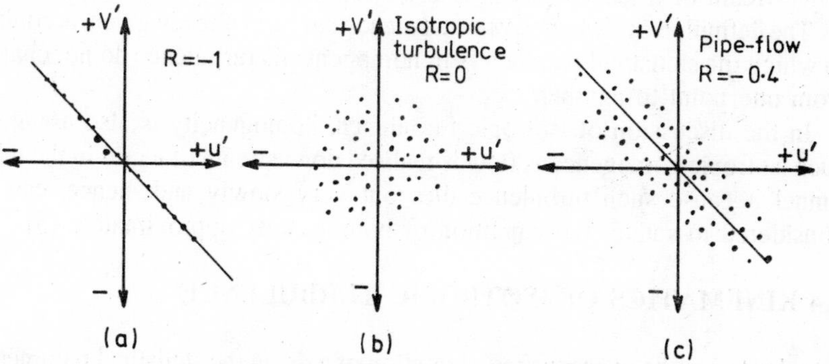

Fig. 5.4 Physical meaning of correlation

In a turbulent flow, it is more difficult to obtain correlation between turbulent fluctuations at two points at the same time. However, it is possible to obtain relatively easily the turbulent fluctuations at a given point at various times. Hence, one defines the Eulerian correlations with respect to time. If one were to consider velocity fluctuations u' at a given point, the *Eulerian autocorrelation coefficient* with respect to time is defined as

$$R_E(t) = \frac{\overline{u'(t' - t)\, u'(t')}}{\overline{u'^2}} \qquad \ldots (5.9)$$

Here t' is any time, and t is the increment in time. The coefficient R_E (t) will be unity when $t = 0$, and it tends to zero as t increases. Lastly, in the analysis of turbulent diffusion one can study the motion of the individual particle, and define the correlation coefficient of velocity fluctuations of a particle as

$$R_L(t) = \frac{\overline{v'(t')\,v'(t'-t)}}{\overline{v'^2}} \qquad \ldots (5.10)$$

The correlation coefficient $R_L(t)$ has the same properties as $R_E(t)$. It is known as *Lagrangian autocorrelation coefficient*.

5.3 ISOTROPIC TURBULENCE AND HOMOGENEOUS TURBULENCE

Earlier, *isotropic turbulence* has been defined as the turbulence in which the statistical parameters related to turbulence remain invariant with respect to rotation of axes or reflection. Even through isotropic turbulence is of hypothetical type, it has been studied in greater detail, firstly because its mathematical analysis is relatively easier, and, secondly, because isotropy is nearly achieved in many practical situations. For example, in a wind tunnel, the turbulence in the central region outside the boundary layer is near isotropic. Similarly the average values of the three components of velocity fluctuations in the central region of the pipe, or two dimensional open channel, are nearly equal even through $\overline{u'^2} > \overline{v'^2} > \overline{w'^2}$. In the atmospheric tubulence also similar conditions are observed. The turbulence generated in the wind tunnel downstream of a lattice grid is also isotropic in nature.

The definition of *homogeneous turbulence* has been already given, according to which the statistical parameters in homogeneous turbulence do not change from one point to another.

In the discussion of isotropic turbulence, homogeneity is also assumed; such isotropic homogeneous flow is realised downstream of a grid in the wind tunnel because such turbulence dies out very slowly and, hence, can be considered to satisfy the condition of homogeneity approximately (5).

5.4 KINEMATICS OF ISOTROPIC TURBULENCE

In order to see the consequence of rotation of axis on the statistical parameters of isotropic turbulence, consider (6) the coordinate frame OX, OY and OZ (see Fig. 5.5), and rotate it through 180° about OX. Let the new axes be OX, OY' and OZ'. Let a $(x_1, 0,0)$ and b $(x, 0,0)$ be the co-ordinates of two points

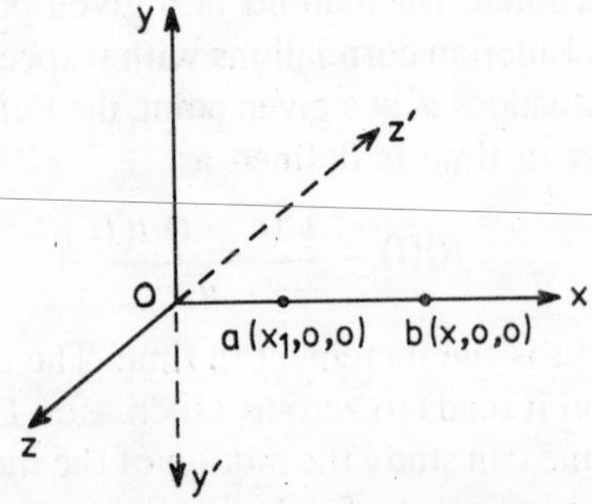

Fig. 5.5 Rotation of axes

on x axis. Consider the quantity $\overline{u'_a v'_b}$ in the orginal and rotated system. Since it must remain invariant in two coordinate systems,

$$\overline{(u)'_a\,(v)'_b} = \overline{u'_a\,(-v'_b)} = -\,\overline{(u)'_a\,(v)'_b}$$

This can be true only if $\overline{(u')_a\,(v')_a} = 0$. Similarly it can be seen that

$$\overline{(u')_a\,(w')_b} = \overline{(u')_a\,(-w')_b} = -\,\overline{(u')_a\,(w')_b}$$

which can be true only if $\overline{(u')_a\,(w')_b} = 0$

Next, the reflection in XZ plane can be considered. Similar reasoning will show that

$$\overline{(v')_a\,(w')_b} = \overline{(-v')_a\,(w')_b} = -\,\overline{(v')_a\,(w')_b}$$

which will be true only if $\overline{(v')_a\,(w')_a} = 0$. As the distance between a and b is reduced to zero, the above quantities give Reynolds stresses. Hence, it is shown that in an istropic turbulence, Reynolds shear stresses are zero.

Karman and Howarth (6) have also shown that the requirements of isotropy and incompressibility dictate that

$$\overline{(p')_a\,(u')_b} = \overline{(p')_a\,(v')_b} = \overline{(p')_a\,(w')_b} = 0$$

Next, consider the double velocity covariance and correlation coefficient. Since for homogeneous isotropic turbulence

$$(\overline{u'^2})_a = (\overline{v'^2})_a = (\overline{w'^2})_a = (\overline{u'^2})_b = (\overline{v'^2})_b = (\overline{w'^2})_b = \text{say } \overline{u'^2}$$

the following correlation coefficients can be defined

$$\left.\begin{aligned} f(r) &= \frac{\overline{u'_a u'_b}}{\overline{u'^2}} \\ g(r) &= \frac{\overline{v'_a v'_b}}{\overline{u'^2}} \end{aligned}\right\} \qquad \ldots (5.11)$$

These are called the longitudinal and lateral correlation coefficients, respectively, which are a function of the distance r between a and b. In the computation of $f(r)$ and $g(r)$, the averaging procedure is to be carried out with respect to time. However, since the turbulence is homogeneous, it is also possible to apply averaging procedure with respect to r; i.e. consider a large number of points at a distance r at a given time and obtain $\overline{u'_a u'_b}$ or $\overline{v'_a v'_b}$. However, the latter is more difficult to measure. The correlation functions $f(r)$ and $g(r)$ can be shown to have the following properties:

(i) $f(0) = g(0) = 1$;

(ii) $f(r)$ and $g(r)$ tend to zero as r becomes large;

(iii) $f(r)$ and $g(r)$ are even functions of r, i.e. $f(r) = f(-r)$, and $g(r) = g(-r)$. Also the odd derivatives of f and g with respect to r are zero.

(iv) f and g have opposite signs beyond a certain value of r;

(v) Karman has shown that as a consequence of incompressibility, f and g are related as

$$g(r) = f(r) + \frac{r}{2}\frac{df}{dr} \qquad \ldots (5.12)$$

In fact, in homogeneous isotropic turbulence, all the correlation functions of the second order can be expressed in terms of single correlation function, say $f(r)$. Typical variation of $f(r)$ and $g(r)$ with r is shown in Fig. 5.6. It can be seen that while $f(r)$ is always positive, $g(r)$ becomes negative for larger values of r.

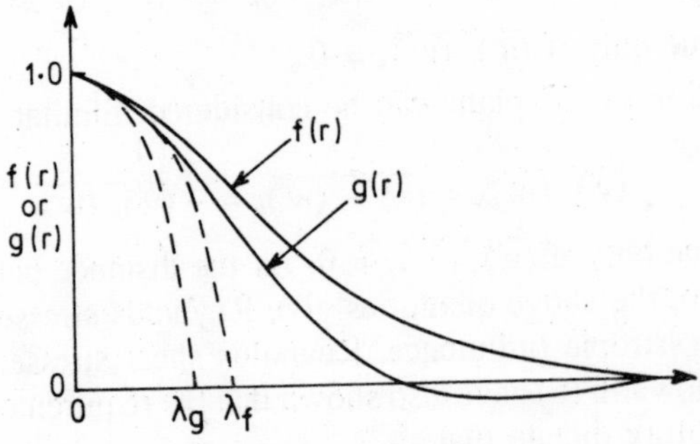

Fig. 5.6 Variation of $f(r)$ and $g(r)$ with r

In many cases at high Reynolds numbers, $f(x)$ and $g(y)$ curves can be approximated by the exponential curves

$$\left.\begin{aligned} f(x) &= e^{-x/L_f} \\ g(y) &= e^{-y/L_g} \end{aligned}\right] \qquad \ldots (5.13)$$

Here L_f and L_g are length scales defined by Eq. 5.17. Even though for large x or y values, Eq. 5.13 approximates the actual variation reasonably well, near the vertex they are a poor representation. Other form of equation which is used is

$$g(y) = e^{-|cy|^m} \qquad \ldots (5.14)$$

where $c = (\text{const.}/\lambda_g)$ and m lies between 1 and 2.

Since f and g are even functions of r, they can be expanded in the neighbourahood of $r = 0$ using Taylor series as under:

$$\left.\begin{aligned} f &= 1 + \frac{r^2}{2}\left(\frac{d^2 f}{dr^2}\right)_{r=0} + \cdots \\ g &= 1 + \frac{r^2}{2}\left(\frac{d^2 g}{dr^2}\right)_{r=0} + \cdots \end{aligned}\right] \qquad \ldots (5.15)$$

If one defines $\left(\frac{d^2 f}{dr^2}\right)_{r=0} = -\frac{1}{\lambda_f^2}$ and $\left(\frac{d^2 g}{dr^2}\right)_{r=0} = -\frac{1}{\lambda_g^2}$, Eq. 5.15 can be written as

$$\left.\begin{aligned} f &\doteq 1 - \frac{r^2}{2\lambda_f^2} \\ g &= 1 - \frac{r^2}{2\lambda_g^2} \end{aligned}\right] \qquad \ldots (5.16)$$

for very small r values. Equations 5.16 define parabolas with vertex at $r = 0$, $f = 1$ and $r = 0$, $g = 1$, and λ_f and λ_g are evidently the radii of curvature of f vs r and g vs r curves at $r = 0$, (see Fig. 5.6).

The two lengths λ_f and λ_g are called the *longitudinal* and *lateral microscales of turbulence*, and are considered to be measures of diameters of small eddies which are primarily responsible for the dissipation of energy through viscous shear. Using definitions of λ_f and λ_g, and Eq. 5.12, it can be easily shown that $\lambda_g = \lambda_f/\sqrt{2}$.

From f vs r and g vs r curves, one can also define two other scales related to area under these curves; they are called *macroscales of turbulence.*

$$\left.\begin{aligned} &\text{Longitudinal integral macroscale} \quad L_f = \int_0^\infty f(r)\,dr \\ &\text{Lateral integral macroscale} \quad L_g = \int_0^\infty g(r)\,dr \end{aligned}\right] \quad \ldots (5.17)$$

L_f and L_g represent the area under $f(r)$ vs r and $g(r)$ vs r curves. Further, since $g = f + \dfrac{r}{2}\dfrac{df}{dr}$, it can be shown that $L_f = 2\,L_g$. The length scales L_f and L_g are a measure of longest correlation distance in longitudinal and lateral directions. They are also a measure of size of large energy containing eddies.

Since turbulence measurements without any mean motion are very difficult, measurements in the wind tunnel downstream of a grid are used to study nearly homogeneous isotropic turbulence which is convected downstream at a constant velocity U. With these data one can study the variation of $R_E(t)$, the Eulerian correlation coefficient, with time t. The $R_E(t)$ vs t curve has the following characteristics.

(i) $R_E(0) = 1$

(ii) $R_E(t) = 0$ at large t values

(iii) $R_E(t)$ is an even function of t

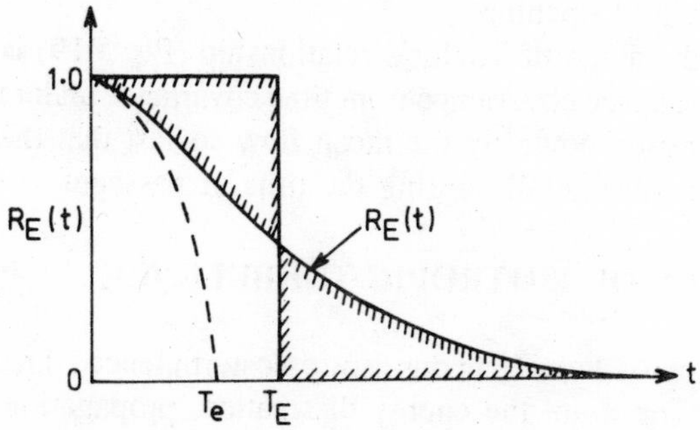

Fig. 5.7 Variation of $R_E(t)$ vs t

Typical variation R_E (t) vs t is shown in Fig. 5.7, which is similar to $f(r)$ vs r curve. For such a curve, two time scales of tubulence can be defined. These are

$$\left.\begin{aligned} &\text{Eulerian micro time scale } T_e \quad && \frac{1}{T_e^2} = -\frac{1}{2}\left(\frac{d^2R_E}{dt^2}\right)_{t=0} \\ &\text{Eulerian integral macro time scale } T_E \quad && T_E = \int_0^\alpha R_E(t)\,dt \end{aligned}\right] \quad \ldots (5.18)$$

The physical interpretation of T_e is that it is a measure of most rapid changes that occur in the velocity field.

Taylor's Hypothesis

According to Taylor's hypothesis, if the turbulent flow field has an average velocity U in the x direction, the turbulent fluctuations at a given fixed point can be approximately interpreted as a result of frozen pattern of turbulence being convected through a point with mean velocity U. If r is replaced by x, according to Taylor's hypothesis

$$\frac{\partial}{\partial t} = -U\frac{\partial}{\partial x} \quad \ldots (5.19)$$

The negative sign is necessary because a positive value of $\partial/\partial x$ at a point in space would correspond to a negative value of $\partial/\partial t$. It has been found (7) that Taylor's hypothesis is valid if $\sqrt{\overline{u'^2}}/U$ is much less than 0.45 for shear flow. It should be of the order 0.01 in isotropic homogeneous turbulence. Taylor's hyothesis is widely used in theoretical and experimental works on microstructure of atmospheric turbulence. As a consequence of Taylor's hypothesis

$$\left.\begin{aligned} &f(x) = R_E(t) \\ \text{and} \quad &L_f = UT_E \end{aligned}\right] \quad \ldots (5.20)$$

Validity of Taylor's hypothesis was confirmed by Favre et al., who obtained a unique curve when $f(x)$ and R_E (t) were plotted against x/M and Ut/M, where M is the grid opening.

The main advantage of Taylor's relationship (Eq. 5.19) is that it allows one to determine space covariance from time covariance under the assumption that they are carried down by the mean flow so fast that they do not have time to change substantially during the time of passage.

5.5 DYNAMICS OF ISOTROPIC TURBULENCE

In the dynamics of homogeneous isotropic turbulence, three aspects need consideration. These are the energy dissipation, propagation of correlation coefficient f (Karman-Howarth's equation) and energy spectrum analysis.

Energy Dissipation

In steady uniform laminar flow in a horizontal plane, the kinetic and potential energies of the fluid per unit volume must remain the same. Therefore, the work done by pressure forces per unit volume of fluid is used in the production of heat at a constant rate due to viscous action. In fact, the work done by the external forces in deforming the fluid elements is converted into heat by its interaction with fluid property of viscosity. For steady two dimensional laminar flow, it can be shown that, even though summation of forces on a fluid element in the flow direction is zero, the shear stresses produce a couple $(\tau\delta x\ \delta z)\delta y$ which performs work causing energy loss E. (see Fig. 5.8)

$$E = (\tau\delta x\ \delta z)\delta y\ \frac{du}{dy}$$

Or the rate of energy loss per unit volume $\varepsilon = \dfrac{E}{\delta x\delta y\delta z} = \tau\dfrac{du}{dy} = \mu\left(\dfrac{du}{dy}\right)^2$

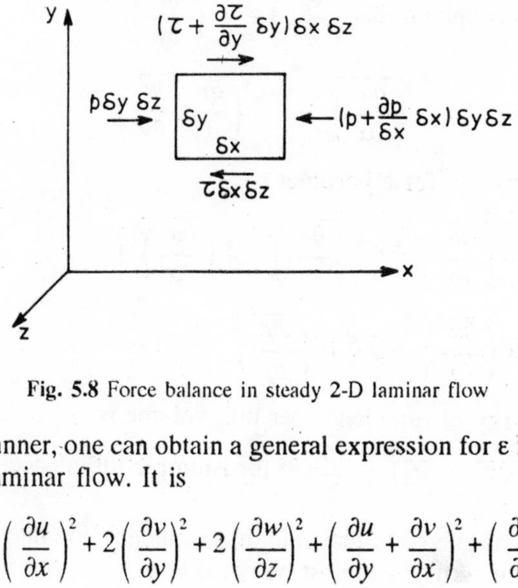

Fig. 5.8 Force balance in steady 2-D laminar flow

In a similar manner, one can obtain a general expression for ε in steady three dimensional laminar flow. It is

$$\varepsilon = \mu\left[2\left(\frac{\partial u}{\partial x}\right)^2 + 2\left(\frac{\partial v}{\partial y}\right)^2 + 2\left(\frac{\partial w}{\partial z}\right)^2 + \left(\frac{\partial u}{\partial y} + \frac{\partial v}{\partial x}\right)^2 + \left(\frac{\partial u}{\partial z} + \frac{\partial w}{\partial x}\right)^2 + \left(\frac{\partial v}{\partial z} + \frac{\partial w}{\partial y}\right)^2\right] \quad \ldots (5.21)$$

The same expression can be used to obtain the rate of energy dissipation in isotropic turbulence except that u, v, w will be replaced by u', v' w', and quantity in each bracket will be time averaged. Hence, for isotropic turbulence

$$\varepsilon = \mu\left[2\overline{\left(\frac{\partial u'}{\partial x}\right)^2} + 2\overline{\left(\frac{\partial v'}{\partial y}\right)^2} + 2\overline{\left(\frac{\partial w'}{\partial z}\right)^2} + \overline{\left(\frac{\partial u'}{\partial y} + \frac{\partial v'}{\partial x}\right)^2} + \overline{\left(\frac{\partial u'}{\partial z} + \frac{\partial w'}{\partial x}\right)^2} + \overline{\left(\frac{\partial v'}{\partial z} + \frac{\partial w'}{\partial y}\right)^2}\right] \quad \ldots (5.22)$$

Taylor (8) has shown that, because of isotropy, the following conditions are satisfied.

(i) $\overline{\left(\frac{\partial u'}{\partial x}\right)^2} = \overline{\left(\frac{\partial v'}{\partial y}\right)^2} = \overline{\left(\frac{\partial w'}{\partial z}\right)^2}$

(ii) $\overline{\left(\frac{\partial u'}{\partial y}\right)^2} = \overline{\left(\frac{\partial u'}{\partial z}\right)^2} = \overline{\left(\frac{\partial v'}{\partial x}\right)^2} = \overline{\left(\frac{\partial w'}{\partial x}\right)^2} = \overline{\left(\frac{\partial w'}{\partial y}\right)^2}$

(iii) $\overline{\frac{\partial v'}{\partial x}\frac{\partial u'}{\partial y}} = \overline{\frac{\partial u'}{\partial z}\frac{\partial w'}{\partial x}} = \overline{\frac{\partial v'}{\partial z}\frac{\partial w'}{\partial y}}$

Substitution of these relations in Eq. 5.22 yields

$$\varepsilon = \mu\left[6\overline{\left(\frac{\partial u'}{\partial x}\right)^2} + 6\overline{\left(\frac{\partial u'}{\partial y}\right)^2} + 6\overline{\frac{\partial v'}{\partial x}\frac{\partial u'}{\partial y}}\right]$$

Taylor has further shown that $\overline{\left(\frac{\partial u'}{\partial x}\right)^2} = \frac{1}{2}\overline{\left(\frac{\partial u'}{\partial y}\right)^2}$

and $$\overline{\left(\frac{\partial u'}{\partial x}\right)^2} = -2\left(\overline{\frac{\partial v'}{\partial x}\frac{\partial u'}{\partial y}}\right)$$

Therefore, expression for ε becomes

$$\varepsilon = \mu\left[6\overline{\left(\frac{\partial u'}{\partial x}\right)^2} + 12\overline{\left(\frac{\partial u'}{\partial x}\right)^2} - 3\overline{\left(\frac{\partial u'}{\partial x}\right)^2}\right]$$

or $$\varepsilon = 15\,\mu\overline{\left(\frac{\partial u'}{\partial x}\right)^2} = 7.5\,\mu\overline{\left(\frac{\partial u'}{\partial y}\right)^2} \qquad \ldots (5.23)$$

The kinetic energy of turbulence per unit volume is $k = \frac{\rho}{2}(\overline{u'^2} + \overline{v'^2} + \overline{w'^2}) = \frac{3\rho}{2}\overline{u'^2}$ for isotropic turbulence. If there is no external supply of energy to maintain the turbulence, the rate of decrease of kinetic energy of turbulence must be equal to ε.

or $$-\frac{\partial k}{\partial t} = 15\,\mu\overline{\left(\frac{\partial u'}{\partial x}\right)^2}$$

or $$\frac{\partial \overline{u'^2}}{\partial t} = -10\,\nu\overline{\left(\frac{\partial u'}{\partial x}\right)^2} = -5\overline{\left(\frac{\partial u'}{\partial y}\right)^2} \qquad \ldots (5.24)$$

Taylor has further shown that $\overline{\left(\frac{\partial u'}{\partial x}\right)^2}$ can be shown to be equal to $\overline{u'^2}/\lambda_f^2$.

Substitution of this value in Eq. 5.24, and use of Taylor's hypothesis yields

$$U \frac{\partial \overline{u'^2}}{\partial x} = 10\, \nu \frac{\overline{u'^2}}{\lambda_f^2} \qquad \ldots (5.25)$$

Equation (5.25) shows that the rate of change of $\overline{u}'^2$ is directly proportional to ν and $\overline{u'^2}$, and inversely proportional to square of microscale of turbulence. In other words, smaller eddies are more efficient in dissipating the energy of turbulence than the larger eddies. The term λ_f^2/ν has the dimension of time and, hence, Rouse prefers to call λ_f^2/ν as the time constant of decay.

Karman–Howarth's Equation

Karman and Howarth (6) have derived an equation which gives information on relationship for propagation of the correlation coefficient f with time and distance r in an isotropic flow. This equations is

$$\frac{\partial}{\partial t}(f\overline{u'^2}) + 2\,(\overline{u'^2})^{3/2}\left(\frac{\partial f}{\partial r} + \frac{4}{r}h\right) = 2\nu\overline{u'^2}\left(\frac{\partial^2 f}{\partial r^2} + \frac{4}{r}\frac{\partial f}{\partial r}\right)$$

which can also be written in alternate form as

$$\frac{\partial}{\partial t}(f\overline{u'^2}) + (\overline{u'^2})^{3/2}\frac{2}{r^4}\frac{\partial}{\partial r}(r^4 h) = 2\nu\,(\overline{u'^2})\frac{1}{4}\frac{\partial}{\partial r}\left(r^4\frac{\partial f}{\partial r}\right) \qquad \ldots (5.26)$$

Here h is the triple velocity correlation coefficient given by $h = \overline{v_a'u_b'}/(\overline{u'^2})^{3/2}$. As mentioned later, this equation is useful for making predictions about the decay of isotropic turbulence at various stages, e.g. when eddies are large and viscous effects negligible, and when eddies become very small and viscous effects predominate over the inertial forces. Since f and h can be measured with the help of hotwire anemometer, the above equation can be verified as has been done by Stewart (10). Here also it can be noticed that there are two unknowns, f and h, and only one equation. Several inferences or important conclusions can be drawn by making some assumption connecting f and h based on similarity, and by statistical and physical considerations.

If one assumes that h, and terms involving h, i.e. the 2nd term on the left hand side of Eq. 5.26, are small and can be neglected, it can be shown (7) that,

$$f(r,t) = e^{-r^2/8\nu t} \qquad \ldots (5.27)$$

Naturally this is valid when viscous effects dominate over the inertial effect, and under such condition, for given r value, $f(r,t)$ curve has a Gaussian error curve shape (see Fig. 5.9). Thus, when the viscous forces dominate, the above equation shows decay of $f(r)$ with t according to Gaussian law. In this range, it is found that

$$\overline{u'^2}\int_0^{\infty} r^4 f(r,t)\, dr = J \qquad \ldots (5.28)$$

where J is a constant with respect to time and is known as Loitsianskii's invariant. According to Loitsianskii, the integral in Eq. 5.28 gives the total amount of disturbance introduced by the turbulence generating system, and,

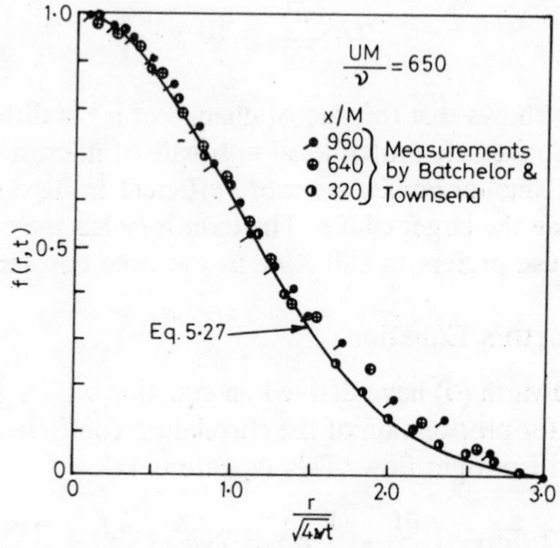

Fig. 5.9 Longitudinal velocity correlation coefficient $f(r,t)$ in the final period of decay

hence, such an amount should remain constant with time. This is again supposed to be true for viscosity dominated condition only. This is obtained on the assumption that all velocity correlations in homogeneous turbulence reduce exponentially as r increases. Batchelor and Proudman (7*) have shown that constancy of J required that $\lim_{r\to\infty} (r^4 k) = 0$, is not true in general case. However, in the final stage of decay of turbulence, where viscous terms in Karman–Howarth equation predominate, J is invariant.

Here $k(r_1 t)$ is triple velocity correlation coefficient

$k = \overline{u_a'^2 u_b'}/(\overline{u'^2})^{3/2}$ and it can be shown that $k(r,t) = -2h(r,t)$.

Karman and Howarth (6) have also shown that the equation for decay of isotropic turbulence, viz.

$$\frac{d\overline{u'^2}}{dt} = -10\nu \frac{\overline{u'^2}}{\lambda_f^2}$$

can be obtained from Karman–Howarth equation under the limitation that the left hand side of Eq. 5.26 can be neglected in preference to the right hand side, and that $f = 1$ when $r = 0$. Analysis of Karman-Howarth equation has led to the following conclusions:

(i) In the initial stages when the effect of viscosity is small compared to inertia, Eq. 5.26 can be reduced to

$$\frac{\partial(f\overline{u'^2})}{\partial t} + 2\,(\overline{u'^2})^{3/2}\left(\frac{\partial f}{\partial r} + \frac{4}{r}\,h\right) = 0$$

the solution of which gives

$$\left.\begin{aligned} \overline{u'^2}\, t^{10/7} &= \text{constant} \\ \text{and} \qquad \lambda_f^2 &= 7\nu t \end{aligned}\right] \qquad \dots (5.29)$$

(ii) In the final stages when $h << f$, the decay of $\overline{u'^2}$ is given by

$$\overline{u'^2} = \frac{\text{const.}}{t^{5/2}} \quad \ldots (5.30)$$

(iii) In the intermediate stages, the exponent n of t in the equations

$$\overline{u'^2} = \frac{\text{const}}{t^n} \quad \text{and} \quad \lambda_f^2 = \frac{10\nu t}{n}$$

is a function of $\sqrt{\overline{u'^2}}\,\lambda_f/\nu$. In this intermediate stage, Taylor found n to lie between 10/7 and 2.0, and can, therefore, be taken as 2.0 approximately.

5.6 SPECTRUM ANALYSIS OF ISOTROPIC TURBULENCE

One method of analysis of isotropic turbulence is the study of correlation coefficients as regards their variation with time and distance as is done through Karman-Howarth's equation. Another method is to analyse velocity fluctuations in terms of their spectrum in much the same ways as a beam of light can be separated into its spectral components. This technique is based on Eulerian concept of velocity. This form of spectral analysis of turbulence was first proposed by Taylor (11). It is not limited to isotropic turbulence; if the turbulence is anisotropic, spectrum analysis must be carried out in three dimensions and, hence, becomes more involved. Here only one-dimensional spectrum will be discussed in brief.

It has already been mentioned that turbulence consists of superposition of eddies of various sizes, strengths, and orientations. Further, eddies of large size cause fluctuations of low frequency, whereas smaller eddies cause fluctuations of high frequency. Hence, observed turbulent fluctuations at a given point with respect to time are caused by the passage of a whole range of sizes of eddies representing a very wide band of frequencies. Each eddy has a certain kinetic energy which is determined by its velocity or corresponding frequency. One can study the manner in which turbulence kinetic energy is distributed according to frequencies. This distribution of energy between different frequencies is known as the *energy spectrum*.

Two special features of specturm analysis applied to turbulence are the following: firstly, turbulence signal is not periodic and, hence, one must have a long enough record such that the Fourier components are averaged and are, therefore, effectively steady state values. Secondly, turbulence signal contains a very broad band of frequencies, since the largest size of the eddy is governed by the size of the flow itself, whereas the smallest size of eddy is determined by viscosity and it decreases with increasing average velocity of flow.

Consider a quasi-steady (i.e. statistically homogeneous with respect to time) turbulent flow field with $\overline{u'^2}$ as mean of squares of turbulent velocity fluctuations in the x direction. Then $\frac{\rho\overline{u'^2}}{2}$ will be the kinetic energy per unit volume. If $\frac{\rho\overline{u'^2}}{2} F_1(n)dn$ represents the contribution of kinetic energy due to fluctuations between n and $(n + dn)$ frequencies, then it is obvious that

$$\int_0^\infty \frac{\rho \overline{u'^2}}{2} F_1(n)\, dn = \frac{\rho \overline{u'^2}}{2}$$

or

$$\int_0^\infty F_1(n)\, dn = 1.0 \qquad \text{. . . (5.31)}$$

The function $F_1(n)$ is known as the normalised one dimensional energy spectrum, first introduced by Taylor (11) in 1938, and by Obukhov (12). The upper limit of n would be $1/Te$, where Te is the Eulerian microscale of time.

It can be logically as well as analytically shown that the normalised one dimensional spectrum function $F_1(n)$ and the longitudinal correlation function $f(x)$, or Eulerian correlation function with respect to time $R_E(t)$ are related. Consider (a) variation in instantaneous velocity u at a fixed point due to passage of small eddies at a constant velocity; and (b) the variation in instantaneous velocity u under the same conditions due to passage of large eddies. In the first case the fluctuations at a fixed point will be much more rapid than they are in the second case, but the energy associated with them will be smaller. In the second case the energy associated with the fluctuations will be larger, but the frequencies will be smaller. Hence, $F_1(n)$ vs. n curves for large and small eddies will be as shown in Fig. 5.10 (a). Furthermore, if larger eddies are present, $f(x)$ vs x curve will extend over a larger distance than when smaller eddies are present as shown in Fig. 5.10 (b). Thus, when $f(x)$ vs x curve has

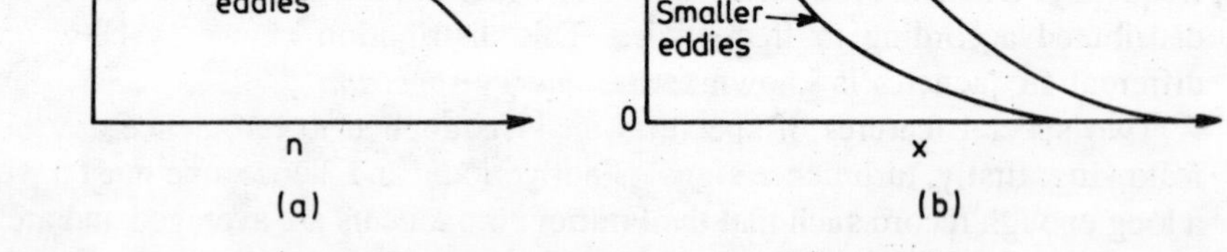

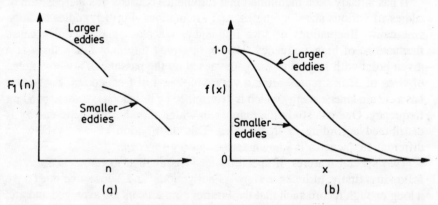

Fig. 5.10 Qualitative relation between $F_1(n)$ and $f(x)$

small spread in the x co-ordinate, $F_1(n)$ vs n curve will extend over larger n values. Similarly, when $f(x)$ vs x curve has a larger spread, $F_1(n)$ vs n curve will be in the smaller n values range and $F_1(n)$ will be large. This shows that $f(x)$ and $F_1(n)$ curves are related. Taylor (11) has shown $f(x)$ and $F_1(n)$ are related as

$$f(x) = \int_0^\infty F_1(n) \cos\left(\frac{2\pi\, nx}{U}\right) dn \qquad \text{. . . (5.32)}$$

where U is the velocity with which turbulence is carried downstream. Alternately

$$F_1(n) = \frac{4}{U}\int_0^\infty f(x) \cos\left(\frac{2\pi nx}{U}\right) dx \qquad \ldots (5.33)$$

In other words $f(x)$ and $F_1(n)$ are Fourier cosine transforms of each other. Therefore, if $f(x)$ vs x curve is known along with U, $F_1(n)$ vs n curve can be obtained by using Eq. 5.33. Alternately, if $F_1(n)$ vs n curve and U are known, $f(x)$ vs x can be obtained from Eq. 5.32. This has been verified by Taylor, using measurements made by Simmons et al. in a wind tunnel behind a grid. Equations 5.32 and 5.33 can be expressed in terms of $R_E(t)$ and t, since $f(x) = R_E t$ and $x/U = t$

$$\left.\begin{aligned} R_E(t) &= \int_0^\infty F_1(n) \cos(2\pi nt)\, dn \\ F_1(n) &= 4\int_0^\infty R_E(t) \cos(2\pi nt)\, dt \end{aligned}\right] \qquad \ldots (5.34)$$

Energy spectrum can also be expressed in terms of wave number

$$K = \frac{2\pi n}{U}. \int_0^\infty F_1(n)\, dn = \int_0^\infty F_1(K)\, dK = 1.0$$

$$\left.\begin{aligned} f(x) &= \int_0^\infty F_1(K) \cos(Kx)\, dK \\ \text{and} \qquad F_1(K) &= \frac{2}{\pi}\int_0^\infty f(x) \cos(Kx)\, dx \end{aligned}\right] \qquad \ldots (5.35)$$

From Eqs. 5.32, 5.33 and 5.34, the following results can be obtained. When $n = 0$, Eq. 5.33 gives

$$F_1(0) = \frac{4}{U}\int_0^\infty f(x)\, dx = \frac{4L_f}{U}$$

Hence,

$$L_f = \frac{UF_1(0)}{4} \qquad \ldots (5.36)$$

Thus, longitudinal integral macroscale of turbulence L_f can be obtained from $F_1(n)$ vs n curve, using $F_1(n)$ value at extremely low values of n. Further, since $f(x) = R_E(t)$, and $L_f = UT_E$, Eqs. 5.36 gives

$$T_E = \frac{F_1(0)}{4} \qquad \ldots (5.37)$$

The same result can be obtained from Eq. 5.34. Also, by differentiating Eqs. 5.32 and 5.34, it can be shown that

$$\left.\begin{aligned}\frac{1}{\lambda_f^2} &= -\frac{1}{2}\left(\frac{\partial^2 f}{\partial x^2}\right)_{x=0} = \frac{2\pi^2}{U^2}\int_0^\infty F_1(n)\,n^2\,dn \\ \frac{1}{T_e^2} &= -\frac{1}{2}\left(\frac{\partial^2 R_E(t)}{\partial t^2}\right)_{t=0} = 2\pi^2\int_0^\infty F_1(n) n^2\,dn\end{aligned}\right] \quad \ldots (5.38)$$

These results indicate that the size of micro eddies, of which λ_f and T_e are a measure, corresponds mainly to the values of $F_1(n)$ for higher values of n.

It was indicated earlier (see Eq. 5.13) that $f(x)$ can be expressed by the relation

$$f(x) = e^{-x/L_f}$$

Substitution of this in Eq. 5.33 yields

$$\begin{aligned}F_1(n) &= \frac{4}{U}\int_0^\infty e^{-x/L_f}\cos\left(\frac{2\pi\, nx}{U}\right)dx \\ &= \frac{4}{U}\,\frac{L_f}{1+\left(\dfrac{4\pi^2 n^2}{U^2}\right)L_f^2}\end{aligned} \quad \ldots (5.39)$$

Hence, a plot can be prepared between $UF_1(n)/L_f$ and nL_f/U, as was done by Favre (see Fig. 5.11). It can be seen that the approximation as expressed by Eq. 5.13 and by Eq. 5.39 seems to be satisfactory. When n approaches zero, $\dfrac{UF_1(n)}{L_f}$ approaches 4.0.

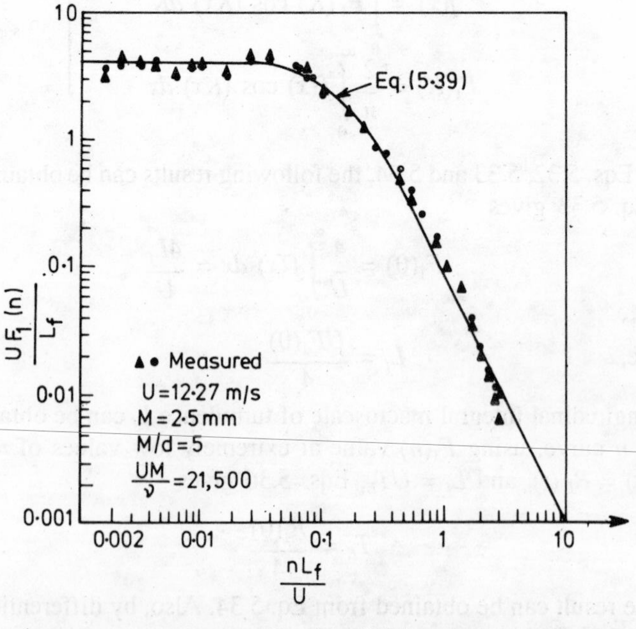

Fig. 5.11 Variation of $UF_1(n)/L_f$ with nL_f/U as obtained by Favre et al.

Two additional comments will be made here. Firstly, one can treat the turbulent fluctiations $v'(t)$ and $w'(t)$, in a similar manner and $F_2(n)$ and $F_3(n)$ can be determined. Secondly, it needs to be emphasised that Taylor's $F_1(n)$ spectrum function is a one dimensional section of what is really a three dimensional spectrum function; this has been studied by Kempe' de Feriet.

5.7 KOLMOGOROV'S THEORY OF LOCAL ISOTROPY

Basic postulate of Kolmogorov's theory (13, 14) is that the boundaries of the flow system cannot affect directly any details of velocity field except those which are typified by lengths of the same order as the boundary scale. As mentioned earlier, larger scale turbulence fluctuations are generated by the mean flow itself through the work it does against the Reynolds stresses. These fluctuations, in turn, give rise to fluctuations of next smaller scale through the same inertial mechanism, and so on until the scale becomes so small that viscous stresses dominate and kinetic energy is directly converted into heat. As the scale decreases, the proportional rate of loss increases and the turbulent motion adjusts itself to some definite state. Further down the scale, the turbulent motion is nearly independent of the large eddies. Thus, at such a small scale, the turbulence will be locally isotropic even though the large scale turbulence may be anisotropic. Under such condition, the turbulence characteristics are solely determined by the rate of energy dissipation per unit volume, ε and kinematic viscosity ν. Kolmogorov's theory can be expressed in the form of following hypotheses.

First Hypothesis: Even though the flow as a whole may be aniostropic, the turbulence in the spatial and temporal domain G, of the general flow system will be essentially isotropic, if G is small enough and Reynolds number of mean flow is large enough. This is known as the hypothesis of local isotropy.

Second Hypothesis: In a locally homogeneous and isotropic domain G, the turbulence characteristics are determined by the forces of friction and inertia (i.e. the mean rate of energy dissipation per unit volume, ε, and the kinematic viscosity).

As a consequence of this hypothesis, Kolmogorov's length and velocity scales are defined as

length scale $l_k = (\nu^3/\varepsilon)^{1/4}$

velocity scale $v_k = (\nu\varepsilon)^{1/4}$

If Reynolds number is formed from these length and velocity scales, one gets

$$\frac{v_k\, l_k}{\nu} = \frac{1}{\nu}\,(\nu\varepsilon)^{1/4}\left(\frac{\nu^3}{\varepsilon}\right)^{1/4} = 1.0$$

This state of flow is known as the universal range. In this range, if one uses

ν and ε for plotting the spectrum at different times, they should collapse on a single curve. The wave number K_d where the viscous effects are very strong is of the order of $K_d = 1/l_k$. Townsend has found that most of the viscous dissipation occurs when $K_d\, l_k$ is less than 0.50.

Third Hypothesis: This hypothesis states that if the Reynolds number of turbulence is very large, there exists a range of wave numbers in the energy spectrum lying between energy containing eddies wave number K_e and Kolmogorov's wave number K_d, where the structure of the spectrum function depends only on the parameter ε and not on ν. This range of wave numbers is called *inertial subrange.*

Energy Spectrum vs K Curve

Some comments on energy spectrum vs K curve for homogeneous isotropic turbulence behind grid are in order. For this a three dimensional spectrum function $E(K)$ is defined as

$$\frac{3}{2}\,\overline{u'^2} = \int_0^\infty E(K)\, dK \qquad \ldots (5.40)$$

It may be noticed that $E(K)$ has the dimensions of L^3/T^2. In fact, $E(K)$ will be a function of x for grid turbulence; and since $x/U = t$, it can be considered to be function of K and t. The dynamic equation for energy spectrum describes the behaviour of energy spectrum as a function of time and space coordinates. It is

$$\frac{\partial E}{\partial t} = W(K,t) - 2\nu K^2 E \qquad \ldots (5.41)$$

where $W(K,t)$ is the energy transfer function representing transfer of energy from larger eddies to the eddies under consideration, while the second term represents the dissipation term.

If for a moment it is assumed for simplicity that the interaction between eddies of different K values is negligibly small, e.g. $W(K,t)$ is zero, solution of Eq. 5.41 yields.

$$E(K,t) = E(K,t_o)\, e^{-2\nu K^2(t - t_o)} \qquad \ldots (5.42)$$

From Eq. 5.42, it can be seen that decrease of kinetic enegy with time occurs at a higher rate for eddies with larger K than for eddies for smaller K, a fact already stated. Further, if one assumes that $f(x,t) = e^{x^2/8\nu t}$ and $g(x,t)$ is related to $f(x,t)$, it can be shown that

$$E(K,t) \sim K^4\, e^{-2K^2\nu t} \qquad \ldots (5.43)$$

According to this relationship $E(K,t)$ increases very rapidly initially according to K^4, reaches a maximum value, and decreases monotonically to zero as K increases. Figure 5.12 shows variation of $E(K,t)$ with t.

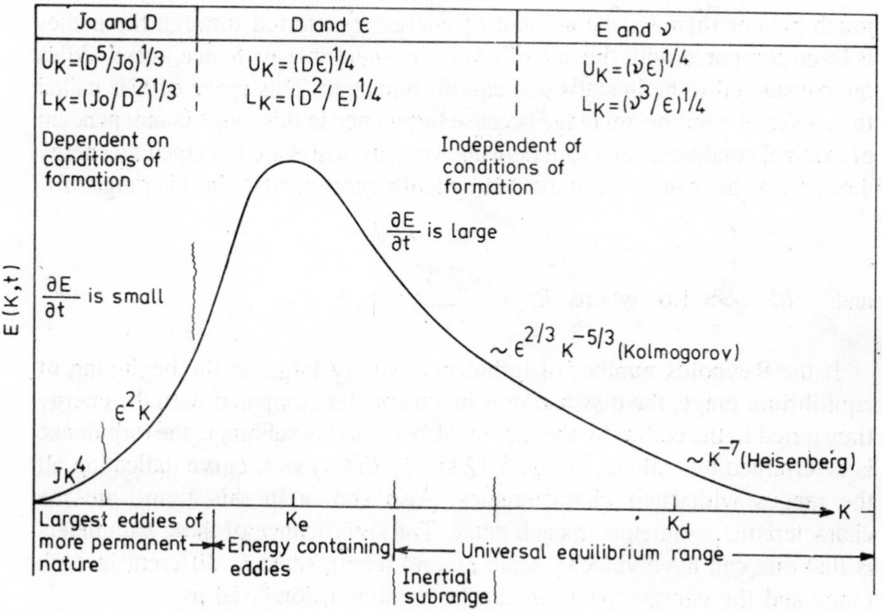

Fig. 5.12 Variation of $E(k,t)$ with K

At very small K values (i.e. when eddies are very large) produced immediately downstream of grid, the turbulence is anisotropic. These large eddies with small wave numbers are of permanent nature, and contain relatively small fraction of total energy, say about 20 percent. Here, energy spectrum is governed by Loitsianskii invariant J. Some distance in the downstream, direction is necessary for larger eddies to break into smaller ones and to change the flow to more isotropic condition. In this range, spectrum is essentially governed by J and eddy diffusivity or the diffusion coefficient D. The range of wave numbers which make the main contribution to the energy of turbulence is called the range of energy containing eddies. In this range, the energy spectrum shows a maxima. One can associate a wave number K_e with the size of eddies providing maximum contribution to the energy. If l_e is defined as $1/K_e$, l_e can be interpreted as the average size of energy containing eddies, and l_e is of the same order of magnitude as integral macroscales L_f or L_g. In the range of energy containing eddies, spectrum is governed by D and ε. Eddy diffusivity can be given by $\overline{u'^2} l_e^2/\nu$.

The energy dissipation by viscous effects increases as the eddies become smaller and smaller (i.e. as K increases) upto a certain size of small eddies associated with wave number K_d. Here, energy dissipation is found to be proportional to $K^2E(K,t)$ and, hence, K_d corresponds to the value of K for which $K^2E(K,t)$ is maximum. This wave number is of the same order of magnitude as $1/l_k$, where l_k is Kolmogorov's length scale.

At high Reynolds numbers, when K_e is much smaller than K_d, and K is much greater than K_e, the amount of energy transferred through the eddies is large compared with the rate of change of their energy; hence, these eddies are considered to be in statistical equilibrium state. This range of K is called the universal equilibrium range because turbulence in this range is independent of external conditions, and its length and velocity scales are governed by energy dissipation rate ε and ν as indicated by Kolmogorov. Also, in this range of K

$$l_e >>> l_d$$

and $\quad R_{e\lambda}^{3/2} >>> 1.0$ where $R_{e\lambda} = \dfrac{\sqrt{\overline{u'^2}}\lambda_g}{\nu}$

If the Reynolds number of turbulence is very large, at the beginning of equilibrium range, the dissipation is much smaller compared with the energy transferred to the eddies by the inertial effects. In this subrange, the turbulence is determined by ε alone. Figure 5.12 shows $E(K,t)$ vs K curve indicating all the ranges with their characteristics. Also shown in this figure are the characteristic parameters in each range. The significance of these parameters is that one can have velocity scale U_k and length scale L_k different in each range and the energy spectrum can be nondimensionalised as

$$E(K,t) = U_k^2 L_k \phi (K L_k)$$

Some attempts have been made to apply the principle of self preservation to energy spectrum for turbulence downstream of grid. The term self preservation is used if it is assumed that the turbulence structure maintains its similarity during decay. This would be true if, for a single choice of U_k and L_k, $E(K,t)/U_k^2 L_k$ vs KL_k would plot as a unique curve. This has not been found to be true. Stewart and Townsend plotted $E\left(K, \dfrac{x}{M}\right) \Big/ \lambda_g^2\sqrt{\overline{u'^2}}$ vs $\lambda_g K$ for grid turbulence data at various x/M values (see Fig. 5.13). It is apparent from this figure that similarity is not achieved in low wave number range; for large wave numbers (i.e. in the inertial subrange and equilibrium range), the curves for various x/M values seem to fall close to one another, indicating at least partial self-preservation. Since spectrum and correlation functions are related, the correlation function also satisfies partial self preservation.

5.8 GRID TURBULENCE AND ITS DECAY

It has been stated earlier that turbulence produced in the central core of wind tunnel by insertion of grid is near isotropic. Since this turbulence decays slowly

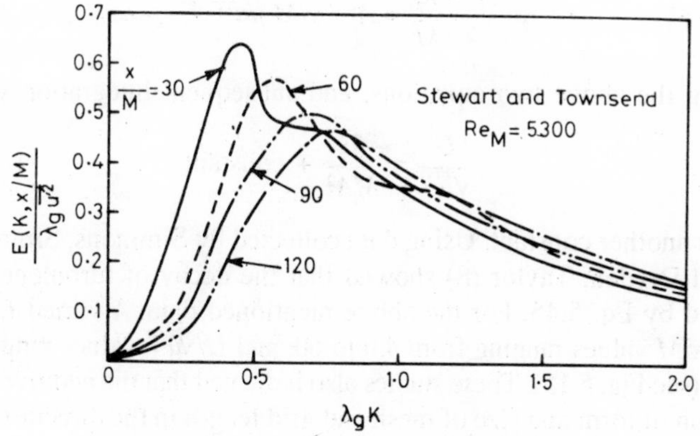

Fig. 5.13 Self-preservation of grid turbulence

because of absence of appreciable shear, one assumes the grid turbulence to be near homogeneous. Various aspects of grid turbulence have been studied experimentally and analytically by Simmons, Dryden, Simmons and Salter, Taylor, Batchelor, Baines and Peterson, and Naudascher and Farrel.

Grids are usually made of round iron bars, or flats, forming a single plane, or double plane square, or rectangular grid; however, square grids are more often used. Typical grids with notations are shown in Fig. 5.14. Wire mesh screens are also often used in wind tunnels. Similarly, hexagonal honeycombs and honeycombs using short circular tubes have also been used. These grids are effective in breaking large energy containing eddies to the size of grid or screen opening, and also in regulating the magnitude of turbulent fluctuations immediately downstream of the grid and its decay process.

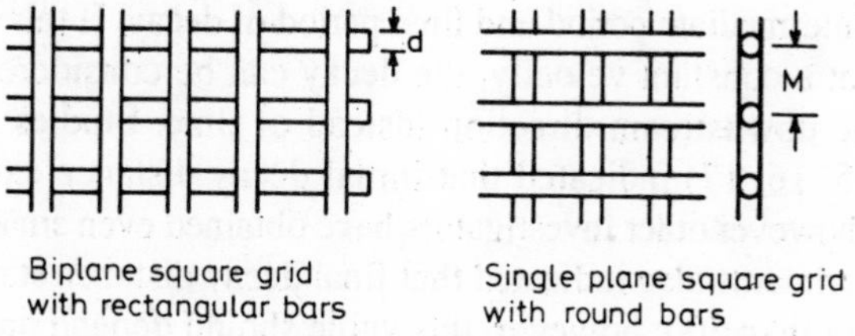

Fig. 5.14 Different grids

Using Taylor's hypothesis, it has been shown earlier that

$$-U \frac{\partial \overline{u'^2}}{\partial x} = 10\nu \frac{\overline{u'^2}}{\lambda_f^2}$$

However, Taylor has shown that λ_f is related to M of the grid as

$$\frac{\lambda_f}{M} = A\sqrt{\nu/M\sqrt{\overline{u'^2}}} \quad \ldots (5.44)$$

Combining the above two equations, and subsequent integration yields

$$\frac{U}{\sqrt{\overline{u'^2}}} = \frac{5x}{MA^2} + \text{constant} \quad \ldots (5.45)$$

where A is another constant. Using data collected by Simmons, Simmons and Salter, and Dryden, Taylor (8) showed that the decay of turbulence can be represented by Eq. 5.45. For the above mentioned data, A varied from 1.95 to 2.0 for x/M values ranging from 4.0 to 60, and $U/\sqrt{\overline{u'^2}}$ values ranging from 4.9 to 110 (see Fig. 5.15). These studies also indicated that the additive constant is a function of form and size of mesh and grid length in the direction of flow.

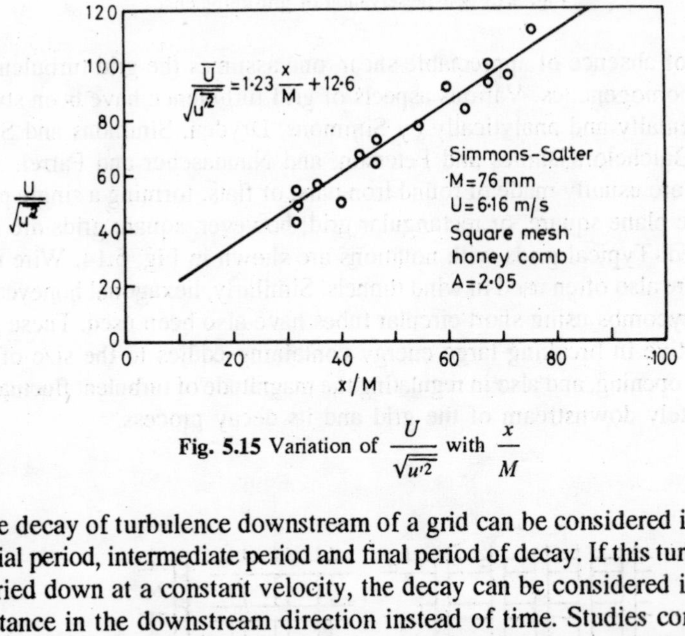

Fig. 5.15 Variation of $\frac{U}{\sqrt{\overline{u'^2}}}$ with $\frac{x}{M}$

The decay of turbulence downstream of a grid can be considered in terms of initial period, intermediate period and final period of decay. If this turbulence is carried down at a constant velocity, the decay can be considered in terms of distance in the downstream direction instead of time. Studies conducted by Townsend (15, 16, 17) indicated that initial decay distance extends upto $x/M = 50$ to 100; however other investigators have obtained even smaller values. Townsend's experiments also indicated that final decay distance starts beyond x/M values greater than 500; however, this value should depend on Reynolds number UM/ν value, which was 650 in case of Townsend's experiments. In the final period of decay, the eddies become very small and viscous effects predominate over the inertial effects. It has been found that in the final period of decay

$$\overline{u'^2} = \text{const } t^{-5/2} \quad \ldots (5.46)$$

Batchelor and Townsend (16) have proposed the following form of relationship for the decay of turbulence downstream of grid.

$$U^2/\overline{u'^2} = \frac{C}{C_D}\left[\frac{x}{M} - \left(\frac{x}{M}\right)_o\right] \qquad \ldots (5.47)$$

where C_D is the drag coefficient for the grid and is given by

$$C_D = \frac{(d/M)\,(2 - d/M)}{(1 - d/M)^4} \qquad \ldots (5.48)$$

for square mesh grid of round bars, and $C = 106$. For square mesh grids often used, i.e. with $M/d = 5$ to 6, C/C_D is about 135. It was also found that, for larger values of M/d, d as a length parameter is more appropriate than M in x/M. Batchelor and Townsend (15) found the turbulence to be isotropic downstream of grid when x/M is greater than 10 to 15. Also, if M/d is equal to two the isotropy would reach earlier. Grant and Nisbet (7*) observed considerable departure from homogeneity up to x/M equal to eighty. They believe these departures to be responsible for difference in the results obtained by various investigators about decay of turbulence in the initial period.

When UM/ν and $\sqrt{\overline{u'^2}}M/\nu$ are large, and if x/M is greater than twenty, linear law between $U^2/\overline{u'^2}$ and x/M is found to be valid for x/M values up to 200 (see Fig. 5.15). Probably transition begins at x/M equal to 200. The virtual origin downstream at which $U/\sqrt{\overline{u'^2}}$ is zero increases as x/M increases. For M/d equal to four or five $(x/M)_o$ was found to be unity.

Figure 5.16 shows the variation of $(U^2/\overline{u'^2})^{2/5}$ with x/M for Batchelor and Townsend's data, which shows that beyond x/M equal to 600 the final period of decay starts where $(U^2/\overline{u'^2})^{2/5}$ varies linearly with x/M. This figure also shows variation of λ_g^2 with x/M. Plot of $(\lambda_g^2 U/10\nu M)$ with x/M in the initial period of decay indicated a linear relationship.

Naudascher and Farrel (18) have presented a unified analysis of the data on grid turbulence. The data used by them include those collected by Dryden, Von Karman, Batchelor and Townsend, Wyatt, Zijnen, Compte-Bellot and Corrsin, Baines and Peterson, and Stewart and Townsend covering the following ranges of pertinent variables:

$Ud/\nu = 123$ to 19750 $\quad M = 1.59$ mm to 80 mm

$d = 0.30$ mm to 24.8 mm $\quad$ Solidity ratio $S = 0.15$ to 0.75

$C_D = 0.30$ to 3.00

for one plane and biplane grids using bars and rods. Their analysis permits estimation of turbulence intensity, integral macroscale L_f, and microscale λ_f for the grid turbulence when U and grid characteristics are known. Their analysis of the above mentioned data indicated that just downstream of the grid, the turbulence is isotropic but farther downstream K_1 in

$$\overline{u'^2} = K_1\,\overline{v'^2} = K_1\,\overline{w'^2}$$

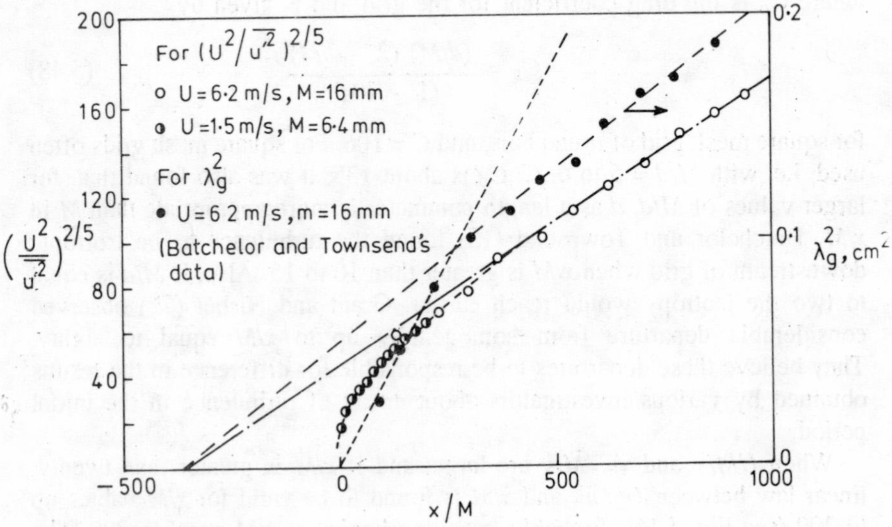

Fig. 5.16 Decay of $U / \sqrt{\overline{u'^2}}$ with x/M in the final period of decay

assumes a constant value between 1.30 and 1.45, the K_1 value depending, among other aspects, on grid construction and grid geometry. According to Naudascher and Farrel, assumption of a constant value of K_1 is incompatible with Kolmogorov's hypothesis of local isotropy. Hence, they argued that Kolmogorov's hypothesis does not apply to grid turbulence at least within the grid-flow Reynolds numbers of practical interest. The analysis of Naudascher and Farrel is based on two assumptions. Firstly, they assumed that

$$\frac{dL_f}{dx} = C_1 \frac{\sqrt{\overline{u'^2}}}{U} \qquad \ldots (5.49)$$

where $\overline{u'^2} = \overline{v'^2} = \overline{w'^2}$. Such an assumption is usually made in the analysis of flow past point and line sources of turbulence energy. They have further postulated that C_1 depends on grid geometry and Ud/ν where d is the diameter or width of grid elements. The second assumption made by them is

$$\frac{dL_f}{dx} = C_2 \frac{d\lambda_f}{dx} \qquad \ldots (5.50)$$

Omitting the subscripts of L_f and λ_f for convenience, Eqs. 5.49 and 5.50 give

$$\frac{\sqrt{\overline{u'^2}}}{U} = \frac{1}{C_1}\frac{dL}{dx} = \frac{C_2}{C_1}\frac{d\lambda}{dx} = C\frac{d\lambda}{dx} \qquad \ldots (5.51)$$

Substitution of Eq. 5.51 in

$$U \frac{d\overline{u'^2}}{dx} = -10 \frac{\overline{u'^2}}{\lambda^2}$$

and subsequent integration with the boundary conditions

$L = L_o$ and $\lambda = 0$ at $x = 0$,

and $L = L_\infty$ and $\lambda = \lambda_\infty$ at $x \to \infty$ yields

$$-5(x - x_o)d\, Re_d = (\lambda_\infty/d)^2 \log\left(1 - \frac{\lambda/d}{\lambda_\infty/d}\right) + \left(\frac{\lambda}{d}\right)\left(\frac{\lambda_\infty}{d}\right) \quad \ldots (5.52)$$

and

$$\frac{\sqrt{\overline{u'^2}}}{U}\frac{Re_d}{C} = 5\left(\frac{1}{\lambda/d} - \frac{1}{\lambda_\infty/d}\right) \quad \ldots (5.53)$$

and

$$Re_\lambda = \frac{\lambda\sqrt{\overline{u'^2}}}{\nu} = 5C\left(1 - \frac{\lambda}{\lambda_\infty}\right) \quad \ldots (5.54)$$

in conjunction with

$$L/d = \left(\frac{L_\infty - L_o}{\lambda}\right)\left(\frac{\lambda}{d}\right) + \frac{L_o}{d} \quad \ldots (5.55)$$

Here $Re_d = Ud/\nu$. Equations (5.53), (5.54) and (5.55) completely describe variations of $\sqrt{\overline{u'^2}}$, λ and L with x. To estimate these quantities one must know C, λ_∞, L_∞ and L_o in addition to U, d and ν. On the basis of analysis of data, Naudascher and Farrel have recommended the following values:

$C = 0.15\,(Re_d)^{1/2}$ for sharp edged grids and grids with large Re_d values.

$\lambda_\infty/d = 24.0$, $L_\infty/\lambda_\infty = 2.0$ and $L_o/d = 1.25$

In addition x_o/M varies with Re_d as follows:

Re_d	100	600	800	1000	1500	2000	4000	6000	10000	20000
x_o/M	15.0	13.8	13.0	12.0	10.0	7.8	5.0	4.0	3.4	3.0

Using the above information one can compute λ, L, $\overline{u'^2}$ in the following manner. For known Re_d, x_o can be determined from the above table and $\lambda_\infty = 24\, d$. Then, λ can be determined for any value of x using Eq. 5.52. With known C and λ for give value of x, Eq. 5.53 will given the magnitude of $\sqrt{\overline{u'^2}}$, Equation 5.55 will give magnitude of L for known value of x, L_∞, L_o and λ.

5.9 TURBULENCE IN STIRRED TANKS (19)

In many unit processes in chemical and environmental engineering, turbulence is generated in a circular tank by rapid rotation of impeller blades. Different types of blades used are shown in Fig. 5.17. The relation of power requirements with the turbulence characteristics can be obtained (19) as follows. The power required P in Nm/s depends on the speed of rotation N in rpm, characteristic length of blades l, diameter of tank D, and mass density ρ and viscosity of the liquid μ. Hence dimensional analysis yields

$$P/\rho N^3 l^5 = f\,(Nl^2\rho/\mu,\ l/D) \qquad \ldots (5.56)$$

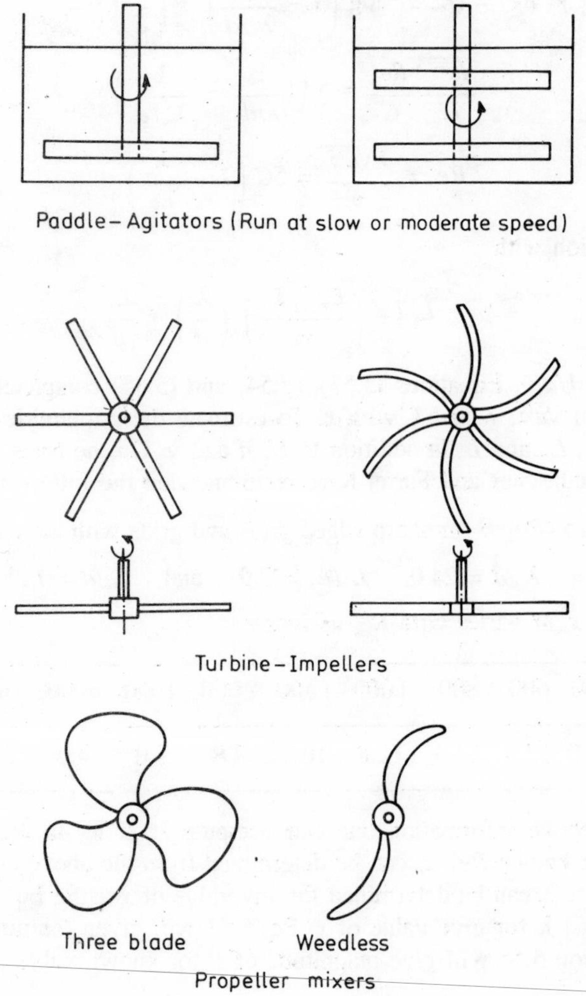

Fig. 5.17 Different types of blades used in stirred tanks

If the Reynolds number $Nl^2\rho/\mu$ is greater than 300, $P/N^3\rho l^5$ will have a constant value which depends on shape and number of blades, l/D and baffling used to inhibit circulatory flow. This power number $P_N = P/\rho N^3 l^5$ varies from 0.3 to 1.0 for propellers and 1.5 to 11.0 for turbine blades and paddles. If H is the depth of flow in the tank

$$\text{Power per unit mass } \varepsilon = \frac{P_N N^3 \rho l^5}{(\pi/4) D^2 H\rho}$$

$$\text{or} \qquad = \frac{4 P_N}{\pi} \frac{N^3 l^5}{D^2 H} \qquad \ldots (5.57)$$

If one assumes that the turbulence in the tank is isotropic, one can obtain expression relating ε with velocity fluctuations and macroscale of turbulence L_g as follows:

According to Kolmogorov

$$v_k = (\varepsilon \nu)^{1/4}$$

and

$$l_k = (\nu^3/\varepsilon)^{1/4}$$

Eliminating kinematic viscosity ν from these two equations, one gets

$$\varepsilon = v_k^3/l_k \qquad \ldots (5.58)$$

Assuming the same form of relationship to be valid if v_k is replaced by $\sqrt{\overline{u'^2}}$ and l_k by L_g, one can write

$$\varepsilon = (\overline{u'^2})^{3/2}/L_g \qquad \ldots (5.59)$$

Equating ε from Eqs. 5.57 and 5.59, one gets

$$\sqrt{\overline{u'^2}} = \left(\frac{4 P_n L_g N^3 l^5}{\pi D^2 H} \right)^{1/3}$$

Experiments have indicated the $L_g = 0.08\ l$.

$$\text{Hence,} \qquad \sqrt{\overline{u'^2}} = \left(\frac{0.32 P_N}{\pi D^2 H} \right)^{1/3} N l^2 \qquad \ldots (5.60)$$

Assuming geometric similarity, the above equation would imply that

$$\sqrt{\overline{u'^2}} \propto Nl$$

Expeimental data have demonstrated the validity of the above equation outside the immediate discharge zone of the impeller blades. Volumetric flow rate Q from an impeller is proportional to N and cross sectional dimension of blade. Hence, from dimensional consideration one can write

$$Q = C_1 N l^3$$

For propeller and turbine impellers, C_1 is in the range of 0.4 to 0.60. If one computes power per unit discharge

$$\frac{P}{Q} \propto \frac{N^3 l^5}{N l^3}, \text{ or } N^2 l^2 \text{ or } \overline{u'^2}$$

REFERENCES

1. Bayazit, M. Boyok Porozloloklo Acik Kanallarda Akim Turkiye Bilimsel Ve Teknik Arastima Kurumu Mühendislik Arastirma Grubu. Proj. No. Mag. 353/A, Istambul, 1975.
2. Landahl, M.J. and E. Mollo-Christensen. *Turbulence and Random Processes in Fluid Mechanics*. Cambridge University Press, Cambridge, 1986.
3. Lin, C.C. *Statistical Theories of Turbulence*. Princeton Aeronautical Paperbacks. Princeton University Press, Princeton (U.S.A.) 1969.
4. Panchev, S. *Random Functions and Turbulence*. Pergamon Press, Oxford, 1971.
5. Batchelor, G.K. *The Theory of Homogeneous Turbulence*. Cambridge University Press, U.K. 1960.
6. Karman, T.V. and L. Howarth. On the Statistical Theory of Isotropic Turbulence. Proc. RSL, A 164, 1936.
7. Hinze, J.O. *Turbulence: An Introduction to its Mechanism and Theory*. McGraw Hill Series in Mechanical Engineering. New York, 1959. Chapter III.
8. Taylor, G.I. Statistical Theory of Turbulence. Pt 1–4. Proc. RSL, A 151. 421–478, 1935.
9. Rouse, H. (Ed), *Advanced Fluid Mechanics*. John Wiley and Sons, Inc. New York, 1959.
10. Stewart, R.L. Triple Velocity Correlation in Isotropic Turbulence. Proc. Cambridge Philosophical Society, Vol. 47, 1951.
11. Taylor, G.I. The Spectrum of Turbulence. Proc. Royal Society of London, A 164, 1938.
12. Obukhov, A.M. On the Spectral Distribution of Energy in Turbulent Flow. Izv. Acad. Sci. USSR, Geogr. and Geophys. Ser. Nos. 4, 5. 1941.
13. Kolmogorov, A.N. The Local Structure of Turbulence in Incompressible Viscous Fluid for Very Large Reynolds Numbers. Comptes Rendus (Doklady) de '1' academy des Science de 'l' U.R.S.S. 32, 1941.
14. Karman, T.V. and C.C. Lin. On the Concept of Similarity in the Theory of Isotropic Turbulence. Reviews of Modern Physics, Vol. 21, 1949.
15. Batchelor, G.K. and A.A. Townsend. Decay of Vorticity in Isotropic Turbulence. Proc. RSL, 190 A, 1947.
16. Batchelor, G.K. and A.A. Townsend. Decay of Isotropic Turbulence in the Initial Period. Proc. RSL, 193 A, 1948.
17. Batchelor, G.K. and A.A. Townsend. Decay of Turbulence in Final Period. Proc. RSL, 194 A, 1948.
18. Naudascher E. and C. Farell. Unified Analysis of Grid Turbulence. Jour of Engg. Mech., Proc. ASCE Vol. 96. No. EM2, April 1970.
19. Davies, J. T. *Turbulence Phenomena*. Academic Press Inc. New York, 1972.

CHAPTER VI

Turbulence Models

6.1 INTRODUCTION

In the discussion of Reynolds equations of motion in Chapter IV it was mentioned that these equations, together with the continuity equation, contain more unknowns than the number of equations; hence they cannot be solved, or their closure cannot be attained. With the availability of high speed computers, one could solve Navier Stokes' equations for instantaneous velocities to obtain solution of turbulent flow problems. However, this would require very small grid, of the order of fraction of millimeter in size, in order that decay of turbulence due to small eddies is properly modelled. This would necessitate a very large storage and computer time. Further, in most of the situations one would like to know only the mean motion. Hence, this approach is considered too expensive and unnecessary from engineering point of view.

Since the time of Boussinesq (1877) attempts have been made to obtain additional equation or equations which, together with the Reynolds equations of motion in their original or simplified form would allow calculation of mean turbulent flows. Since the turbulent fluid motion has a great ability for diffusion, it can transport with it any quantity associated with the turbulent fluid motion, such as momentum, vorticity, heat, suspended matter, etc. These transport processes were modelled in the earlier investigations on turbulent motion by Boussinesq, Prandtl, Karman, Taylor and Reichardt. Essentially, these studies provide a relationship, albeit empirical, between mean flow characteristic, such as velocity gradient, and turbulent stress. These theories are known as phenomenological or semi-empirical theories of turbulence. It may be mentioned that these theories do not necessarily describe the mechanism underlying the transport process correctly; however, since some of these theories give useful and practical empirical relations, their importance cannot be underestimated.

Since 1945 additional equations for closure of Reynolds equations are proposed on the premise that quantities such as turbulent kinetic energy, or eddy size, or turbulent stress are transported by turbulent motion and, hence additional differential equation/s are obtained from equations of motion. They also contain some empirical constants. These equations, together with Reynolds equations, are then solved numerically. These are known as turbulence models.

All these approaches are briefly discussed in this chapter. For a more complete description of the turbulence models and related aspects, the reader is advised to refer to excellent texts by Launder and Spalding (1), Rodi (2) and Hinze (3).

6.2 PHENOMENOLOGICAL THEORIES OF TURBULENCE

Boussinesq (4) in 1877 introduced eddy dynamic viscosity μ_t to express turbulent shear $\bar{\tau}$ as

$$\bar{\tau} = \mu_t \frac{d\bar{u}}{dy} \qquad \ldots (6.1)$$

where $\bar{u}$ is the time averaged velocity at distance y from the boundary and $\bar{\tau}$ is the time averaged turbulent shear $(-\rho\,\overline{u'v'})$ in the xy plane in x direction. It may be noted that from here onwards in this chapter, bars over the quantities will be dropped for convenience. Some comments about turbulence viscosity μ_t are in order. Firstly, μ_t is found to be much larger than the dynamic viscosity μ of the fluid. Secondly, whereas μ is constant for a given fluid at a given temperature, μ_t is found to depend on flow conditon and fluid characteristic ρ. Lastly, unless one knows before hand how μ_t varies with the flow conditions, Equations 6.1 cannot be integrated to obtain velocity distribution even in the case of simple flows such as in circular pipes and in wide open channels.

Mixing Length Hypothesis

Prandtl (5) in 1925 proposed a hypothesis to connect the eddy viscosity μ_t to the flow conditions which is known as mixing length hypothesis. In kinetic theory of gases, the viscosity of a gas is given by

$$\mu = \frac{1}{3} \rho c \lambda \qquad \ldots (6.2)$$

where λ is the mean free path of the molecules and c is the root mean square value of molecular velocity. Probably taking clue from this relationship, Prandtl hypothesised that

$$\frac{\tau}{du/dy} = \mu_t = \rho l_m v_t \qquad \ldots (6.3)$$

where l_m is the mixing length and v_t is the turbulence velocity in the y direction, both of which were supposed to vary from one point to the other. Mixing length was considered to be such a lateral length that lumps of fluid would move from one layer to the other, distance l_m apart, without losing their momentum. He further hypothesised that

$$v_t = l_m \left| \frac{du}{dy} \right| \qquad \ldots (6.4)$$

$$\text{Hence, } \tau = \rho l_m^2 \left| \frac{du}{dy} \right| \frac{du}{dy} \qquad \ldots (6.5)$$

$$\text{or} \quad \mu_t = \rho l_m^2 \left| \frac{du}{dy} \right| \qquad \ldots (6.6)$$

Yet, to solve Eq. 6.5 one needs to specify the variation of l_m. Since for flow in pipes

$$\tau = \tau_0 \left(1 - \frac{y}{R}\right) \qquad \ldots (6.7)$$

where τ is turbulent shear at distance y from the pipe boundary, τ_o is the shear at the wall and R is pipe radius,

$$\tau_o \left(1 - \frac{y}{R}\right) = \rho l_m^2 \left(\frac{du}{dy}\right)^2$$

or

$$\frac{l_m}{R} = \frac{u_* \sqrt{1 - y/R}}{R\,(du/dy)} \qquad \ldots (6.8)$$

Here $u_* = \sqrt{\tau_o/\rho}$. Thus, for known R, u_* and du/dy at various values of y obtained from experimental data, one can obtain variation of l_m/R vs. y/R. Analysis of carefully collected data on turbulent flow in pipes by Nikuradse gave the following relationship for l_m/R for large Reynolds numbers

$$\frac{l_m}{R} = 0.14 - 0.08\left(1 - \frac{y}{R}\right)^2 - 0.06\left(1 - \frac{y}{R}\right)^4$$

which reduces to

$$\left.\begin{aligned} \frac{l_m}{R} &= 0.40\,\frac{y}{R} \\ \text{or} \quad l_m &= \kappa y \end{aligned}\right] \qquad \ldots (6.9)$$

for small y, where κ is known as *Karman constant* and has the value of 0.40. As an example of use of above relations, consider turbulent flow in a pipe.

$$\tau = \tau_o \left(1 - \frac{y}{R}\right) = \rho \kappa^2 y^2 \left(\frac{du}{dy}\right)^2$$

$$\frac{u_*^2}{\kappa^2 y^2}\left(1 - \frac{y}{R}\right) = \left(\frac{du}{dy}\right)^2$$

$$\frac{du}{dy} = \frac{u_*}{\kappa y}\sqrt{1 - \frac{y}{R}}$$

which on integration yields

$$\frac{u}{u_*} = \frac{2}{\kappa}\left[-\frac{1}{2}\ln\left(1 + \sqrt{1 - \frac{y}{R}}\right) + \frac{1}{2}\ln\left(1 - \sqrt{1 - \frac{y}{R}}\right) + \sqrt{1 - \frac{y}{R}}\,\right] + C$$

When $y = R$, $u = u_m$ can be used as the boundary condition Hence, $C = u_m/u_*$, and

$$\frac{(u - u_m)}{u_*} \frac{1}{\kappa} \cdot \left[2\sqrt{1 - y/R} + \ln \frac{1 - \sqrt{1 - y/R}}{1 + \sqrt{1 + y/R}} \right] \qquad \ldots (6.10)$$

If one makes an assumption that near the boundary $\tau = \tau_o$, one can write

$$\tau_o = \tau = \rho l_m^2 \left(\frac{du}{dy} \right)^2 \qquad \ldots (6.11)$$

or $\dfrac{u_*}{\kappa y} = \dfrac{du}{dy}$

which can be integrated with the boundary condition, $u = u_m$ at $y = R$ to obtain

$$\frac{(u - u_m)}{u_*} = \frac{1}{\kappa} \ln \left(\frac{y}{R} \right) \qquad \ldots (6.12)$$

Equations 6.10 and 6.12 are known as velocity defect laws for turbulent flow in pipes, and are valid for smooth as well as rough pipes. In passing, it may be mentioned that if boundary condition near the wall is specified, one must use different boundary conditions for smooth and rough walls. The equations then obtained from Eq. 6.11 are

Smooth boundary $\qquad \dfrac{u}{u_*} = \dfrac{2.3}{\kappa} \log \left(\dfrac{u_* y}{\nu} \right) + 5.50 \qquad \ldots (6.13)$

Rough boundary $\qquad \dfrac{u}{u_*} = \dfrac{2.3}{\kappa} \log \left(\dfrac{y}{k} \right) + 8.5 \qquad \ldots (6.14)$

where k is average height of roughness. The constants in the above equations have been obtained by Karman and Prandtl from the analysis of Nikuradse's data.

Prandtl's mixing length theory is also applied to free turbulent shear flows. From the analysis of extensive data collected by Reichardt on free turbulent shear flows, Prandtl proposed a simpler model for μ_t in which he assumed that the turbulent eddy viscosity is given by

$$\mu_t = \kappa_1 \, b \, (u_m - u_{min}) \qquad \ldots (6.15)$$

where κ_1 is a constant, b is the width of mixing zone, and u_m and u_{min} are maximum and minimum velocities.

Two major defects were noticed in the mixing length hypothesis of Prandtl. The first was that when $du/dy = 0$, one gets $\mu_t = 0$ as at the centre of the pipe. However, at this point turbulent mixing is still there and μ_t will have a finite value there. Realising this limitation, Prandtl proposed the following formula for μ_t.

$$\mu_t = \rho l_m^2 \left[\left(\frac{du}{dy} \right)^2 + l'^2_m \left(\frac{d^2u}{dy^2} \right)^2 \right]^{1/2} \qquad \ldots (6.16)$$

Because Eq. 6.16 has two lengths l_m and l'_m available for adjustment, it is likely to give better results. However, it has not been used much. The other objection (3) is raised concerning interpretation of mixing length hypothesis as linked to the momentum transport. In this interpretation, no account is taken of the effect of pressure fluctuations on the momentum transport. Because of pressure fluctuations to which each fluid lump is subjected during its path over l_m, the momentum of the lump will not remain constant and, hence, it is not preserved.

Karman's Similarity Hypothesis

Karman assumed that the turbulence mechanism is independent of viscosity in the whole flow field except in the immediate vicinity of the wall, and that the mixing length is determined by the local flow conditions. Therefore, mixing length would be given by

$$l_m \frac{\partial^2 u/\partial y^2}{\partial u/\partial y} = \kappa = l_m \frac{\partial^3 u/\partial y^3}{\partial^2 u/\partial y^2}, \text{ etc.} \quad \ldots (6.17)$$

Taking the first two terms, one gets

$$l_m = \kappa \frac{\partial u/\partial y}{\partial^2 u/\partial y^2}$$

and hence,

$$\mu_t = \rho l_m^2 \left(\frac{\partial u}{\partial y} \right)$$

or

$$\mu_t = \rho \kappa^2 \frac{(\partial u/\partial y)^2}{(\partial^2 u/\partial y^2)^2} \frac{\partial u}{\partial y} \quad \ldots (6.18)$$

This equation gives $\mu_t = \infty$, when $\partial^2 u/\partial y^2 = 0$ and $\frac{\partial u}{\partial y}$ is not zero, which is not very reasonable. Equation 6.18 can be used to obtain velocity defect law for turbulent flow in pipes.

$$\tau = \tau_o \left(1 - \frac{y}{R} \right) = \mu_t \frac{\partial u}{\partial y} = \rho \kappa^2 \frac{(\partial u/\partial y)^4}{(\partial^2 u/\partial y^2)^2}$$

or

$$(\partial u/\partial y)^2/(\partial^2 u/\partial y^2) = \frac{-u_*}{\kappa} \sqrt{1 - \frac{y}{R}} + C \quad \ldots (6.19)$$

It may be noted that a –ve sign is introduced on the right hand side of Eq. 6.19 since throughout the region of integration (i.e. $y = 0$ to R), $\partial^2 u/\partial y^2$ is negative. Integration of the above equation yields

$$1/(\partial u/\partial y) = \frac{-2\kappa R}{u_*} \sqrt{1 - y/R + C} \quad \ldots (6.20)$$

The constant of integration C can be evaluated if one assumes that near the boundary $\partial u/\partial y$ is very large so that as $y \to 0$, $1/(\partial u/\partial y) \to 0$, so that

$$C = \frac{2\kappa R}{u_*}$$

$$\frac{\partial u}{\partial y} = \frac{u_*}{2\kappa R} \frac{1}{1 - \sqrt{1 - y/R}}$$

which on integration and substitution of boundary condition $u = u_m$ at $y = R$ yields

$$\left(\frac{u - u_m}{u_*}\right) = \frac{1}{\kappa}\left[\ln\left(1 - \sqrt{1 - \frac{y}{R}}\right) + \sqrt{1 - \frac{y}{R}}\right] \qquad \ldots (6.21)$$

It may be mentioned that the value of Karman constant in Eqs. 6.19, 6.20 and 6.21 is also 0.40 as in earlier equations. Figure 6.1 shows variation of

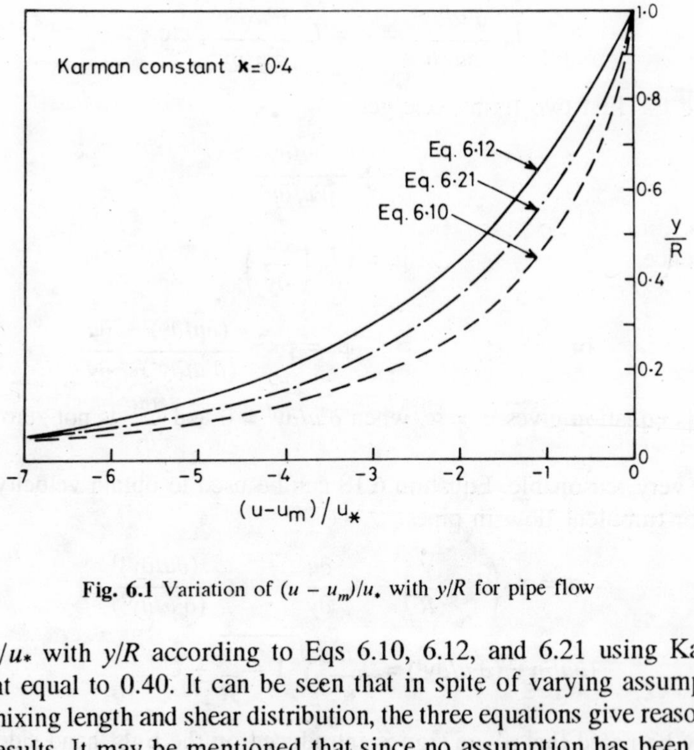

Fig. 6.1 Variation of $(u - u_m)/u_*$ with y/R for pipe flow

$(u-u_m)/u_*$ with y/R according to Eqs 6.10, 6.12, and 6.21 using Karman constant equal to 0.40. It can be seen that in spite of varying assumptions about mixing length and shear distribution, the three equations give reasonably close results. It may be mentioned that since no assumption has been made about the nature of boundary, these equations are valid for smooth as well as rough boundaries.

Vorticity Transport Theory (6)

In Taylor's Vorticity Transport theory, it is assumed that (i) the flow is two dimensional, and (ii) viscosity is negligible, so that the vortices will follow Helmholtz's law and vorticity of each fluid particle remains constant. Since viscosity is negligible, for two dimensional flow, one can start with Euler's equation of motion in x direction

$$\frac{\partial u}{\partial t} + u\frac{\partial u}{\partial x} + v\frac{\partial u}{\partial y} = -\frac{1}{\rho}\frac{\partial p}{\partial x} \quad \ldots (6.22)$$

By adding and subtracting $v\dfrac{\partial v}{\partial x}$ on the left hand side, Eq. 6.22 can be written as

$$\frac{\partial u}{\partial t} + \frac{\partial}{\partial x}\left(\frac{u^2 + v^2}{2}\right) + 2\,v\,\omega = -\frac{1}{\rho}\frac{\partial p}{\partial x} \quad \ldots (6.23)$$

where $\omega = \dfrac{1}{2}\left(\dfrac{\partial u}{\partial y} - \dfrac{\partial v}{\partial x}\right)$ is vorticity about z axis. Now consider the mean motion given by

$$\bar{u} = \bar{u}(y) \text{ and } \bar{v} = 0$$

This can represent flow between parallel plates, or approximately the boundary layer flow. Consequently,

$$\partial/\partial x = 0 \text{ and } \partial/\partial t = 0;\text{ but } \frac{\partial p}{\partial x} \neq 0.$$

Hence, Eq. 6.23 becomes

$$2\,v\,\omega = -\frac{1}{\rho}\frac{\partial p}{\partial x} \quad \ldots (6.24)$$

However $v = \bar{v} + v'$, $p = \bar{p} + p'$ and $\omega = \bar{\omega} + \omega'$, and $\bar{v} = 0$. Substitution of these relations in Eq. 6.24, and subsequent application of Reynolds rules of averages, yields

$$2\,\overline{v'\omega'} = -\frac{1}{\rho}\frac{\partial p}{\partial x} \quad \ldots (6.25)$$

For parallel flows $\dfrac{\partial \bar{v}}{\partial x} = 0$; hence, $\bar{\omega} = \dfrac{1}{2}\dfrac{\partial \bar{u}}{\partial y}$

Taking clue from Prandtl, assume

$$\left.\begin{aligned} \omega' &= L'\frac{\partial \bar{\omega}}{\partial y} = \frac{L'}{2}\frac{\partial^2 \bar{u}}{\partial y^2} \\ \text{and} \quad v' &= L\frac{\partial \bar{u}}{\partial y} \end{aligned}\right] \quad \ldots (6.26)$$

where L and L' are mixing lengths.

$$\therefore \frac{\partial \bar{p}}{\partial x} = -2\,\rho\overline{v'\omega'} = -2\,\rho\frac{LL'}{2}\left(\frac{\partial \bar{u}}{\partial y}\right)\left(\frac{\partial^2 \bar{u}}{\partial y^2}\right)$$

or

$$\frac{\partial \bar{p}}{\partial x} = l_\omega^2\,\frac{\partial \bar{u}}{\partial y}\frac{\partial^2 \bar{u}}{\partial y^2} \quad \ldots (6.27)$$

where l_ω is mixing length for vorticity. Since $\partial^2\bar{u}/\partial y^2$ is negative, the sign in Eq. 6.27 has been changed to positive sign. Remembering that for parallel flow $\frac{\partial \bar{p}}{\partial x} = \frac{\partial \bar{\tau}}{\partial y}$, one gets

$$\frac{\partial \bar{\tau}}{\partial y} = \frac{\rho l_\omega^2}{2}\left(\frac{\partial \bar{u}}{\partial y}\right)^2 \qquad \ldots (6.28)$$

Comparison of Eq. 6.28 with Eq. 6.5 shows that

$$\sqrt{2}\, l_m = l_w \qquad \ldots (6.29)$$

In general, Prandtl's mixing length theory and Taylor's vorticity transport theory give different results.

Reichardt's Inductive Theory of Turbulence (6*)

In his inductive theory, Reichardt tried to dispense with hypothesis. Analysis of a large volume of experimental data on free turbulent flows led him to the conclusion that velocity distribution can, in most of the cases, be represented by Gaussian distribution. Starting with this premise, he attempted to cover all the cases of free turbulent shear flows by a set of formulae. Momentum equation for time averaged two dimensional frictionless flow with the aid of continuity equation can be written as

$$\frac{\partial}{\partial x}\left(\frac{p}{\rho} + \overline{u^2}\right) + \frac{\partial}{\partial y}(\overline{uv}) = 0$$

In the case of free turbulent flow, p is constant; hence,

$\frac{\partial p}{\partial x} = 0$. He expressed $\overline{uv}$ as

$$\overline{uv} = -A\,\frac{\partial \overline{u^2}}{\partial y} \qquad \ldots (6.30)$$

where $A(x)$ has a dimension of length, and must be determined empirically. Since $\overline{uv}$ is shear stress, the above equation can be interpreted as an empirical law for momentum transport. The coefficient A is called the *momentum transfer length.* If $A(x)$ were constant, this equation would have been identical to one dimensional heat conduction equation if x corresponds to t and y corresponds to x. Thus, distribution of momentum in free turbulent shear flows, according to Reichardt, is governed by heat conduction equation, and its solution yields Gaussian distribution. According to Hinze (3), Reichardt's theory is found to be incorrect if it is considered in all its consequences.

6.3 GENERAL COMMENTS ON TURBULENCE MODELS

As mentioned in the beginning of this chapter, one can attempt to solve the N.S. equations for instantaneous u,v,w, and p values for given boundary conditions. However, this would entail a very large computer time and storage, and such detailed information may not be required in most of the problems. One would normally need time averaged description of the flow. However, in the averaging process enter the statistical correlations, involving fluctuating velocities, temperatures, etc. Since their magnitudes are not known, they must be modelled in terms of the quantities that one can determine. This involves modelling, and phenomenological theories were the simplest models. Thus, by turbulence model, one means a set of equations which, when solved with the mean flow equations, allows calculation of relevant correlations and, hence, helps in the solution of equations simulating the behaviour of real fluids in important aspects.

Turbulence Models can be classified according to the number of differential equations used. Models in which only algebraic equation is used for determining turbulent viscosity do not use any additional differential equation; hence, they are called zero equation models. Those which use additional equation or equations are called one equation, or two equation models. The additional equations may be for transport of turbulence energy or other such quantities.

Phenomenological theories discussed earlier could provide solutions to simple problems, such as flows in straight pipes and channels, boundary layers on flat plates, and jets and wakes. However, complex flows, either two or three dimensional, involving change of direction, recirculatory flow and complex boundary conditions, could not be analysed by phenomenological theories. Similarly, problems involving heat exchange, chemical reactions, and species transport in turbulent flow cannot be solved by phenomenological theories and need turbulence modelling. Compressibility and buoyancy effects further complicate the analysis.

6.4 ZERO EQUATION MODELS

These are constant eddy viscosity/diffusivity models (2). This type of modelling is adopted for large bodies of water and the value of eddy viscosity ν_t, or diffusivity Γ is obtained experimentally or by trial and error. They are primarily used for solving hydraulic problems; in some cases different values of Γ have to be used in horizontal and vertical directions. It may be mentioned that when heat or mass transport takes place, eddy diffusivity is defined as

$$-\overline{u_i' \phi'} = \Gamma_{\phi,t} \frac{\partial \bar{\phi}}{\partial t} \quad \ldots (6.31)$$

where ϕ is the scalar quantity such as temperature or concentration, and $-\rho \overline{u_i' \phi'}$ is the turbulent diffusion of flux ϕ having kg/m^{2}s times ϕ units. This is Fick's law of diffusion.

The eddy diffusivity $\Gamma_{\phi,t}$ for heat or mass can be expressed in the same manner as μ_t in terms of mixing length $l_{m,\phi}$ and v_t as

$$\Gamma_{\phi,t} = l_{m,\phi}\, v_t \qquad \ldots (6.32)$$

Here Prandtl number and Schmidt number for heat and mass transfer can be defined as ratio of ν_t and $\Gamma_{\phi,t}$

$$\sigma_{\phi,t} = \nu_t/\Gamma_{\phi,t} = \frac{l_m}{l_{m,\phi}} \qquad \ldots (6.33)$$

The parameter $\sigma_{\phi,t}$ is found to be nearly constant across the flow and from one flow to other. For free turbulent flows, for heat as well as mass transfer, following average values of $\sigma_{\phi,t}$ have been obtained.

Plane mixing layers:	$\sigma_{\phi,t} = 0.50$	(heat and mass transfer)
Plane jets in stagnant surrounding:	$\sigma_{\phi,t} = 0.50$	(heat and mass transfer)
Round jets in stagnant surroundings:	$\sigma_{\phi,t} = 0.70$	(heat and mass transfer)
For pipe flow:	$\sigma_{\phi,t} = 0.85$	(heat transfer)
For boundary layer flow:	$\sigma_{\phi,t} = f\left(\frac{u_* y}{\nu}\right)$	(heat transfer)

Mixing length hypothesis has been used for studying free turbulent flows. However, in such flows the mixing length distribution has to be specified. Analysis of experimental data has given the following information:

Plane mixing layer:	$\frac{l_m}{\delta} = 0.07$	δ is mixing layer width
Plane jet in stagnation surroundings:	$\frac{l_m}{\delta} = 0.09$	
Fan jet in stagnation surroundings:	$\frac{l_m}{\delta} = 0.125$	. . . (6.34)
Round jet in stagnation surroundings:	$\frac{l_m}{\delta} = 0.075$	δ is half the jet width
Plane wake:	$\frac{l_m}{\delta} = 0.06$	δ is the wake width

In all these cases, since δ increases along the length, l_m also increases along the length; however l_m is found to vary very little across the jet width and, hence, can be taken as constant.

In the case of boundary layers on plane walls, Escudier (7), after analysis of large volume of data, found that mixing length distribution can be specified as follows:

$$\left.\begin{aligned} \frac{l_m}{\delta} &= \kappa \frac{y}{\delta} \text{ for } \frac{y}{\delta} < \frac{\lambda}{\kappa} \\ \text{and} \quad \frac{l_m}{\delta} &= \lambda \text{ for } \frac{y}{\delta} > \frac{\lambda}{\kappa} \end{aligned}\right] \quad \ldots (6.35)$$

Here δ is the nominal boundary layer thickness, $\kappa = 0.4$, and $\lambda = 0.09$. This is valid for variable as well as constant density flows. For boundary layers with positive or negative pressure gradients $\kappa = 0.435$ gives better prediction of velocity distribution. For wall jets in stagnant surroundings $\kappa = 0.6$ and $\lambda = 0.075$ give better conformity with the experimental data. On the other hand, shear variation on the wall is better predicted by $\kappa = 0.435$ and $\lambda = 0.09$.

For stable stratification (density decreasing away from the wall), i.e. when $Ri > 0$, Monin–Obukhov relation

$$\frac{l_m}{l_{mo}} = 1 + \beta_1 \, Ri \quad \ldots (6.36)$$

gives the mixing length relation. Here l_{mo} is mixing length without any buoyancy effect and $\beta_1 = 7.0$. For unstable condition when $Ri < 0$, l_m is given by

$$\frac{l_m}{l_{mo}} = (1 - \beta_2 Ri)^{-1/4} \quad \ldots (6.37)$$

where $\beta_2 = 14.0$ and $Ri = -\dfrac{g}{\rho} \dfrac{\partial\rho/\partial y}{(\partial u/\partial y)^2}$ is Richardson number.

From the above discussion it can be concluded that mixing length hypothesis is a simple one, requiring no additional differential equation. If proper choice is made for the mixing length distribution, it allows realistic predictions of velocity and shear distributions and general behaviour of turbulent boundary layers. Mixing length hypothesis is not suitable where convective and diffusive transport of turbulence are important, e.g. in rapidly developing flows, circulatory flows etc.

6.5 ONE EQUATION MODELS

Since limitations of mixing length hypothesis were soon realised, Kolmogorov (8) and Prandtl (9) proposed that the turbulent viscosity should be determined by means of a differential equation rather than an algebraic equation. Instead of relating turbulent velocity v_t to the mean velocity gradient, they suggested that it should be related to the square root of time averaged turbulence kinetic energy $k = \dfrac{1}{2}\left(\overline{u_1'^2} + \overline{u_2'^2} + \overline{u_3'^2}\right)$

Hence,

$$\mu_t = C'_\mu \sqrt{k}\, l \quad \ldots (6.38)$$

where C'_μ is an empirical constant and l is the mixing length. The turbulent kinetic energy term k is to be obtained from the solution of convective transport equation for k which was obtained in Chapter IV. This equation is

$$\frac{\partial k}{\partial t} + u_i \frac{\partial k}{\partial x_i} = -\frac{\partial}{\partial x_i}\left[\overline{u'_i\left(\frac{u'_i\, u'_j}{2} + \frac{p'}{\rho}\right)}\right] - \overline{u'_i u'_j}\,\frac{\partial u_i}{\partial x_j} - \nu\,\overline{\frac{\partial u'_i}{\partial x_j}\,\frac{\partial u'_i}{\partial x_i}} \qquad (6.39)$$

(Local rate of change of k) + (Convective transport of k) = (Diffusive transport of k) + (Production of k by shear) + (Viscous dissipation ε of k)

Various terms occurring in the above equation need to be modelled. This has been done by Prandtl, Kolmogorov, and others in the following manner.

The diffusive transport of k is modelled as

$$-\overline{u'_i\left(\frac{u'_i\, u'_j}{2} + \frac{p'}{\rho}\right)} = \frac{\nu_t}{\sigma_k}\,\frac{\partial k}{\partial x_i} \qquad \ldots (6.40)$$

where σ_k is an empirical diffusion constant. The viscous dissipation term ε is expressed as

$$\varepsilon = C_D\, k^{3/2}/l \qquad \ldots (6.41)$$

where C_D is another empirical constant and ε is rate of dissipation of k. And finally one can write

$$\overline{u'_i u'_j} = \nu_t\left(\frac{\partial u_i}{\partial x_j} + \frac{\partial u_j}{\partial x_i}\right) \qquad \ldots (6.42)$$

Substitution of Eqs. 6.40, 6.41 and 6.42 in Eq. 6.39 gives

$$\frac{Dk}{Dt} = \frac{\partial}{\partial x_i}\left(\frac{\nu_t}{\sigma_k}\,\frac{\partial k}{\partial x_i}\right) + \nu_t\left(\frac{\partial u_i}{\partial x_j} + \frac{\partial u_j}{\partial x_i}\right)\frac{\partial u_i}{\partial x_j} - C_D\, k^{3/2}/l \qquad \ldots (6.43)$$

where $\dfrac{Dk}{Dt} = \dfrac{\partial k}{\partial t} + u_i\,\dfrac{\partial k}{\partial x_i}$

For two dimensional boundary layers, it can be reduced to

$$\frac{Dk}{Dt} = \frac{\partial}{\partial y}\left(\frac{\nu_t}{\sigma_k}\,\frac{\partial k}{\partial y}\right) + \nu_t\left(\frac{\partial u}{\partial y}\right)^2 - C_D\, k^{3/2}/l$$

Equations 6.43 has to be solved with the continuity and other equations of motion. The constants σ_t and C_D must be assigned appropriate values, and mixing length distribution has to be prescribed. For diffusion of other matter, Eq. 6.43 will have additional term on the right hand side, namely $-\beta\, g_i\, \overline{u_i'\phi'}$, which represents the buoyancy production/destruction term. Here β is an empirical constant. This term is negative in stable stratification; hence, k is reduced and turbulence is damped while potential energy is increased. This term is positive in unstable stratification; hence, k is increased at the expense of potential energy. The term $-\beta\, g_i\, \overline{u'_i\, \phi'}$ henceforth is not considered.

The constants in Eq. 6.43 are determined in the following manner. Near the wall, convection and diffusion terms are negligible. Hence, production and dissipation must be equal. This gives.

$$\nu_t \left(\frac{\partial u}{\partial y}\right)^2 = C_D\, k^{3/2}/l \qquad \ldots (6.44)$$

Also $\nu_t = C'_\mu \sqrt{k}l$

Elimination of k from these two equations yields

$$\nu_t = (C'_\mu/C_D)^{1/2}\, l^2 \left|\frac{du}{dy}\right| \qquad \ldots (6.45)$$

which can be compared with Prandtl's mixing length relation

$$\nu_t = l_m^2 \left|\frac{du}{dy}\right|$$

Hence, $l_m = l\,(C'_\mu/C_D)^{1/4}$ $\qquad \ldots (6.46)$

Also from Eq. 6.44 one can obtain the expression for shear τ_o as

$$\tau_o/\rho k = \sqrt{C'_\mu\, C_D} \qquad \ldots (6.47)$$

The values of empirical constants obtained from analysis of data are $C'_\mu = 1.0$, $\sigma_k = 1.0$ and $C_D = 0.08$.

Hence, $\tau_o = (0.08)^{1/2}\, \rho k$. For axisymmetric flows, the constants obtained are $C'_\mu = \sigma_k = 1.0$ and $C_D = 0.07$.

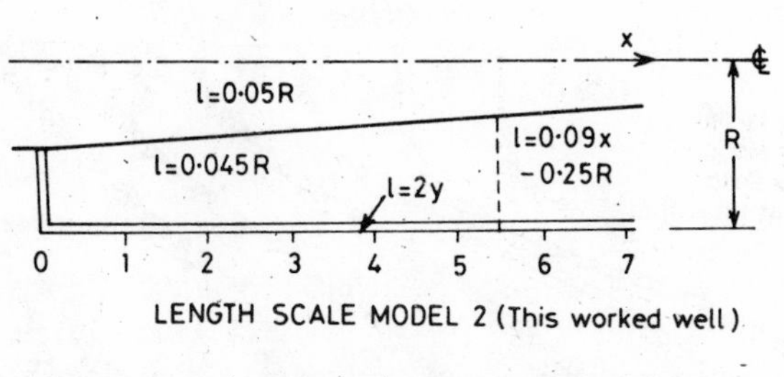

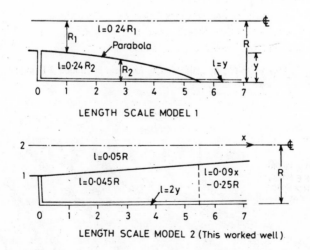

Fig. 6.2 Runchal's length scale models for flow in sudden expansion in axisymmetric flow

The determination of the length scale l has to be done. The parameter l/δ depends on y/δ and type of flow. However, prescribing l/δ is not as easy in one-equation models as in the mixing length hypothesis. Spalding suggests that the length scale may be made proportional to the distance from the wall y for small values of y, and for flows away from the wall, it may be considered constant. Runchal (10), while modelling the flow in axisymmetric sudden expansion, tried two distributions shown in Fig. 6.2 out of which the second was found to produce results in good agreement with the experimental data. Mehta (11) studied two dimensional sudden expansion, and adopted mixing length distributions similar to those adopted by Runchal (see Fig. 6.3). Further,

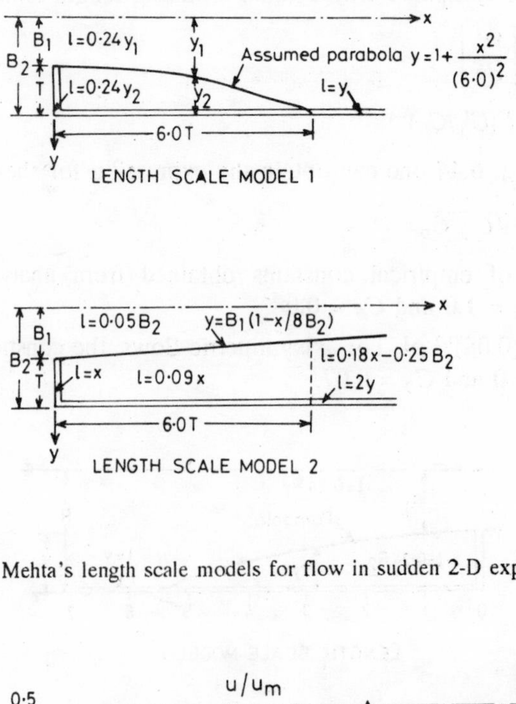

Fig. 6.3 Mehta's length scale models for flow in sudden 2-D expansion

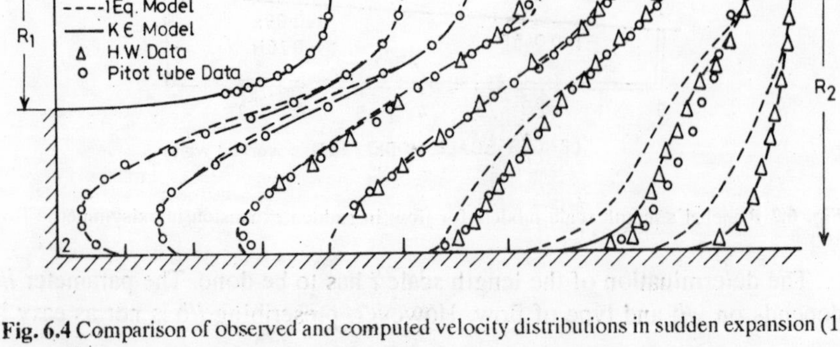

Fig. 6.4 Comparison of observed and computed velocity distributions in sudden expansion (12)

Runchal adopted the constants as $C'_\mu = 0.20$, $\sigma_k = 1.54$ and $C_D = 0.13$. One equation model has been used to determine the velocity distribution in sudden expansion in a pipe by Ha Minh and Chassaing (12). The agreement is quite good immediately downstream of expansion (see Fig. 6.4); however, farther away, two equation model gives better results. The authors have not mentioned the details about mixing length distribution adopted.

Experience with one equation models has indicated that, one equation models are superior to the mixing length hypothesis because they take into account the convective and diffusive transport of turbulent velocity scale. This is important in nonequilibrium boundary layers, rapidly changing free stream conditions, in boundary layers with free stream turbulence, and in recirculatory flows. This model has been restricted mainly to shear layers, since for other flows it is difficult to specify the mixing length distribution empirically.

6.6 TWO EQUATION MODELS OF TURBULENCE

It is argued that the length scale l characterising the average size of energy containing eddies is subject to transport process in a similar manner to turbulence energy k. Thus, eddies generated by grid are convected downstream so that their size at a point depends much on their initial size. It is influenced by dissipation and vortex stretching. Hence, in two equation models, the turbulence energy k and the length sacle l are determined by way of transport equations.

Two equation models use eddy viscosity concept and Kolmogorov-Prandtl expression for it, namely

$$\nu_t = C'_\mu \sqrt{k}\, l.$$

The length scale to be determined can occur explicitly as l as used by Rotta (13) and Spalding (14), or a combination of k and l in the form of variable Z. The variable Z is expressed as

$$Z = k^m l^n \qquad \ldots (6.48)$$

with m and n as constants. Once k is known, l can be determined. The following table given by Launder and Spalding (1) gives various proposals for dependent variable Z in Eq. 6.48.

TABLE 6.1
Proposals for adoption of Z (1)

Proposer	Z	*Symbol*
Kolmogorov	$k^{1/2}/l$	f (frequency)
Chu, Davidov Harlo-Nakayama Jones-Launder	$k^{3/2}/l$	ε
Rota, Spalding	l	l
Rota, Rodi-Spalding	kl	kl
Spalding	k/l^2	ω (vorticity)

The earliest proposal was Kolmogorov's who chose quantity proportional to the mean frequency of the most energetic motion. The turbulence dissipation rate ε has been favoured by many workers; the reason for this partly lies in the relative ease with which the exact equation for ε can be derived, and partly in the fact that ε appears directly as an unknown in the equation for k. Rotta (13) is the only investigator who has derived a partial differential transport equation for length scale l. Spalding (14) provided an ordinary differential equation for l, which he used to calculate development of a number of shear flows. But this has not been successful because l just does not diffuse at the rate proportional to $\partial l/\partial y$. Spalding's transport equation involves vorticity ω which is just square of Kolmogorov's f.

Various investigators have derived the second differential equation for Z in different manners. However, when their end results are compared, they bear a remarkable conformity. Most of them possess the following common form for nonbuoyant flows.

$$\underbrace{\frac{\partial Z}{\partial t}}_{\text{Temporal changes}} + \underbrace{u_i \frac{\partial Z}{\partial x_i}}_{\text{Convective transport}} = \underbrace{\frac{\partial}{\partial x_i}\left(\frac{\sqrt{k}\, l}{\sigma_z} \frac{\partial Z}{\partial x_i}\right)}_{\text{Diffusion}} + \underbrace{C_{Z1} \frac{Z}{k} P}_{\text{Production}} - \underbrace{C_{Z2} \frac{\sqrt{k}}{l}}_{\text{Destruction}} + \underbrace{S}_{\text{Source}} \quad \ldots (6.49)$$

Here σ_z, C_{z1} and C_{z2} are emperical constants, P is the production of kinetic energy term defined by $\overline{u_i' u_j'} \dfrac{\partial u_i}{\partial x_j}$, and S is the secondary source term which differs according to choice of Z. The ε equation does not require S term, and $k - \varepsilon$ model is more popular. The equations for complete $k - \varepsilon$ model are given below.

$$\nu_t = C'_\mu \frac{k^2}{\varepsilon}$$

$$\frac{\partial k}{\partial t} + u_i \frac{\partial k}{\partial x_i} = \frac{\partial}{\partial x_i}\left(\frac{\nu_t}{\sigma_k} \frac{\partial k}{\partial x_i}\right) + \nu_t \left(\frac{\partial u_i}{\partial x_j} + \frac{\partial u_j}{\partial x_i}\right) - \varepsilon \quad \ldots (6.50)$$

$$\frac{\partial \varepsilon}{\partial t} + u_i \frac{\partial \varepsilon}{\partial x_i} = \frac{\partial}{\partial x_i}\left(\frac{\nu_t}{\sigma_\varepsilon} \frac{\partial \varepsilon}{\partial x_i}\right) + C_{\varepsilon 1} \nu_t (P) - C_{\varepsilon 2} \frac{\varepsilon^2}{k} \quad \ldots (6.51)$$

where $P = \left(\dfrac{\partial u_i}{\partial x_j} + \dfrac{\partial u_j}{\partial x_i}\right) \dfrac{\partial u_i}{\partial x_j}$

The constants in Eq. (6.51) are specified as $\sigma_\varepsilon = 1.30$, $C_{\varepsilon 1} = 1.45$, $C_{\varepsilon 2} = 1.92$ while those in Eq. (6.50) as $C'_\mu = 0.09$, $\sigma_k = 1.0$

It has also been found that the calculations are most sensitive to the values of $C_{\varepsilon 1}$ and $C_{\varepsilon 2}$. For example (1) five per cent change in $C_{\varepsilon 1}$ or $C_{\varepsilon 2}$ results in twenty per cent change in spreading of jet.

$k - \varepsilon$ model with the above constants has been successfully used to study two dimensional boundary layers, duct flows, free shear flows, recirculatory flows and three dimensional wall boundary layers, confined flows and jets.

Boundary Condition (2)

While considering the boundary conditions, one must take into account three types of boundaries,, namely a solid wall, a free surface, and non-turbulent flow, e.g. free boundary of a jet or wake. Here the boundary conditions near the wall rather than at the wall are specified, because in turbulent flow, the turbulence properties near the wall are function of $u_* y/\nu$. Hence, most economical and straight forward practice is to apply experimentally established boundary condition at some point near the wall; e.g.

$$\frac{u}{u_*} = \frac{1}{\kappa} \ln \left(\frac{E u_* y}{\nu} \right) \text{ for } 30 < \frac{u_* y}{\nu} < 100 \qquad \ldots (6.52)$$

where E is a constant and is taken as 9.0. In this region, production equals dissipation and shear stress equals approximately the wall shear. These conditions lead to the equality

$$\frac{k}{u_*^2} = \frac{1}{\sqrt{C'_\mu}} \qquad \ldots (6.53)$$

This is used as the boundary condition for k. The condition for ε is

$$\varepsilon = u_*^3/\kappa y \qquad \ldots (6.54)$$

Relations 6.52 and 6.54 are valid for both smooth and rough walls.

At the free boundaries, the velocities and the scalar quantities must be equal to their free stream values. Since many times the free stream or ambient stream is assumed to be entirely free of turbulence, the turbulent stresses, fluxes and ε are zero at the free boundary. At the symmetry planes or lines, the normal gradients are zero for all the quantities with symmetrical behaviour, such as the scalar quantities ϕ, k, $\overline{\phi'^2}$, ε, normal stresses etc. Velocity components normal to the symmetry planes or lines, and shear stresses as well as scalar fluxes are themselves zero.

The free surface can be considered as a plane of symmetry if there is no wind induced shear stress there. But presence of free surface certainly reduces the scale of turbulence.

Other Two Equation Models

Even through $k - \varepsilon$ model is the most popular, other models have also been tried. The $k - l$ model was used by Rodi and Spalding (15), who predicted velocity, shear and energy profiles in a plane turbulent jet in stagnant surrounding, and compared the results with experimental results of Bradbury and Robbins; they found good agreement between the two (see Fig. 6.5). $k - \omega$ models has been used by Runchal (10) to predict the flow in pipe expansion and by Wolfsthein (16) to predict the characteristics of a jet of fluid emerging from a pipe and impinging on a nearby wall.

General Comments

Considerable effort has gone into the development of two-equation models, and they have been subjected to many checks and their predictive capacities have been established for boundary layer flows. Some studies have been made about their use to predict the recirculatory flows. However, as pointed out by Rodi (2), for weak shear layers and axisymmetric jets, the constants have to be replaced by functions. Further, the standard $k - \varepsilon$ model is based on the assumption of the same eddy viscosity for all Reynolds stresses $\rho\overline{u'_i u'_j}$, e.g. isotropic eddy viscosity. This assumption is rather crude and can give inaccurate results when shear stresses other than $-\rho\overline{u'v'}$ are important.

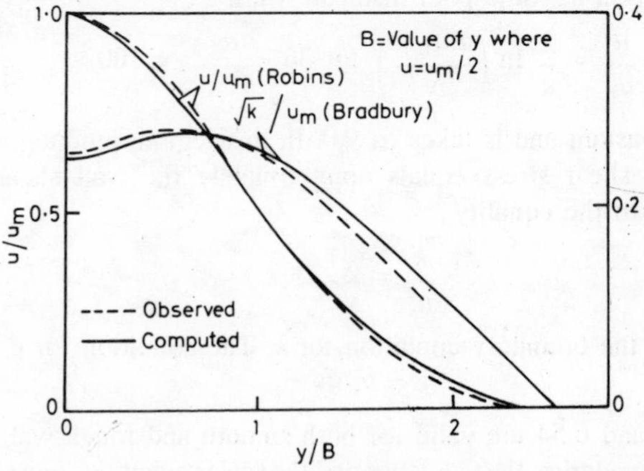

Fig. 6.5 Observed and computed variation of u/u_m and $\sqrt{k}/u_m$ in plane jet (15)

6.7 MULTI-EQUATION MODELS

Even though two equation models are much superior to one equation model, they have the limitations mentioned earlier. Hence, attempts have been made by different investigators by proposing models which use stresses rather than turbulent viscosity. This leads to models which need solutions of more than two differential equations. Some proposals for multi-equation models are listed below (1).

TABLE 6.2

Proposals for multi-equation models (1)

Author	Equations proposed for				
	l	$\overline{u'v'}$	$\overline{u'^2}$ etc.	$\overline{u'v'^2}$ etc.	Duffusion of l
Chou (1945)	√	√	√	√	×
Rotta (1951)	√	√	√	×	×
Davidov (1961)	√	√	√	√	√
Donaldson (1968)	×	√	√	×	×
Harlow et al. (1970)	√	√	√	×	×
Hanjalic (1970)	√	√	×	×	×

With such models, the number of differential equations to be solved becomes very large and, hence, computations become very expensive. Therefore, little progress has been made in exploring their utility.

REFERENCES

1. Launder, B.E. and D.B Splading. *Lectures in Mathematical Models of Turbulence*. Academic Press, London, 1972.
2. Rodi, W. *Turbulence Models and their Application in Hydraulics*, Int. Asso. for Hyd. Research, Delft. 1980.
3. Hinze, J.O. *Turbulence: An Introduction to its Mechanism and Theory*. McGraw Hill Series in Mechanical Engineering, New York, 1959. Chapter 5.
4. Boussinesq. T.V. Theorie de l'Ecoulement Tourbillant. Me'm. Press. Par. Div. Sav. XXIII, Paris, 1877.
5. Prandtl, L. Über die Ausgebildete Turbulenz. ZAMM, 5, 1925.
6. Schlichting H. *Boundary Layer Theory*. McGraw Hill Series in Mechanical Engineering, New York, 1st Ed. 1955.
7. Escudier, M.P. The Distribution of Mixing Length in Turbulent Flow Near Walls. Heat Transfer Section Rep. JWF/TN/1, Imperial College, London, 1966.
8. Kolmogorov A.N. Equations of Turbulent Motion of an Incompressible Turbulent Fluid. Izv. Akad. Nauk SSSR Ser Phys. VI, No. 1–2, 1942.
9. Prandtl, L. Uber Ein Neues Formelsystem für die Ausgebildete Trubulenz. Nachrichten van der Akad der Wissenschaft in Gottingen, 1945.
10. Runchal, A.K. Transfer Processes in Two-Dimensional Separated Flows. Ph. D. Thesis Imperical College, London, 1969.
11. Mehta, P.R. Characteristics of Mean and Turbulent Flow at Sudden Two Dimensional Expansion. Ph. D. Thesis, Civil Engg. Deptt. I.I.T., Bombay, 1977.
12. Ha Minh and P. Chassaing. Perturbations of Turbulent Flow. Proc. Symposium on Turbulent Shear Flows, Pennsylvania State University, Apr. 1977.
13. Rotta, J. Statistische Theorie Nichthomogener Turbulenz. Zeitsch für Physik, Vol. 129 and 131, 1951.
14. Spalding D.B. The Calculation of the Length Scale of Turbulence in Some Turbulent Boundary Layers Remote from Walls. Heat Transfer Section, Rep. TWF/TN/31, Imperial College London, 1967.
15. Rodi, W. and D.B. Splading. A Two Parameter Model of Turbulence and its Application to Free Jets. Wärme and Stoffübertragung No. 3, 85, 1970.
16. Wolfshtein, M. Numerical Solution of the Problem of a Turbulent Impinging Jet. Heat Transfer Section. Rep. EF/TN/A/171, 1969.

CHAPTER VII

Measurement of Turbulence

7.1 INTRODUCTION

Since the turbulent flows are very complex, the possibility of completely analytical treatment of such flows seems remote. Therefore, most of the analytical/semi analytical treatments of turbulent flows need inputs from carefully collected data. It is through the judicious use of such data that the knowledge of turbulent flows can be advanced, and our design methods improved. The quantities that need to be measured include *rms* values of turbulent fluctuations in velocity, density, pressure, force and concentration of suspended matter, the energy spectra, velocity correlations, turbulent shear, etc. Various techniques of measurement of turbulence can be broadly classified into three categories, namely (i) methods in which tracers are introduced in the flowing fluid to make the fluid motion visible by suitable detecting apparatus (such as photographic camera) outside the flow field; (ii) methods in which the detecting element is introduced in the flow field and the turbulence quantities are measured by recording the mechanical, physical or chemical change that occurs in the element; and (iii) Laser technique, in which no tracer or detecting element is introduced in the flow, but change in doppler frequency is recorded outside. The second group methods have been more often used to measure velocity, pressure and force fluctuations in a turbulent flow. In such a case, certain restrictions have to be imposed on the detecting element, often known as the transducer, so that turbulence measurements are accurate. These (1) are: (i) The detecting element is sufficiently small so that it causes minimum disturbance in the flow. The size of the detecting element should be such that instantaneous velocity distribution over this size is uniform. This means that its size should be smaller than the microscale of turbulence. For all practical flows encountered in low speed fluid mechanics and hydraulic engineering, detecting element size should not exceed about 1.0 mm to 2.0 mm. (ii) It should have low intertia, so that it can respond almost instantaneously to most rapid fluctuations which can be of the order of 5000 Hz. (iii) The transducer must be sufficiently sensitive to record small differences in fluctuations. (iv) It should be sufficiently rigid and strong to exclude possibility of its vibrations due to turbulent motion, and (v) It should be stable so that its calibration does not change, at least during one test run.

It is not the intention of the author to give complete theory and mechanical details of the measuring techniques; the intention is to give a general description of some commonly used techniques and indicate their limitations. Also, only techniques for measurement of velocity, pressure and force fluctuations have been discussed here.

7.2 TRACER TECHNIQUES

In the first category of methods, very small particles that are insoluble in the flowing fluid are introduced in the flow, and photographs are taken at short intervals and at known exposure times. From the analysis of these photographs, it is possible to find the direction and displacement of particles; hence, components of velocity parallel to photographic plane can be determined. Kalinske and Robertson (2) used a mixture of carbon tetrachloride and benzene containing powdered anthracite to measure turbulence in water flowing in the laboratory flume. Alternately, a mixture of ethylene dibromide and olive oil can be used. These particles or droplets have to be small in relation to the microscale of turbulence. Further, the mixtures used should have the same density as the flowing fluid. For the measurement of turbulence in low velocity air flow, tiny soap bubbles can be used as was done by Kampe' de Fe'riet (1*). For visualisation of flow in wind tunnel, microaluminium flakes were used by Bourot (1*).

Instead of using discrete particles, one can also use a steady source from which diffusable material such as salt solution or dye or vapours of ammonia, etc. is introduced in the flow. The concentration distribution profiles can be measured at various sections in the down stream direction. These data can be analysed to determine turbulent fluctuations, as well as the diffusion coefficient.

Another method that is increasingly used in the study of structure of turbulence in flowing liquids is the hydrogen bubble technique. This method was earlier used by Schraub and Kline (3), and later, by Kemp and Grass (4), and Clark and Markland (5). In this method, a platinum wire of a very small diameter, e.g. 0.025 mm, is used which acts as the negative electrode. The wire is speck insulated by passing it through a rectangular comb which masks alternate sections of wire at regular intervals of 1.5 mm to 2.0 mm. A d.c. supply is used for the electrolysis process, which accumulates hydrogen ions on the cathode. Through electrical arrangement, the current is pulsed so that hydrogen bubbles are formed and released at regular intervals. A transformer is used to give a working voltage range of 0 to 200 V, the actual voltage required for optimum bubble formation being dependent on the quality of water. When the experiments are conducted in a glass walled flume, the bubbles can be illuminated by a beam projector bulb of 1000 W, or from a mercury vapour lamp. The beam can be condensed to the required size by using a lens. A cine camera running at high speed of the order of 250 frames per second can be used to photograph the motion. Kemp and Grass (4) collected data in an open channel rectangular flume, while Clark and Markland (5) studied the turbulent boundary layer on a plate. Even though Clark and Markland did not report the computations of turbulent velocity components, they studied vortex formation in the inner portion of turbulent boundary layer with $u_* y/\nu$ less than 120.

7.3 METHODS USING TRANSDUCERS

There are a number of methods of measuring turbulence in which a detecting element is introduced in the flow. One of the characteristics of the detecting element (or transducer) is changed in the presence of turbulence, the measure of which can be related to the flow conditions. Hot wire, current meter, total head tube using capacitance or resistance variation, electromagnetic transducer and electrokinetic transducer, which are used in measurements of turbulent velocity components, are briefly discussed below.

Hot Wire and Hot Film Anemometers (1, 6, 7, 8, 9, 10)

Hot wire anemometer is one of the most perfected and often used instrument for measurement of various quantities related to turbulence in air and gases; hence, this is discussed in somewhat greater detail here. The detecting element of HWA consists of a very fine and short wire of platinum, tungsten, or platinum-tungsten alloy which is heated by an electric current. When heated wire is exposed to gas flowing at right angles to the wire, the wire is cooled and it loses heat, as a consequence of which its temperature and resistance are reduced. The total amount of heat loss takes place due to heat conduction, free and forced convection and by radiation. Experiments have shown that free convection effect can be neglected if Reynolds number of the wire is greater than 0.50, and $GrPr$ is less than 10^{-4} (1); here Gr is Grashof number and Pr is the Prandtl number. Radiation effect in heat transfer to the ambient air is negligibly small if wire temperature does not exceed 300°C. Further, effect of compressibility can be neglected if Mach number is less than 0.30. Under these conditions the nondimensional heat transfer coefficient, known as Nusset number Nu, is related to Prandtl number and Reynolds number Re by the relation.

$$Nu = 0.42\, Pr^{0.20} + 0.57\, Pr^{0.33}\, Re^{0.50} \qquad \ldots (7.1)$$

which can be written as

$$Nu = A + B\, Re^{0.50} \qquad \ldots (7.2)$$

By relating the rate of heat loss to temperature difference between wire and air, and using other relations, one gets from the above equation

$$\frac{I^2 R_w}{R_w - R_o} = a + b\sqrt{\bar{u}} \qquad \ldots (7.3)$$

where I is the current, R_w is the wire resistance, R_o is resistance at reference temperature, $\bar{u}$ is the velocity to which wire is exposed, and a and b are constants which depend on the dimensions and physical properties of wire, and on the properties of gas. It may be mentioned that in 1914 King derived and experimentally verified expression similar to Equation 7.3 over a limited range, assuming potential flow around the wire. However, structure of the constants a and b in the two approaches is different, except that in both the cases heat loss is proportion to $\sqrt{\bar{u}}$.

The probe wire for turbulence measurements is very fine. It should be easily available and strong. For low speed work, platinum wire using silver coating is used. The silver coating increases its diameter, and it is easier to work with it and solder it to the support. Later, the silver is removed by etching in nitric acid. Platinum is used in clear air upto about 70 m/s velocity, tungsten wire upto sonic velcoity, and platinum-iridium, or platinum-rhodium alloys in supersonic flows. Wire diameter varies from 2 to 10 microns. For such diameter, length/diameter ratio is about 200 to 300, so that the time constant of the wire is small. The response of a hot wire to fluctuations of fixed amplitude declines (as well as shifts in phase) at some high frequency due to the wire being unable to exchange heat rapidly enough with its surroundings. This is known as the thermal inertia of the wire. This attenuation of the signal and phase shift can be corrected by passing it through a RC circuit with opposite characteristics. Such wires with appropriate electronics can respond to turbulent fluctuations upto 300,000 Hz when exposed to fluctuating velocity, as in the case of turbulent flow in air or gases.

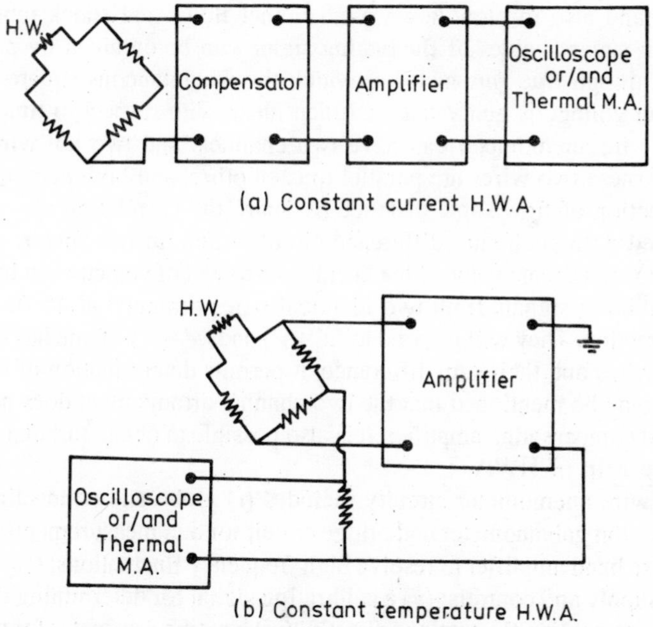

Fig. 7.1 Hot wire anemometer circuits

In practice, hot wire forms one arm of the Wheatstone bridge which, when combined with other electronic equipment, gives the desired turbulence quantities. The hot wire anemometer can be operated either under constant temperature (or resistance) condition, or under constant current condition. Figure 7.1 shows the HWA circuits for both these cases. In constant temperature system, as the wire gets cooled due to exposure to air velocity, the heating current is automatically varied so as to maintain the wire temperature (or resistance) constant. In such a case, the heating current is the measure of velocity. The variable current is obtained by using a variable current source controlled by a feedback amplifier. In the constant current anemometer, the heating current is kept constant. Because of the exposure to air current, the wire is cooled and the wire resistance and, hence, the voltage drop across the wire is changed, which is a measure of velocity or velocity fluctuation. The constant current HWA has the following advantages:

(i) It resolves the velocity fluctuations 20 times smaller than those resolved by constant temperature HWA;
(ii) even though it does not operate at very low frequencies (about 1 Hz). its overall frequency range is much larger;
(iii) it is more stable; and
(iv) it can be used for measurement of rapid temperature fluctuations at frequencies upto 12000 Hz. However, it must be emphasised that the electronic equipment required for constant temperature operation is much more complicated than that for constant current operation. Hence, normally constant current HWA is used more often, and constant temperature HWA is used when large fluctuations are to be measured, and also in blow down wind tunnel flows and shock tubes.

Mean square values of fluctuating signal can be obtained by passing the signal through true *rms* meter in which the instantaneous square values of the input voltage is generated, and then mean with respect to time is taken.

Hot wire anemometer can have two channels and two hot wires can be used. If these two wires are parallel to each other, and both perpendicular to the direction of flow some distance Δx apart, the correlation $\overline{u'_x u'_{x+\Delta x}}$ can be measured with such sum difference circuitry and an *rms* meter.

Root mean square value of the lateral component of velocity can be obtained by combining signals from two identical wires arranged at 45° to the main flow direction. They will respond to $(u' + v')$ and $(u' - v')$. If one has the signals of two wires and their sum difference, it permits determination of $\overline{u'^2}$, $\overline{v'^2}$ and $\overline{u'v'}$. It may be mentioned that the two channel arrangement does not require a second compensating amplifier. It is also possible to obtain turbulence spectra with the help of HWA.

Hot wire anemometer circuity includes (i) probe and connecting cables; (ii) precision galvanometer and bridge circuit for d.c. measurements; (iii) very low noise band amplifier to resolve high frequency fluctuations; (iv) low noise power supply and controls; (v) a calibrating signal for determining the system response to cooling fluctuations directly; (vi) a.c. true *rms* meter for turbulence intensity and correlation measurements; (vii) provision of additional channels for sum differnce and correlation measurements; and (viii) a recording device for *a.c.* and *d.c.* signals, such as an oscilloscope or wave analyser.

Hot wire anemometer is unsuitable for measurement of turbulence in water because of the contamination of the wire by dissolved gases, dirt and chemicals in water, electrolysis of wire, and conductivity of wire. Hence Ling (11) introduced hot film probe which consists of a thin platinum film as a sensing element. The film, about 0.10 mm long and 0.20 mm wide is fused on the wedge shaped glass or ceramic support. The coating eliminates stability problems caused by electrolysis and conductivity through fluid medium. Richardson and Mcquivey (12) used constant temperature hot film anemometer for measurement of turbulence in water in a laboratory flume. Figure 7.2 shows different types of hot film probes that are commonly used.

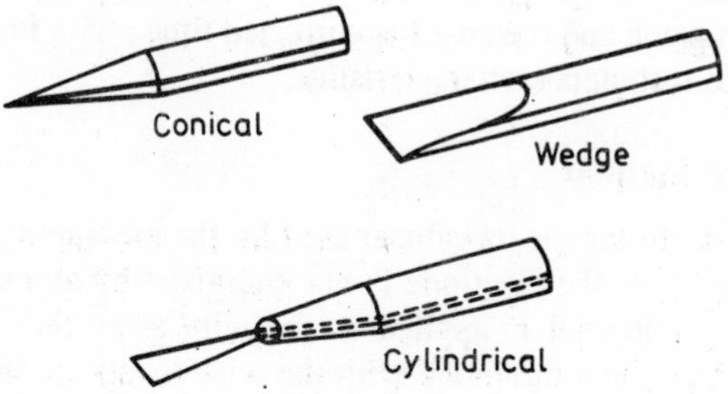

Fig. 7.2 Different types of hot-film probes

Many investigators, however, have faced difficulties in application of hot film anemometer (HFA) for measurements in water. The main problems include probe selection, operating temperature/overheat ratio selection, calibration methods, temperature compensation, and cable length compensation. The recent research work undertaken at Central Water and Power Research Station (CWPRS), Poona (India) has helped in providing guidelines in order to take reliable and accurate measurements of velocity fluctuations in turbulent flows. Regarding the operating temperature of the sensor when used in liquid media, the boiling point of the liquid plays an important role. The operating temperature of the order of 52° to 60°C is recommended for water applications. Correspondingly the overheat ratios in the range of 1.04 to 1.06 are recommended, the lower overheat ratio for velocity range from 20 mm/s to 2.0 m/s and higher overheat ratio for higher velocities upto 7.0 to 8.0 m/s. Proper care is required to be taken to compensate for resistances of cables and fixtures connecting the actual sensor at the measurement site to the anemometer instrument at the laboratory. This is especially important as hot film sensors have resistances of the order of 4 to 6 Ohms at ambient temperature while cables and fixtures have resistances of the order of 0.1 to 0.5 Ohms.

As the HFA works on the principle of energy transfer which in turn depends upon the fluid velocity, temperature and density, it is important to control the temperature and density during the measurements to give reliable velocity measurements. If environmental temperature changes during the measurement, the compensation for the change in temperature is essential. CWPRS has developed algorithms for temperature compensation by obtaining calibration curves between velocity and voltage for various temperatures.

HFA is very sensitive at low velocities in the range of 10 mm/s to 0.25 m/s. The probe is calibrated at low velocities in a special calibration tank fabricated in CWPRS and another calibrator used for high velocities. A fourth degree polynomial is fitted for the calibration curve at velocities above 0.30 m/s while at lower velocities piecewise linearisation is done.

However, use of HFA for two phase or bubbly flows still poses a problem as the density of fluid during calibration and actual measurement is different. If the bubble contents in the fluid are not very high, special software techniques are used to remove the portions of the record indicating appearance of bubble on the hot film patch and then use the corrected time series for further analysis to find out the turbulence characteristics.

Electrokinetic Method

In the case of electrokinetic transducer used for the measurement of turbulence in water, the wire or the electrode is not energised by an external source of energy, e.g. no potential is applied to the wire as in the case of hot-wire anemometer. Yet when the probe with the wire is introduced into a suitable turbulent flow and the active or so called "hot" lead, which is nothing but extension of sensing wire, is connected to an oscilloscope through an electrometer and an amplifier, a random signal is obtained. The probe used by Chuang and Cermak (13), and Binder (14) are shown in Fig. 7.3. The following qualitative description of the phenomenon taking place at the wire surface is offered by Binder (14).

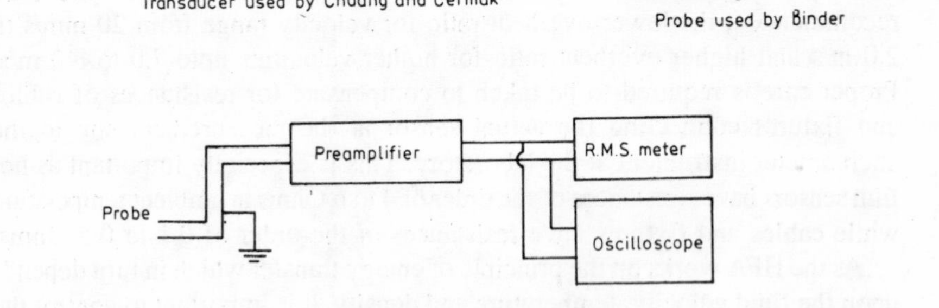

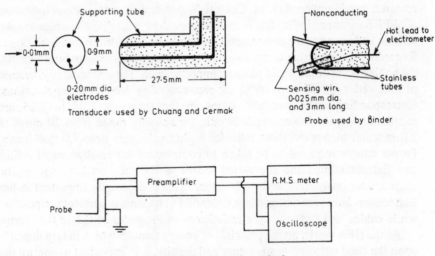

Fig. 7.3 Eletrokenetic method

When water or an aqueous solution is in contact with a solid, conductor or insulator, due to preferential adsorption of some ion species, an electric double layer exists at the solid-liquid interface. This double layer on the whole has zero charge, but consists of an adsorbed layer, positive or negative, depending on the solute whose charges are bound to the solid and of a layer in the fluid of opposite sign to the former, which is diffused because of molecular agitation (see Fig. 7.4). The flow of a fluid causes entrainment of

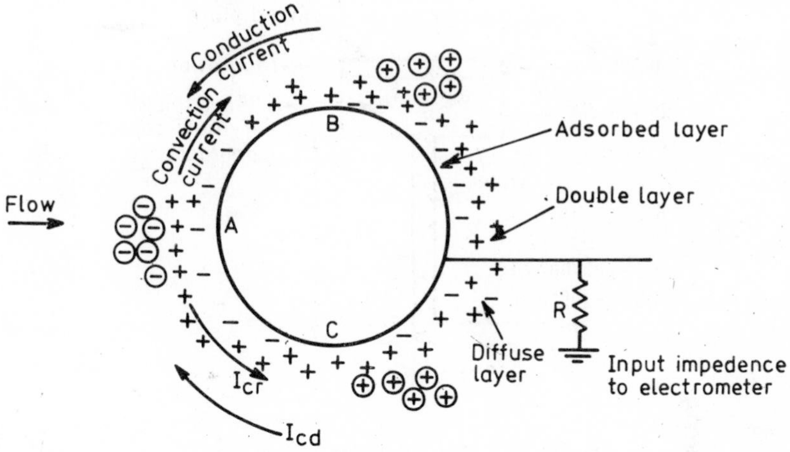

Fig. 7.4 Schematic representation of the interface process

charges of the diffuse layer along its interface. This produces an electrical "convection" current whose density depends on the flow field and on the charge distribution in the diffuse layer. Due to the flow, the diffuse layer slides from A towards B and C (as shown in Fig. 7.4) and, hence, at these points there is concentration of positive charges. At A there is excess negative charge. Under steady state condition, positive and negative charge concentration at B, C and A causes a conduction current from B, C to A, and a convection current from A to B, C. In unsteady or turbulent flow, these positive or negative charge accumulations fluctuate with incident velocity and hence, electric potential of the wire with respect to ground also fluctuates. These potential fluctuations appear across the input impedence of the electrometer.

The electronic measurement equipment used consisted of a low level amplifier, a true *rms* meter and an oscilloscope. The pre-amplifier had a selective gain and frequency response range that were set at 1000x and 1 to 10,000 Hz, respectively. The *rms* meter had a voltage range from 1 to 1000 volts and frequency range from 1 to 200,000 Hz. The equipment was tested for flows in open channels, in the wake caused by a flow past circular cylinder (13), jet impinging on a flat plate, and flow behind grid in turbulent flow in a pipe. Bergmann and Hodgson (15) have given theoretical explanation of the phenomenon and have shown that in an oscillating poiseuille flow in a pipe, electrode positioned in the wall detects a signal with a phase angle of $\pi/4$ ahead of the fluctuating velocity as measured by a hot film probe in the pipe centre.

Total Head Tube-Pressure Cell Combination

Ippen and Raichlen (16), and Eagleson and Perkins (17) have described a total head tube-pressure cell combination type of instrument for measurement of

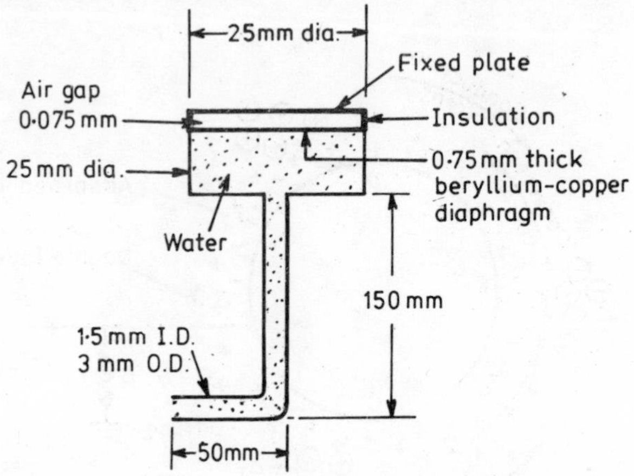

Fig. 7.5 Total head capacitance type transducer

turbulence in water. Figure 7.5 gives the schematic drawing of the total head tube-pressure cell combination used by Ippen and Raichlen (16). It consists, essentially, of the total head tube inserted in the flow facing the direction of flow and the pressure transducer mounted on the other end of the tube. At one end of pressure chamber is a flexible diaphragm clamped to the chamber walls firmly. A fixed electrode is placed at small distance away from the membrane separated by a thin air gap. Change in the instantaneous total head at the tip of tube causes change in the pressure at the diaphragm, which in turn deflects the membrane. Deflection in membrane causes change in air gap and, hence, change in capacitance which is electronically recorded. By proper static calibration, the change in the capacitance can be directly related to the instantaneous change in total head. In such a case one can write

$$H_t = H + h_s = \frac{(\bar{u} + u')^2}{2g} + \frac{p'}{r} + h_s \qquad \ldots (7.4)$$

where H_t is the instantaneous total head recorded, h_s is the mean hydrostatic head on the tube tip, $\bar{u}$ is the mean velocity in front of the tube tip, u' is the turbulent velocity fluctuation both in x direction, and p' is the turbulent local pressure fluctuation. The values of H and H_t are fluctuating in random manner. Hence,

$$H = H_t - h_s = \frac{\bar{u}^2}{2g} + \frac{\bar{u}u'}{g} + \frac{u'^2}{2g} + \frac{p'}{r} \qquad \ldots (7.5)$$

It may be noted that since $\bar{u}$ is much larger than u', $\bar{u}u' >> u'^2$. Further, even though $\frac{u'^2}{2g}$ and $\frac{p'}{r}$ are of the some order of magnitude, they are normally not in phase. Therefore, $\frac{u'^2}{2g}$ and $\frac{p'}{r}$ can be neglected. Therefore, if the total head tube is statically calibrated, it is possible to know $\bar{u}u'$ since $\bar{u} = \sqrt{2gH}$ is known from calibration, and then u'. Making some realistic assumptions, Ippen and Raichlen have determined the natural frequency of the system as

$$f_m = \frac{1}{2} \frac{A_1}{A_2} \sqrt{\frac{K}{\rho A_1 L_1}} \qquad \ldots (7.6)$$

here A_1 and A_2 are cross sectional areas of total head tube and diaphragm well, respectively, L_1 is the length of total head tube, and K is the spring constant of the diaphragm which for a circular diaphragm clamped at its edge is given by

$$K = \frac{16\pi\, Eh^3}{a^2\,(1 - \xi^2)}$$

Here, a is radius of diaphragm, h is its thickness, E is the volume modulus of elasticity of the diaphragm and ξ is Poisson's ratio, equal to 0.30. They tested initially a gauge whose natural frequency was 15.4 Hz and hence, a modified version was tested by them with the dimensions given in Fig. 7.5. This had a natural frequency of 255 Hz. This device was used by them to measure $\overline{u'^2}$, mean intensity spectrum and autocorrelation function in turbulent flow in open channels. They also measured these quantities in the wake of the cylinder. The major advantages of this device seem to be ease in calibration, ease in determining its natural frequency, and its utilisation in clear water as well as sediment laden flows.

This device was later modified by Eagleson and Perkins (17) by using a small piezoelectric ceramic pressure transducer at the tip of total head tube. This device was found to be satisfactory upto a frequency of 1200 Hz. Sushil Kumar et al. (18) used a device similar to that used by Ippen and Raichen, but using a strain gauge type of transducer. They studied the turbulence in open channels with spot roughnesses on the bed. Arndt and Ippen (19) studied in detail the performance of such a device and discussed its design aspects. As regards low frequency response, it is emphasised that, since most of energy of the eddies in turbulent flow in open channels is contained in eddies with frequencies of the order of 200 Hz, relatively low frequency response of total head tube type of device is not a serious drawback.

Electromagnetic Induction Method

Electromagnetic induction method for measurement of turbulent velocity fluctuations in liquids which are slightly ionised depends (1) on the fact that a velocity component perpendicular to an electromagnetic field induces an electric field whose strength is given by Maxwell-Faraday law

$$V = \frac{\mu_1}{C} Bu \qquad \ldots (7.7)$$

where u is the velocity component perpendicular to the electromagnetic field of strength B, μ_1 is the relative magnetic permeability, C is the velocity of light in the medium, and V is the strength of induced electrical field. The direction of the induced field is perpendicular to both the velocity component u and the magnetic field B, and its direction is obtained by left-hand rule. If two electrodes with a gap l are introduced into the flowing liquid in the direction of induced electric field, the potential difference between the electrodes E is given by

$$E = Vl = \frac{\mu_1 Bul}{C} \qquad \ldots (7.8)$$

Two important characteristics about the above phenomenon are that the voltage E is a linear function of u, and that it is direction sensitive since it is affected only by the velocity component perpendicular to the direction of the magnetic field and the line joining the electrodes. It is also worth noting that the relation between E and u is unaffected by density, viscosity, temperature and electric conductivity of the liquid. Electrodes should be made of materials that are unaffected by acids and cannot be polarised. One can use either of the two methods in the application of electromagnetic field: (i) expose the entire flow to an electromagnetic field; or (ii) introduce in the flow field a relatively small electromagnet connected to a pair of electrodes in a line perpendicular to the line through the poles.

In the first case, a stronger electromagnetic field is required, depending on width or depth of flow. Further, since fluid motion is not uniform throughout the flow field, one must take into consideration the effect of locally induced currents, which will be important near the boundaries. In the second case, the detecting element consisting of the electromagnet and electrodes must be quite small if one is to measure small scale turbulence. Kolin (1*) used a system with 23 mm as overall length of magnet, 1 mm distance between the magnet poles, and 1 mm as diameter of poles. In most of laboratory work, the first method has been used; the second can be used in large dimensioned flows for measurement of macroturbulence.

Grossman et al. (2) have used this method to measure turbulent velocity fluctuations in fully developed flow in a circular pipe. These measurements were made in a 50 mm diameter lucite pipe. The magnetic field was provided by direct current electromagnet with 140 mm diameter circular pole pieces, and a 65 mm gap. Maximum flux intensity was about 9000 gauss, which was nearly uniform over the measuring region. d.c. current was obtained from a d.c. generator, and it was ensured that ripple voltage was eliminated by use of a condenser across the coils of the magnet. Probes were constructed from 0.40 mm diameter insulated copper wires enclosed in a 1.3 mm outer diameter copper tube, bent into an L shape, so that the wire tips were 18 mm upstream from the stem. Tips of wires were separated by 0.60 mm distance. Signal from the probe was fed to the differential input of a low level preamplifier. The output signal of the preamplifier was fed to a cathode ray oscilloscope. In addition, the instantaneous signal of preamplifier was amplified once again by a signle stage amplifier and fed to a squaring amplifier to obtain time average of the signal. Two complete channels were provided so that two input signals from the probe could be measured at the same time (see Figs. 7.6 & 7.7).

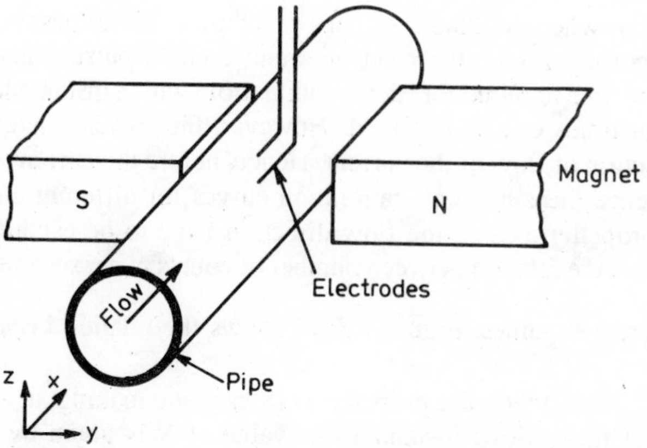

Fig. 7.6 Definition sketch for electromagnetic induction method

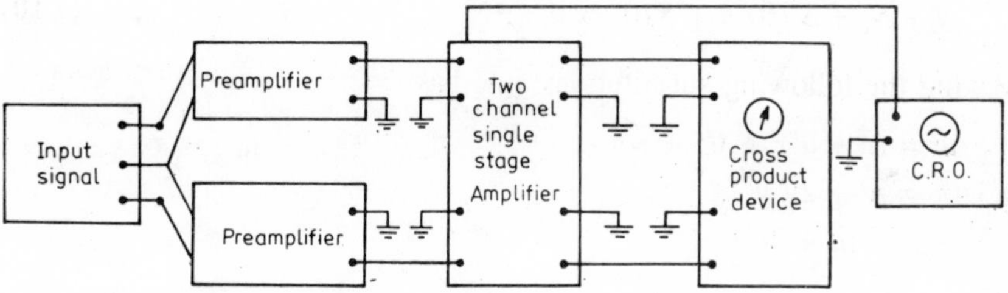

Fig. 7.7 Block diagram: Electronic equipment for electromagnetic induction method

Pande (21) has also used this method for measurement of turbulence characteristics in a pipe.

This method, intrinsically most perfect, has the disadvantage that a uniform magnetic field of large strength is required across the flow. Such a field can be maintained only if the flow field is small. As such, it cannot be easily used for field measurements unless the magnet forms an integral part of the probe.

Turbulence Measurement Using Current Meters

Knowing the limitations of hot-wire anemometer, and hot film anemometer, attempt has been made to use the current meter in measurement of turbulence in water. The major problem with hot wire/film aneometer is the uncontrollable influence of wire contamination and the formation of air bubbles on the surface of heated element. These problems make the calibrations of probe unstable to such an extent that the interpretation of the measured data becomes doubtful. These problems do not exist in the use of a small propeller type current meter for the measurement of turbulence. This has been studied and used by Schyuf

(22), Hansen and Christensen (23), Tiffany (24), Iwasa (25), and Ishihara and Yokosi (26).

Normally, when the blades or cups of a current meter pass a certain point on the suporting frame, the electronic unit emits a pulse. This information can give the time required for each rotation, from which instantaneous velocity at different times can be obtained. However, the current meter is sensitive to the direction of flow of the current. Hence, before the current meter is used for turbulence measurement, calibration curves for different angles of yaw, θ of the propeller axis to the flow direction have to be established. Figure 7.8 (a) shows the relation between number of counts N per second and velocity u for different θ values. Figure 7.8 (b) shows the relation between $\frac{dN}{du} = S$ and θ (23). Now, when the propeller is in xy plane making an angle θ to the direction of mean flow, instantaneous value of N is given by

$$N = N_o + S(\theta + \alpha)u \qquad \ldots (7.9)$$

where N_o is the number of rotations per second when θ is zero, and α is as shown in Fig. 7.8. If α is much smaller than θ, Eq. (7.9) can be written as

$$N + N_o + S(\theta)\, u + S'(\theta)\, \alpha\, u \qquad \ldots (7.10)$$

Making the following substitutions, one has

$$\begin{aligned} u^2 &= (\bar{u} + u')^2 + v'^2 + w'^2 \\ &= \bar{u}^2 + 2\bar{u}\, u' \\ &= \bar{u}\, (\bar{u} + 2u') \end{aligned}$$

or

$$u = \bar{u} + u'$$

and

$$\alpha u = v'$$

since for small α , $\alpha \approx \sin \alpha$. Therefore, Eq. 7.10 becomes

$$N = N_o + S(\theta)\, \bar{u} + S(\theta)\, u' + S'\, (\theta)\, v' \qquad \ldots (7.11)$$

or $N' = S(\theta)\, u' + S'(\theta\ )\, v'$ $\qquad \ldots (7.12)$

From this equation one obtains $\overline{N'^2}$ as

$$\overline{N'^2} = \overline{S'^2}\, \overline{u'^2} + S'^2(\overline{v'^2}) + 2SS'\, \overline{u'v'} \qquad \ldots (7.13)$$

Thus, it becomes evident from the above equation that, by measuring $\overline{N'^2}$ at a point for three different directions, say at +30°, 0° and –30°, one can obtain $\overline{u'^2}$, $\overline{v'^2}$ and $\overline{u'v'}$.

Tiffany (24) has reported measurements of turbulence in axial direction in the Mississippi river by standared Price current meter. The measurements were made at a point for 3 hours, when readings were recorded at intervals of 1 minute. For a depth of 10.67 m, when measurements were taken at 0.55 m above bed, it was found that

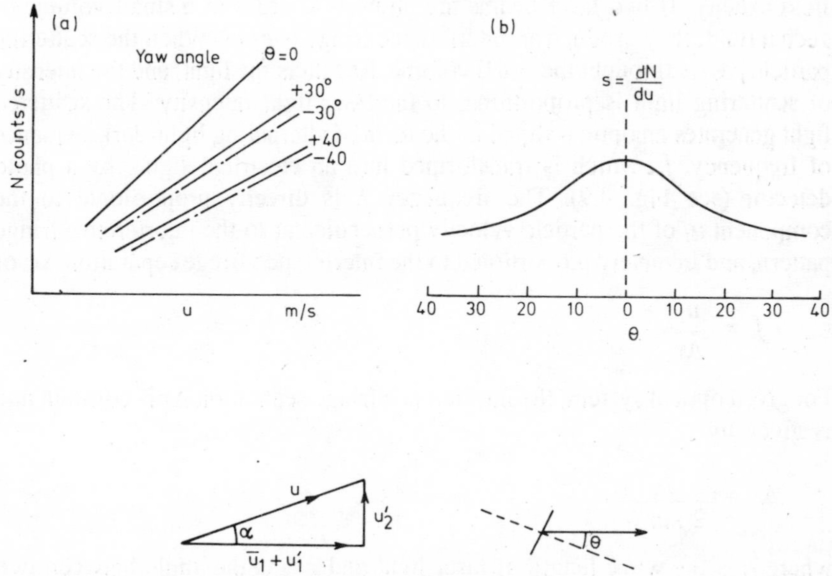

Fig. 7.8 Calibration of current meter for turbulence measurement

average velocity = 1.0 m/s
maximum velocity = 1.47 m/s
minimum velocity = 0.70 m/s

Further, it was found that there were cycles of periodic fluctuations of approximately 4 minutes, 20 minutes and 2.5 to 3 hours, superimposed on one another. Ishihara and Yokosi (26) measured the one dimensional spectrum in the Uji river in Japan, which was 100 m wide and 2 m deep at the sampling station. The current meter was 0.14 m in diameter.

Certain limitations of use of current meter in measurement of turbulence must be noted. Firstly, because of finite size of the current meter of the order to 10 mm to 0.15 m in laboratory and field respectively, only relatively large scale turbulence can be measured. Secondly, due to inertia of the current meter, it may not respond to low intensity turbulence, and further, a real lag of time will be produced between rotation and fluctuating velocity. Statistical analysis of data by Iwasa (25) has indicated that if

$$\frac{\text{Sampling time interval}}{\text{Eulerian integral time scale}} \leq 10^{-1}$$

and

$$\frac{\text{Sampling time}}{\text{Eulerian integral time scale}} \geq 10^{2}$$

the inertia effect will be negligible.

7.4 LASER DOPPLER ANEMOMETER (27, 28)

In liquid and gas flows, there are often sufficient number of scattering particles present (and if they are not present, they can be added) which follow the flow field exactly. If two laser beams are allowed to cross in a small volume of such a fluid, they produce an interference fringe pattern. When the scattering particle passes throught the small volume, it scatters the light, and the intensity of scattering light is proportional to the laser light intensity. The scattered light generates an optical signal in the form of alternating light-dark sequence of frequency, f_s, which is transformed into an electrical signal by a photo detector (see Fig. 7.9). The frequency f_s is directly proportional to the component u_1 of the particle velocity perpendicular to the interference fringe pattern, and inversely proportional to the interference fringe separation Δx; or

$$f_s = \frac{u_1}{\Delta x}$$

For given optical system, the interference fringe separation Δx is constant and is given by

$$\Delta x = \frac{\lambda}{2 \sin \delta}$$

where λ is the wave length of laser light and 2 ϕ is the angle between two laser beams. When both laser beams are of identical frequency, the interference fringe pattern in the measuring volume is fixed in space. If these two frequencies differ, the interference fringe pattern moves through the measuring volume with a frequency f_{SH} which corresponds to the difference in frequencies between the two beams. From the principle of superposition, one can write

$$u_1 = \Delta x(f_s \pm f_{SH}) = \Delta x f_{so}$$

where f_{so} is the frequency generated by the moving particle. Thus, u_1 can be known by knowing the movement of fringe pattern.

If the test section of the flow system, such as flume or wind tunnel, has both sides transparent, the transmitter and receiver optical elements can be located on two separate base plates on either side of the test section. This is known as the forward scattering mode. Appropriate rotation of the optical system permits the measurement of required velocity component. If the test section is accessible from only one side, it being transparent, the transmitter and receiver elements of the optical system are kept on the same side of the test section, and this is known as the backward scattering mode.

Two optical systems are normally used for getting the laser beam, these are Helium-Neon system, and Argon system. The Helium-Neon system is economical and is mainly used for single channel measurements in liquid and low velocity gas flows. The Argon system has a higher laser beam power and, hence, it is possible to take measurements in high flow velocities and in backward scattering mode. Further, with Argon system, simultaneous measurements of two velocity components with a two colour optical system is possible. Naturally, Argon system is more expensive.

The signal processing unit contains filter elements for signal conditioning before the active tracking loop circuit. The instrument follows the particular signal frequency of the photodetector unit, and delivers a voltage level which

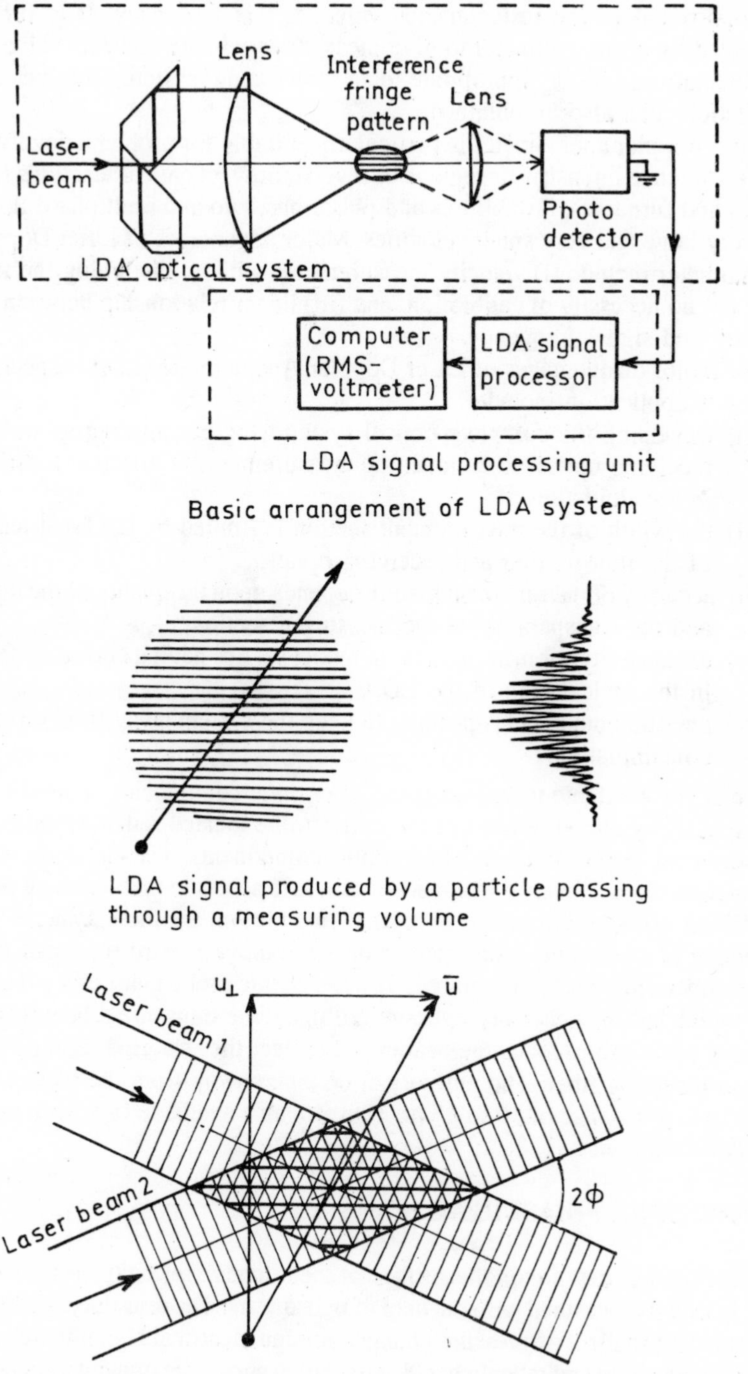

Basic arrangement of LDA system

LDA signal produced by a particle passing through a measuring volume

Interference fringe pattern of the measuring volume

Fig. 7.9 Interference in LDA

is proportional to the instantaneous value of this frequency. It is further processed by a *rms* voltmeter to give mean flow velocity and *rms* values of its fluctuations. With appropriate data processing system, the velocity correlations can also be obtained.

Laser doppler anemometer is particularly suitable for problems involving high turbulence intensity, or regions of reverse flow, chemical reactions (e.g. flames and furnaces), instabilities and pulsations, two and multiphase flows, and very large and very small velocities. Major advantages of Laser Doppler Anemometer include (i) velocity measurements without disturbing the flow field; (ii) no necessity of calibration; and (iii) linear relationship between the velocity and signal frequency.

The major disdvantages of Laser Doppler Anemometer (LDA) especially for water application include:

(i) traversing the complete optical system (for transmitting as well as receiving optics) to conduct the measurements at different locations in the fluid flow.
(ii) the width of the measurement section is limited by the focal length of the transmitting and receiving optics;
(iii) accuracy of measurement greatly depends upon the quality of the liquid and the transparancy of the measuring section;
(iv) unsuitability for measurement in model set ups having curved surfaces in the optical path of the LDA;
(v) environmental care, specially to keep the system away from dirt and contamination.

Recently the available technique using fiber optic components enhances the use of LDA by allowing most of the optics to be located remotely from the measurement point. With the fiber optic components, velocity/turbulence information can be obtained in applications where access to measurement point is difficult or the environment is hostile. In addition, for these systems, the traversing is much more simplified as only the movement of the small light weight fiber optic probe is required. The fiber optic probe features a compact head which has the necessary optics to collimate the transmitted beams from the fiber optic cable, focus the beams and collect the scattered light on to a built-in receiving fiber. The probes can be taken away from the main LDA optics upto a distance of 10 to 15 m. They are also available in a wide range of focal distances.

7.5 PRESSURE MEASUREMENT

The fluctuating pressure on the surface of the boundary or body, in turbulent flow, is usually measured with the help of transducers. The transducer converts the input in the form of pressure changes into an electrical signal which can be more accurately and reliably handled by the electronic instrumentation units. The electrical output from the transducer can be in the form of fluctuating voltage, current, resistance, capacitance or inductance. Commonly used transducers can be of resistive, inductive, capacitive, voltage or current, photovoltaic, thermocouple or piezoelectric type. These are discussed by

Rangan et al. (29). Of these, three types are briefly discussed below. Transducers are classified into active and passive transducers. An active transducer does not need external excitation; electrokinetic transducer used for velocity measurement and discussed earlier is an active transducer. A passive transducer needs external supply of d.c. or a.c.; hot wire anemometer is a passive transducer.

Strain gauge type transducer is a passive transducer which transforms elongation or compression caused in a wire due to pressure variation into a resistance change. Change in resistance occurs because the length and cross sectional area of the gauge wire change. Gauge factor of a strain gauge is defined as

$$\text{Gauge factor} = \frac{\Delta R/R}{\Delta l/l}$$

where R and ΔR are the initial resistance and change in it, and l and Δl are wire length and change in it, respectively. This factor normally varies from one to three. If the strain gauge is pasted on the underside of a flexible diaphragm on the other side of which varying pressure is applied, change in pressure will cause change in the length of strain gauge and, hence, change in resistance. Materials used for strain gauges include copper-nickel, nickel-chromium, platinum-irridium, iron-chromium-aluminium, and iron-nickel-chromium alloys. If the strain gauge forms one arm of the Wheatstone bridge, this change in resistance can be measured and related to change in pressure.

The capacitance type of transducer is essentially of the same type as used by Ippen and Riachlen (16) and shown in Fig. 7.5 for measurement of total head at the tip or total head tube. The capacitive transducers are small in size, have good linearity and resolution.

Piezoelectric pressure transducer works on the principle that when a stress is applied to certain materials, an electrostatic charge or voltage is generated. An opposite effect is also generated when electrostatic charge or voltage is applied to this material, resulting in the mechanical deformation of the device.

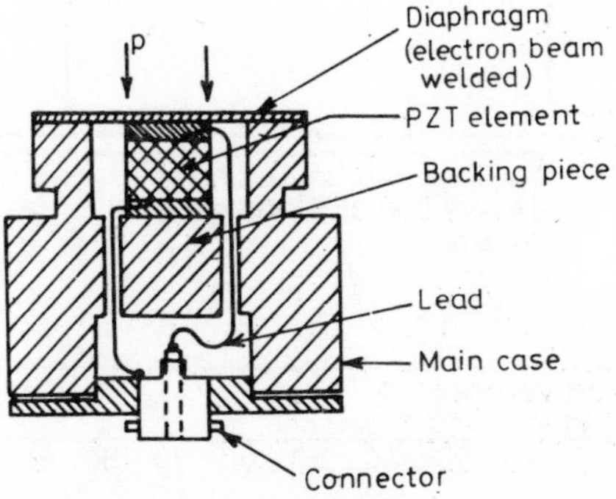

Fig. 7.10 Assembly of a piezoelectric pressure transducer

This property of piezoelectricity has been utilized in the design of pressure-transducers. The most popular piezoelectric materials are natural quartz, Rochelle salt, and synthetic ceramic materials like barium-titanate, and lead-zircoate-titanate. Of these, the natural quartz is the most stable device, since it has lower temperature sensitivity and high resistivity, thus giving an inherently long time constant which permits static calibration. Further, it exhibits linearity over a wide range of stress level. Piezoelectric pressure transducers are widely used for measurement of rapidly varying pressure as well as shock pressure. They provide a flat freauency response from 1 Hz to 20 kHz, the natural frequency being of the order of 50 kHz. If piezoelectric crystal is in the form of a disc of thickness t mm and is subjected to normal stress, its natural frequency is $923/t$, kHz. A typical complete assembly of a normal compressive mode trarducer is shown in Figure 7.10.

Tiffnay (24) measured the pressure distribution near the bed in the case of the Mississippi river. The full scale equipment consisted of a circular cast iron disc of 1.52 m diameter, 100 mm thickness and weighing about 1250 kg. The edge of the disc was formed on a 2:1 quadrant of an ellipse to minimise the effects of surface discontinuity. A 3-cable bridle suspension was found necessary to handle the rig. Vertical and horizontal stabilizers, designed from small scale tests, were provided for proper orientation of flow lines. In the centre of the disc was a transformer type pressure cell for the measurement of pressure fluctuations on the surface of the disc. Attached to, and suspended immediately above the disc was a Price current meter for measurement of velocity fluctuations. Figure 7.11 shows the assembly and data obtained. It can be seen that high instantaneous velocity is associated in general with low instantaneous pressure.

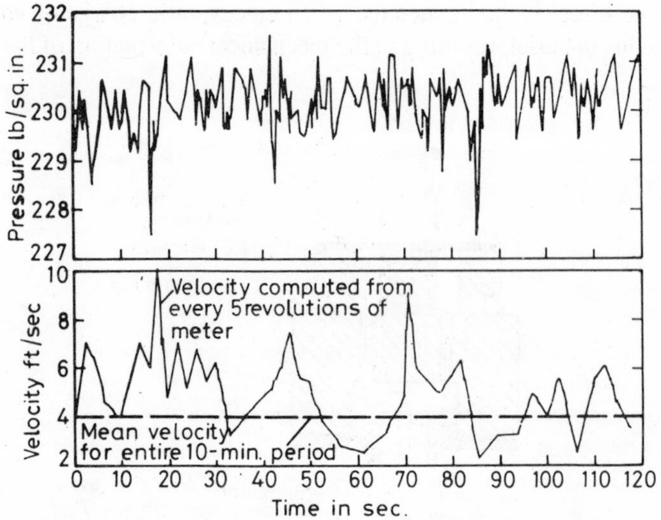

Fig. 7.11 Set-up used in Mississippi river for measurement of velocity and pressures, and typical data

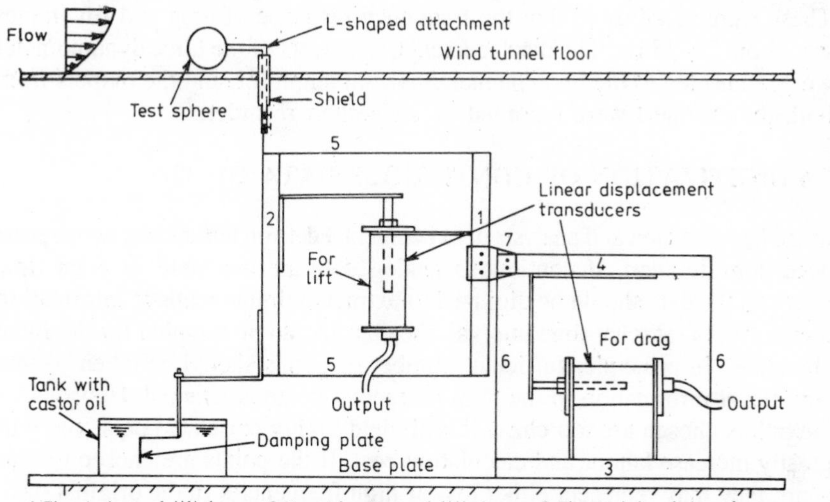

1 Fixed strap of lift frame; 2 Movable strap of lift frame; 3 Fixed plate of drag frame
4 Movable strap of drag frame; 5 Flat springs of lift frame; 6 Flat springs of drag frame;
7 Revet plate

Fig. 7.12 Force dynamometer (30)

The transducers can also be utilised for measurement of force, drag as well as lift, on a body. The arrangement used by Patnaik (30) with linear displacement transducer is described below. Figure 7.12 shows the spring activated parallel link force dynamometer used for simultaneous measurement of drag and lift on a sphere placed near the boundary in a wind tunnel. The dynamometer comprised of two frames, one for the measurement of drag 3–4, 6–6, and the other for the lift 1–2, 5–5. Each frame had a pair of aluminium straps 1–2 and 3–4 with a pair of flat springs connected at their ends as shown in Fig. 7.12. The lift frame which supported the test sphere, was itself supported on the drag frame. The whole unit was mounted on a heavy base plate keeping the lift and drag frames in a single plane. With this arrangment, a vertical force caused a deflection of the lift frame while a horizontal force actuated the drag frame. The displacement of either of the frames was sensed by linear displacement transducers along with the associated circuitry. The mutual non-interference between the lift and drag frames was verified by the application of static loads at the centre of test sphere. The test sphere was attached to the dynamometer through an L-shaped attachment, which was properly shielded upto the tunnel floor in order to eliminate the force on the exposed vertical limb of the attachment.

For the purpose of damping, the whole system was connected to two aluminum plates, one in the vertical plane and the other in the horizontal plane, dipped in a closed tank filled with castor oil. By varying the size of the plates the system damping could be varied. The stiffness of the dynamometer springs and the size of the damping plates were adjusted by trial and error to obtain optimum peformance under dynamic conditions. While stiffer springs increase the frequency response of the system, the sensitivity goes down considerably. These were so adjusted that the frequency response of drag and lift frames was about 20–25 Hz. The displacement transducers of the force dynamometer were connected to the two channel universal amplifier, and the outputs from both the channels were recorded on strip chart recorder.

7.6 DIGITIZATION OF CONTINUOUS DATA (31, 32)

In the interpretation and analysis of experimental data on turbulence, two aspects need consideration. If continuous analog data are available, at what time intervals the data should be digitized to extract optimum realistic information therefrom; or, at what time interval, the data should be sampled for the same objective? In principle, sufficient number of points should be taken so that significant information in the high frequency range is retained. However, if the points chosen are too close, it will yield highly correlated data, and will greatly increase labour and calculation cost. If the points are chosen too far apart, they may represent either low or high frequencies in the original data. This property is called *aliasing*, and can be a source of error which does not occur in analog processing of data. Figure 7.13 shows how sampling interval can affect the interpretation of data. If data were read at points O, one could trace the curve shown by the solid line. On the other hand, data taken at point x would trace the curve shown by the dotted line. It can be seen that the frequency of the second curve is one seventh of the first one. Thus, one can appreciate the importance of judiciously choosing the time interval for digitizing the analog data. Figure 7.13 shows how equally spaced data at different time intervals would appear to have come from each of many cosine curves.

If sampling is done at an interval of Δt seconds, one will be collecting samples at the rate of $1/\Delta t$ samples per second. Assuming that a minimum of two points are needed per frequency, this would correspond to a frequency $n_c = 1/2\ \Delta t$. Thus, useful data will be in the frequency range 0 to $1/2\ \Delta t$ Hz,

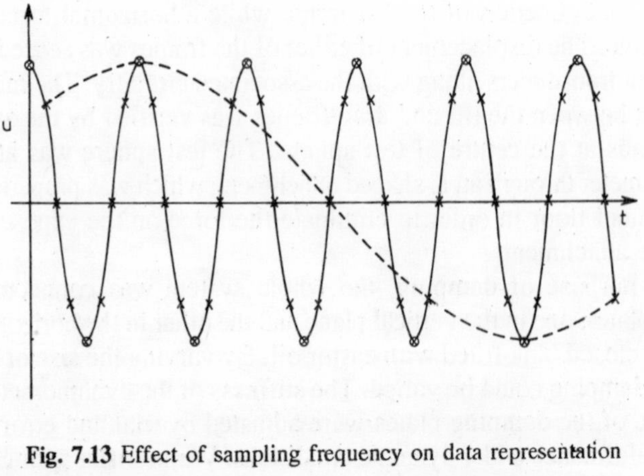

Fig. 7.13 Effect of sampling frequency on data representation

since frequencies in the data which are higher than n_c Hz will be folded into lower range. The frequency n_c is known as Nyquist frequency. For frequency n, ranging between $0 \le n \le n_c$, the higher frequencies that are aliased are $(2Kn_c \pm n)$, where κ = 1, 2, 3, 4, ... This is because all the data at frequencies $(2Kn_c \pm n)$ have the same cosine function as the data at frequency n, when data are sampled at $1/2\ n_c$ time interval apart. If, for example, n_c = 100 Hz, data at n = 30 Hz frequency will aliase with data at 170, 230, 370, 430 ..... Hz. Similarly, power spectra at these frequencies are aliased with spectra in the lower frequencies.

To deal with the problem of aliasing, two methods can be used. The first method is to choose Δt sufficiently small, so that it is physically unreasonable for data to exist above the cutoff frequency n_c. Thus, if information is needed below a freqnency of 1000 Hz for a certain phenomenon, $\Delta t = 1/2\ n_c = 1/2 \times 1000 = 0.0005$ seconds is adequate. In general, it is a good rule to select n_c to be 1.5 to 2.0 times the maximum frequency of interest. The second alternative is to filter the data prior to sampling so that information above the maximum desired frequency is no longer contained in the filtered data. Then choosing n_c equal to the maximum frequency of interest will give accurate results for frequencies below n_c. For accurate correlation function measurements, where the correlation function has frequencies near n_c, one should choose $\Delta t = 1/4n_c$. If power spectra are the prime consideration, choosing $\Delta t = 2/5n_c$ is considered sufficient (32). For correlation lag values, the maximum lag number m is retated to the desired equivalent resolution band width Be of the power spectrum calculation and normalised standard error by the relations

$$m = 1/Be\Delta t$$

and

$$\in = \sqrt{m/N}$$

where N is the total number of observations. Thus, for given values of n_c, m and $\in$ as 2000 Hz, 50 and 0.10, respectively, one can calculate Δt, Be and N as

$$\Delta t = \frac{1}{2n_c} = \frac{1}{2 \times 2000} = 0.25 \times 10^{-3} \text{ s.}$$

$$Be = \frac{1}{m\Delta t} = \frac{1}{50 \times 0.25 \times 10^{-3}} = 80 \text{ Hz}$$

$$N = \frac{m}{E^2} = \frac{50}{(0.1)^2} = 5000$$

Sampling time = $N\Delta t = 5000 \times 0.25 \times 10^{-3} = 1.25$ s. Usually, m should be much smaller than N. As a thumb rule, m should be less than $N/10$. If m is large, high resolution will result; On the other hand, small m tends to avoid certain instabilities that can occur in autocorrelation estimation.

Normally the HFAs and LDAs come with *rms* voltmeters and counters respectively for indicating the fluctuations in fluid flow representing turbulence. They do not provide direct facility for getting continuous time series records which are essential for digital signal processing of the data. Due to advancement in electronics and computer techniques, it is now possible to connect these system to Personal Computers (PCs) which act as data acquisition, analysis and display systems. A user friendly, menu driven, interactive software is available for performing various activities such as calibration, acquisition of data at prescribed sampling intervals, conversion of data from voltage to velocity using calibration curves display of turbulence time series record, computation of mean, *rms*, *rms* of fluctuating components, turbulence intensity, power spectral density, auto and cross correlations etc. These systems have the facilities for extensive graphics at every stage of analysis for ease of interpretation.

REFERENCES

1. Hinze O. *Turbulence: An Introduction to its Mechanism and Theory*. McGraw Hill Series in Mechanical Engineering, McGraw Hill Book Company, Inc. New York, 1959, Chapter 2.
2. Kalinske, A.A. and J.M. Robertson. Turbulence in Open Channels.Engineering News Record, Apr. 10, 1941.
3. Schraub, F.A. and S.J. Kline. A study of the Structure of Turbulent Boundary Layers with and without Longitudinal Pressure Gradients. Rep. MD. 12, Dept. of Mechanical Engg., Stanford University, March 1965.
4. Kemp, P.H. and A.J. Grass. The Measurement of Turbulent Velocity Fluctuations Close to a Boundary in Open Channel Flow. Proc. of 12th Congress of IAHR, Fort Collins, (USA), Vol. 2, 1967.
5. Clark, J.A. and E. Markland. Flow Visualisation in Turbulent Boundary Layers. JHD, Proc. ASCE, Vol. 97, No. HY10, Oct. 1971.
6. Bradshaw, P. and R.F. Johnson. Turbulence Measurements with Hot Wire Anemometers. Notes on Applied Science No. 33, NPL, HMS Office, London 1963.
7. Bradshaw P. *An Introduction to Turbulence and its Measurement*. Pergamon Press, Oxford (U.K.) 1971.
8. The Hot Wire Anemometer. Flow Corporation, Cambridge Massachusetts (USA) Bulletin 53.
9. The Hot Wire Anemometer. Flow Corporation, Cambridge Massachusetts, (USA), Bulletin 36 C.
10. Technical Memorandum: X-Wire Turbulence Measurements. Flow Corporation, Cambridge Massachusetts, (USA) Bulletin 68.
11. Ling, S.C. Measurement of Flow Characteristics by Hot Film Technique. Ph. D. Thesis, Iowa State University, Iowa City (USA) 1965.
12. Richardson, E.V. and R.S. McQuivey. Measurement of Turbulent Flow in Water. JHD, Proc. ASCE, Vol. 94, No. HY-2, Mar. 1968.
13. Chung. H. and J.E. Cermak, Turbulence Measurement by Electrokinetic Transducers. JHD, Proc. ASCE, Vol. 91 No. HY-6, Nov. 1965.
14. Binder, J. Potential Fluctuations Generated by Turbulence on a Small Wire, Proc. 12th Congress of IAHR, Fort Collins (USA), Vol. 2, 1967.
15. Bergmann, B.M. and T.H. Hodgson. The Unpolarised Electrode in Pulsating Poiseuille Pipe Flow. Proc. of Symposium on Turbulence in Liquids. University of Missouri—Rolla (USA), 1971.
16. Ippen, A.T. and F. Raichlen. Turbulence in Civil Engineering—Measurements in Free Surface Stream. JHD. Proc. ASCE, Vol. 83, No. HY-5, Oct. 1957.

17. Eagleson, P.S. and F.E. Perkins. Total Head Tube for the Broad Band Measurement of Turbulent Velocity Fluctuations in Water. Proc. of 9th Congress of IAHR, Dubrovnik, 1961.
18. Sushil Kumar, P.K. Pande and R.J. Garde. Turbulence Characteristics of Rough Open Channels. Proc. of 44th Annual Session, CBIP (India), Jan., 1975.
19. Arndt, R.E.A. and A.T. Ippen. Turbulence Measurements in Liquids Using an Improved Total Pressure Probe. JHR, IAHR Vol. 8, No. 1, 1970.
20. Grossman, L.M., H.Li and H.A. Einstein. Turbulence in Civil Engineering: Investigations in Liquid Shear Flow by Electromagnetic Induction. JHD, Proc. ASCE, Vol. 83, No. HY-5, Oct. 1957.
21. Pande, P.K. Measurement of Liquid Turbulence Spectra by Electromagnetic Induction Method. Ph. D. Thesis, Civil Engineering Deptt., University of Roorkee, 1966.
22. Schuyf, J.P. Measurement of Turbulent Velocity Fluctuations with a Propeller-Type Current Meter. JHR, IAHR, Vol. 4, No. 2, 1966.
23. Hansen, E. and O. Christensen. On the Measurement of Turbulence in Water by Means of a Small Propeller—Type Current Meter. Basic Reserach Rep. No. 14, Coastal Engg. Laboratory, Hydraulic Laboratory, Technical University of Denmark, July 1967.
24. Tiffany, J.B. Turbulence in the Mississippi River, Proc. 12th Congress of IAHR, Fort Collins (USA), Vol. 2, 1967.
25. Iwasa, Y. Turbulence Measurement by Means of Small Current Meter in Free Surface Flows. Proc. 12th Congress of IAHR, Fort Collins (USA), Vol. 2, 1967.
26. Ishihara, Y. and S. Yokosi. The Spectra of Turbulence in a River Flow. Proc. 12th Congress of IAHR, Fort Collins (USA), Vol. 2, 1967.
27. Laser Doppler Anemometer. Opto-Electronische Instrumente GmbH and Co. Karlsruhe (Germany).
28. OET-Counter Signal Processor LD-E-10. Opto-Electronische Instrumente GambH and Co. Karlsrune (Germany)
29. Rangan, C.S., G.R. Sharma and V.S. Mani. ***Instrumentation: Devices and Systems***. Tata McGraw Hills Publishing Co. Ltd., New Delhi, First Ed. 1983, Chapter VII.
30. Patnaik, P.C. Fluid Dynamic Forces on a Sphere Submerged in Turbulent Boundary Layer. Ph. D. Thesis, Civil Engineering Dept., University of Roorkee, Roorkee, 1983.
31. Bendat, J.S. and A.G. Piersol. ***Measurement and Analysis of Random Data***. John Wiley and Sons. Inc. New York, 1958. Chapter VII.
32. Blackman, R.B. and J.W. Tuckey. ***The Measurement of Power Spectra***. Dover Publications Inc., New York, 1958.

CHAPTER VIII

Description of Turbulent Flows-I: Wall Turbulent Shear Flows

8.1 INTRODUCTION

As discussed in earlier chapters, turbulence can be classified depending on where it is generated. If turbulence is generated at the wall due to high shear, it is termed as wall turbulent shear flow. On the other hand, when it is generated at the interface between two layers having large velocity difference, i.e. high shear, it is called free turbulent shear flow. This chapter is devoted to description of some wall turbulent shear flows which include boundary layers, atmospheric boundary layers, flow past a flat plate and internal flows.

TURBULENT BOUNDARY LAYERS

8.2 GOVERNING EQUATIONS

Turbulent boundary layer is a typical example of turbulent flow in which, on one side, the flow of a real fluid is restricted by the wall, and on the other side, the flow is infinite in extent; such flows have been classified as external flows. The basic premises of boundary layer theory on a flat plate have been discussed in Chapter I. The study of turbulent boundary layers has given impetus to the advancement of knowledge about flow in pipes, diffusers, open channels, boundary layers in atmosphere, flow past aerofoils and blades, spillways, etc. From the hydraulic engineer's point of view, more emphasis is needed on the discussion about basic concepts, velocity distribution, boundary drag and turbulence characteristics of the boundary layer, rather than the boundary layer calculations. It may be mentioned that the turbulence in the turbulent boundary layer is anisotropic and nonhomogeneous. If one starts with Navier-Stokes equations and the continuity equation for two dimensional steady flow of an incompressible fluid, and substitutes $u = \bar{u} + u'$, $v = \bar{v} + v'$, $w = w'$, and $p = \bar{p} + p'$, the boundary layer equations can be obtained (1) in the manner similar to that described in Chapter 1. These equations are

$$\bar{u}\frac{\partial \bar{u}}{\partial x} + \bar{v}\frac{\partial \bar{u}}{\partial y} + \frac{\partial \overline{u'^2}}{\partial x} + \frac{\partial \overline{u'v'}}{\partial y} = -\frac{1}{\rho}\frac{\partial \bar{p}}{\partial x} + \nu\frac{\partial^2 \bar{u}}{\partial y^2} \qquad \text{. . . (8.1a)}$$

$$\frac{\partial \overline{v'^2}}{\partial y} = -\frac{1}{\rho}\frac{\partial \bar{p}}{\partial y} \qquad \text{. . . (8.1b)}$$

$$\frac{\partial \bar{u}}{\partial x} + \frac{\partial \bar{v}}{\partial y} = 0 \qquad \ldots (8.1c)$$

The boundary conditions to be satisfied are

$$\left.\begin{aligned} &\text{at } y = 0,\ \bar{u} = \bar{v} = 0,\ u' = v' = w' = 0 \\ &\text{at } y \to \infty,\ \bar{u} = U_o(x) \end{aligned}\right\} \qquad \ldots (8.2)$$

where U_o is the free stream velocity which varies with x in the presence of $\partial\bar{p}/\partial x$. Integration of Equation 8.1 (b) yields

$$\overline{v'^2} = -\frac{\bar{p}}{\rho} + C_1$$

where the constant of integration C_1 can the determined from the condition that at $y \geq \delta$, $\overline{v'^2} = 0$ and, hence, $\bar{p}$ is equal to the outside pressure p_o. Therefore,

$$\frac{\bar{p}}{\rho} = \frac{p_o}{\rho} - \overline{v'^2} \qquad \ldots (8.3)$$

Obtaining $\partial\bar{p}/\partial x$ from Eq. 8.3 and substituting it in Eq. 8.1 (a) gives

$$u\frac{\partial \bar{u}}{\partial x} + \bar{v}\frac{\partial \bar{u}}{\partial y} = -\frac{1}{\rho}\frac{\partial p_o}{\partial x} + \frac{\partial}{\partial y}\left(-\overline{u'v'} + \nu\frac{\partial \bar{u}}{\partial y}\right) - \frac{\partial}{\partial x}(\overline{v'^2} - \overline{u'^2}) \qquad \ldots (8.4)$$

This equation can be compared with Eq. 1.29 for laminar boundary layer which differs from it in additional terms involving derivatives of Reynolds stresses $\overline{u'v'}$, $\overline{v'^2}$ and $\overline{u'^2}$. It can be seen from Eq. 8.4 that mean shear and boundary shear are given by $\bar{\tau} = \rho\left(-\overline{u'v'} + \nu\frac{\partial \bar{u}}{\partial y}\right)$ and $\tau_o = \mu\left(\frac{\partial \bar{u}}{\partial y}\right)_{y=0}$. Since outside the boudary layer

$$U_o\frac{dU_o}{dx} = -\frac{1}{\rho}\frac{\partial p_o}{\partial x}$$

Eq. 8.4 can be modified to

$$\bar{u}\frac{\partial \bar{u}}{\partial x} + \bar{v}\frac{\partial \bar{u}}{\partial y} = U_o\frac{dU_o}{\partial x} + \frac{\partial}{\partial y}\left(-\overline{u'v'} + \nu\frac{\partial \bar{u}}{\partial y}\right) - \frac{\partial}{\partial x}(\overline{v'^2} - \overline{u'^2}) \qquad \ldots (8.5)$$

It may be mentioned that the term $\frac{\partial}{\partial x}(\overline{v'^2} - \overline{u'^2})$ is usually small and can be neglected, except near the point of separation. Equations 8.5 and 8.1 are indeterminate since there are more unknowns than the number of equations. Hence, some empirical relations or hypotheses are required for their solution.

Karman's momentum integral equation for the case of laminar boundary layer is

$$\frac{d\theta}{dx} + \theta\,(2 + H)\,\frac{dU_o/dx}{U_o} = c_f/2 \tag{8.6}$$

where θ is the momentum thickness of boundary layer and H is equal to δ^*/θ. This equation modifies to

$$\frac{d\theta}{dx} + \theta\,(2 + H)\,\frac{dU_o/dx}{U_o} = \frac{c_f}{2} - \frac{1}{U_o^2}\int_0^\infty (\overline{u'^2} - \overline{v'^2})\,dy \tag{8.7}$$

in case of turbulent boundary layers with pressure gradient, i.e. where U_o depends on x. Again as in the case of Eq. 8.5, the turbulence term $\int_0^\infty (\overline{u'^2} - \overline{v'^2})\,dy$ can be neglected in many cases; however, it becomes important near the point of separation.

8.3 STRUCTURE OF WALL FLOW

The flow near the wall in turbulent boundary layer has been studied in detail by Kline et al. (2, 3) by flow visualisation technique using hydrogen bubble method, and by Kim et al. (4*) who used hydrogen bubble technique along with hot film anemometry. These studies have revealed that, near the wall, the flow pattern consists of "islands of hesitation" and longitudinal vortices which impart wispy appearance to the flow; these are interspersed with areas of faster moving fluid. These islands of hesitation appear as long stretched filaments in the direction of flow which move more slowly than the surrounding fluid. The vortices apparently originate as a breakdown, or roll up along the edges of the islands of hesitation. The primary orientation of the vortex elements is longitudinal, i.e. in the direction of flow; however, each vortex stands at a slight angle to the wall so that its distance from the wall increases as it moves downstream. After the vortex element reaches a critical distance from the wall, it breaks up into typical turbulence "hash". This happens in the region $5.0 < \frac{u_* y}{\nu} < 30$. This bursting appears to form distinct streaks whose lateral spacing Z_o is approximately given (4) by $u_* Z_o/\nu = 100$. These streaks are not permanent and shift at random when observed for a long time. Figure 8.1 shows a sketch of the flow model near the wall as given by Kline and Runstadler (3). The instability of the flow near the outer edge of the sublayer and the action of the outer flow are considered to be responsible for such a flow pattern. Einstein and Li (5) have proposed a model to predict the mean velocity distribution, shear stress at the wall, and the longitudinal velocity fluctuations. In this model the flow is considered as an inherently unsteady process, in which the laminar sublayer builds up and decays. At any given $u_* y/\nu$, the above quantities depend only on $U_o^2 T/\nu$ where U_o is the velocity at time $t = 0$, and T is time over which the process of growing continues, the decay being very rapid. As pointed out by Rotta (1*), the model is not in full accord with actual behaviour of the sublayer flow; the mean velocity distribution even violates the continuity equation. The most important defect of the model, however, is the complete disregard to the three dimensional character of the flow. Still the model attracts the attention because of surprisingly good agreement between calculated and observed turbulent intensity distribution.

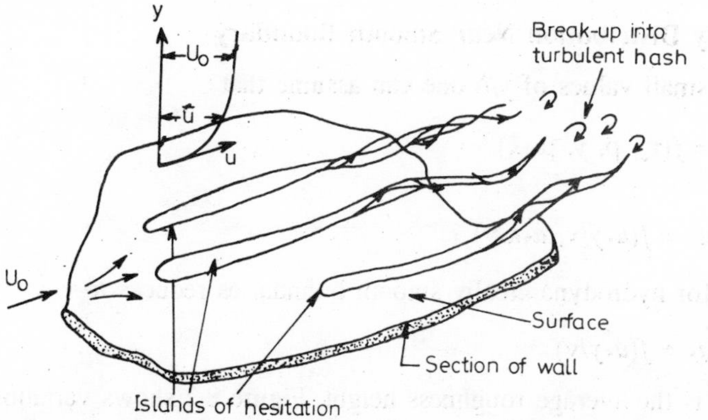

Fig. 8.1 Flow model of wall layer in turbulent boundary layer (3)

8.4 VELOCITY DISTRIBUTION IN BOUNDARY LAYER

The velocity profiles in laminar boundary layer over a flat plate have similar shape at all the distances from the leading edge of the plate if the velocity distribution is plotted as u/U_o vs. η where $\eta = y\sqrt{U_o/\nu x}$; hence $u/U_o = f(\eta)$. In earlier development of analysis of turbulent boundary layers, it was assumed that velocity distribution in turbulent boundary layer on a flat plate can be expressed in the form

$$\bar{u}/U_o = (y/\delta)^n \qquad \ldots (8.8)$$

However, it was soon found that n is not constant but varies with Reynolds number $U_o\delta/\nu$. In fact, n was found to reduce from 1/3 to 1/10 as $U_o\delta/\nu$ increased. This probably occurred because, with change in Reynolds number and the wall roughness, the wall shear also changed; this change in shear plays an important role in supplying energy to turbulence. Clauser (6) has suggested that the behaviour of turbulent boundary layer can be interpreted as that of a laminar layer having a thin sublayer next to the wall, having a much smaller viscosity. Major changes in the velocity from wall to the edge of the boundary layer take place in this thin layer. The viscosity of the outer layer corresponds to eddy dynamic viscosity if Reynolds stress $\rho\overline{u'v'}$ is expressed in terms of Boussinesq's concept. Here ε will be proportional to U_o and δ; hence, $\dfrac{\varepsilon}{\nu}$ changes with $\dfrac{U_o\delta}{\nu}$, and so does the velocity profile. In addition, the wall roughness will also be greatly influential as far as velocity distribution near the wall is concerned. For these reasons, variation of n in Eq. 8.8 with Reynolds number is understandable.

In addition to Eq. 8.8, two other laws which have been used to describe velocity distribution in turbulent boundary layer, are the 'law of the wall' for smooth boundaries as proposed by Prandtl, and the 'velocity defect law' proposed by Karman.

Velocity Distribution Near Smooth Boundary

For small values of y/δ one can assume that

$$\bar{u} = f(\tau_o, \rho, y, \mu, k)$$

or

$$\bar{u}/u_* = f(u_* y/\nu, u_* k/\nu)$$

which for hydrodynamically smooth boundaries reduces to

$$\bar{u}/u_* = f(u_* y/\nu)$$

Here k is the average roughness height. Figure 8.2 shows variation of $\bar{u}/u_*$ with $u_* y/\nu$ for data collected by Nikuradse, and Reichardt for turbulent flow

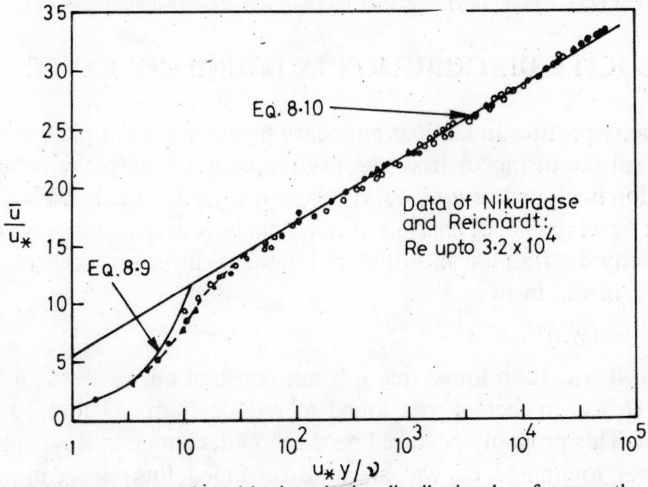

Fig. 8.2 The universal logarithmic velocity distribution law for smooth pipes

in hydrodynamically smooth pipes (which is a limiting case of turbulent boundary layer where boundary layer reaches the centre line). Very close to the wall, the velocity distribution in laminar sublayer is given by

$$\bar{u}/u_* = u_* y/\nu \qquad \ldots (8.9)$$

and it is valid upto $u_* y/\nu$ equal to 10 to 12. Beyond this limit, the velocity distribution is given by the logarithmic law

$$\frac{\bar{u}}{u_*} = 5.75 \log \frac{u_* y}{\nu} + 5.5 \qquad \ldots (8.10)$$

Figure 8.3 shows verification of law of the wall for boundary layer flow without and with pressure gradients. It may be mentioned that the constants in Eq. 8.10 for boundary layer development on a flat plate are 6.0 and 4.0 respectively. It can be seen that the law of the wall is valid even for flows with pressure gradient because near the wall the force due to shear is greater than that produced by the pressure gradient. Therefore, correlation between $\bar{u}/u_*$ and $u_* y/\nu$ for both the laminar flow (Eq. 8.9) and turbulent flow (Eq. 8.10) is unaffected by the pressure gradient for relatively small values of $u_* y/\nu$, namely upto 1000 except that the additive constant is changed to 5.20. However, for larger values of $u_* y/\nu$, there is deviation from logarithmic relationship. Study of Fig. 8.3 reveals that when $\left(\frac{\partial \bar{p}}{\partial x}\nu\right) / \tau_* u_*$ is large, i.e. near the point of separation, Eq. 8.10 with constant equal to 5.20 is not valid. It may be mentioned that slight variations in the constants in Eq. 8.10 have been obtained by various investigators. Clauser (6, 7*) has obtained the coefficient and constant as 5.60 and 4.90 while Coles (8) has obtained the values as 5.75 and 5.10. However, such variations are considered as minor. The law of the wall is valid approximately upto 15 to 20 per cent of the boundary layer thickness.

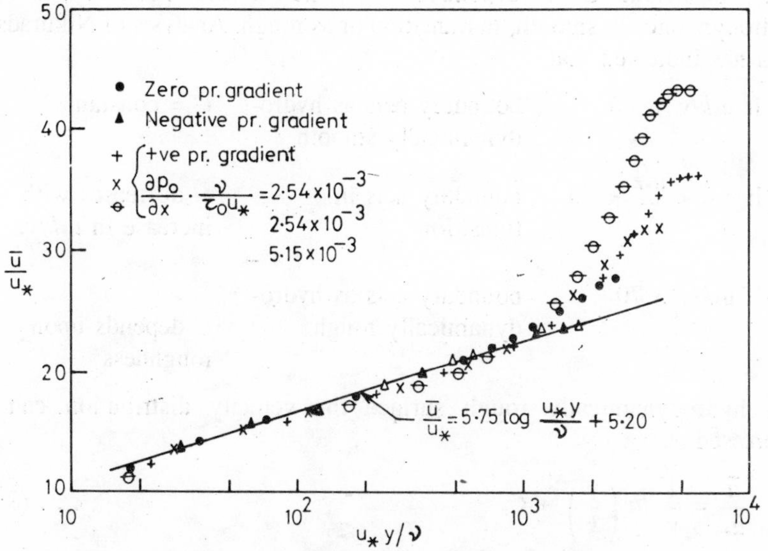

Fig. 8.3 Law of the wall for boundary layer flow

In this region, turbulence intensities are constant multiples of u_* and triple correlations constant multiples of $u_*^{3/2}$. The scale of larger eddies is proportional to the distance from the wall, while the longitudinal scale of larger eddies is governed by outside flow conditions rather than by the wall similarity. In the wall region the turbulence is anisotropic; however, as distance from the wall increases, local isotropy is approached. This happens for $u_* y/\nu$ values greater than 2.5×10^4

Velocity Distribution Near Rough Boundary

With roughness elements on the bed, part of tangential stress on the surface can be due to viscous shear stress $\mu\left(\frac{\partial \bar{u}}{\partial y}\right)_{y=0}$, and the remainder due to form drag caused by pressure distribution on the roughness elements. Because of separation around these roughness elements, they act as vortex generators causing additonal turbulence which diffuses in the main flow. Roughness effect also counteracts the damping effect of viscous forces produced by proximity of wall. Since size, shape and concentration of roughness elements can vary considerably, very little is known about their effect on structure of turbulence and on velocity distribution. However, adequate information is available about roughness caused by closely packed uniform sand particles on the bed surface which has been extensively studied by Nikuradse.

The velocity distribution near the wall on surface with roughness elements can be described by the equation

$$\frac{\bar{u}}{u_*} = \frac{1}{\kappa} \ln \frac{u_* y}{\nu} + C(u_* k/\nu) \qquad \ldots (8.11)$$

The coefficient C is dependent on whether the boundary acts as hydrodynamically smooth, in transition or as rough. Analysis of Nikuradse's data has indicated that

i. If $u_* k/\nu < 5.0$	boundary acts as hydrodynamically smooth	C = constant
ii. If $5.0 < \frac{u_* k}{\nu} < 70$	boundary acts in transition	C decreases with increase in $u_* k/\nu$.
iii. If $u_* k/\nu > 70$	boundary acts as hydrodynamically rough	C depends upon roughness

For hydrodynamically rough surface the velocity distribution can be expressed as

$$\frac{\bar{u}}{u_*} = \frac{1}{\kappa} \ln\left(\frac{y}{k}\right) + C_r \qquad \ldots (8.12)$$

Comparison of Eqs. 8.11 and 8.12 yields

$$C_r - \frac{1}{\kappa} \ln\left(\frac{u_* k}{\nu}\right) = C\left(\frac{u_* k}{\nu}\right) \qquad \ldots (8.13)$$

Figure 8.4 shows variation at C with $\frac{u_* k}{\nu}$ for various types of roughnesses, as shown by Clauser (6). In this figure, the effect of type and concentration of roughness on C is apparent. This relation between C and $u_* k/\nu$ has been determined experimentally for each type of roughness. One can also assume that the action of roughness is equivalent to a reduction of viscous sublayer.

In that case, the velocity distribution law

$\bar{u}/u_* = f(u_* y/\nu)$

is still applicable if the reference plane is shifted beneath the surface by Δy_r. The plane of reference moves with the velocity $\bar{u}(\Delta y_r)$ opposed to main flow. Hence, one gets

$\Delta y_r = k e^{-\kappa C_r}$

or $\Delta y_r = 0.035\ k$ for uniform sand grain roughness.

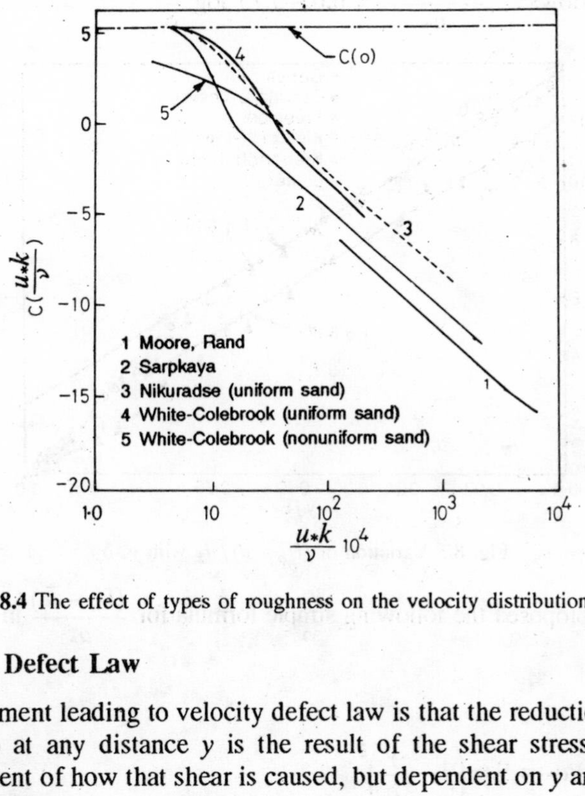

Fig. 8.4 The effect of types of roughness on the velocity distribution, Eq. 8.11

Velocity Defect Law

The argument leading to velocity defect law is that the reduction in velocity $(U_o - \bar{u})$ at any distance y is the result of the shear stress at the wall, independent of how that shear is caused, but dependent on y and U_o. Hence, one can write

$$(U_o - \bar{u}) = f(u_*, y, \nu, U_o)$$

or $$\left(\frac{U_o - \bar{u}}{u_*}\right) = f\left(\frac{y}{\delta}, \frac{U_o}{u_*}\right) \quad \ldots (8.14)$$

In the initial studies, variation of U_o/u_* in the avaiable data was not much; hence, Eq. 8.14 was in the form

$$\left(\frac{U_o - \bar{u}}{u_*}\right) = G\left(\frac{y}{\delta}\right) \quad \ldots (8.15)$$

Whereas Eq. 8.10 is valid for smooth boundaries, Eq. 8.15 is valid for both smooth as well as rough boundaries. However, Eq. 8.15 is not valid for very small values of y. Figure 8.5 shows variation of $(U_o - \bar{u})/u_*$ vs y/δ for several sets of data for channel and boundary layer flows from which some observations can be made. Firstly, the function $G(y/\delta)$ is significantly affected by the turbulence in the free stream and pressure gradient. Secondly, for y/δ less than about 0.15, the velocity defect law can be expressed in logarithmic form

$$\text{Boundary layers:} \quad \frac{(U_o - \bar{u})}{u_*} = 2.35\text{–}5.75 \log \frac{y}{\delta} \quad \ldots (8.16)$$

$$\text{Channel flows:} \quad \frac{(U_o - \bar{u})}{u_*} = 0.65\text{–}5.75 \log \frac{y}{\delta} \quad \ldots (8.17)$$

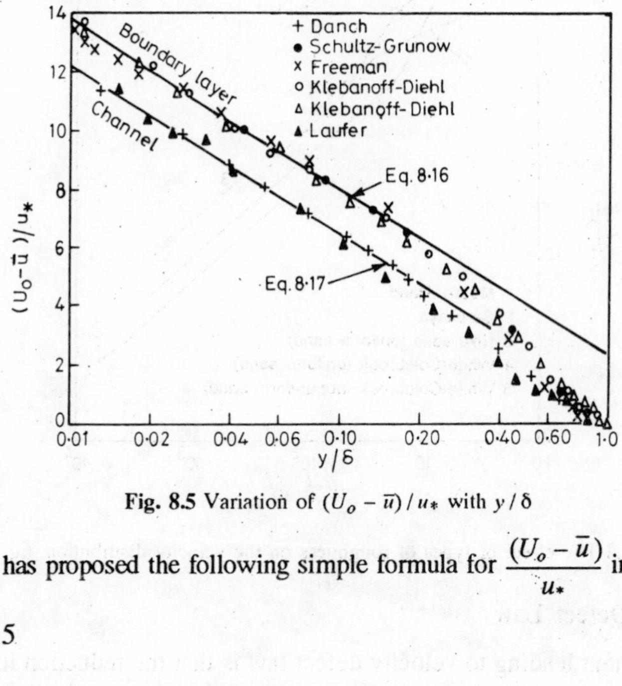

Fig. 8.5 Variation of $(U_o - \bar{u})/u_*$ with y/δ

Hamma has proposed the following simple formula for $\frac{(U_o - \bar{u})}{u_*}$ in the region $\frac{y}{\delta} > 0.15$

$$(U_o - \bar{u})/u_* = 9.6 \left(1 - \frac{y}{\delta}\right)^2 \quad \ldots (8.18)$$

For boundary layer flow detailed studies have revealed that the omission of effect of the parameter U_o/u_* in Eq. 8.14 is not completely justified. Hence, Rotta (1*) suggested use of dimensionless distance $\left(\frac{y}{\delta^*}\frac{u_*}{U_o}\right)$ in place of $\frac{y}{\delta}$. The abscissa scale is so fixed that the area under $\frac{(U_o - \bar{u})}{u_*}$ vs $\left(\frac{y}{\delta^*}\frac{u_*}{U_o}\right)$ curve is unity in the light of the definition of displacement thickness δ^* The variation of $(U_o - \bar{u})/u_*$ vs $\left(\frac{y}{\delta^*}\frac{u_*}{U_o}\right)$ can be represented by the logarithmic relation

$$\frac{(U_o - u)}{u_*} = \frac{1}{\kappa} \ln \left(\frac{y}{\delta^*} \frac{u_*}{U_o} \right) + \text{constant}$$

upto a certain value of $\left(\frac{y}{\delta^*} \frac{u_*}{U_o} \right)$; however, over the entire range, it indicates a unique curve. In the case of boundary layer flows with pressure gradient, velocity profile outside the wall region is affected by pressure gradient, and hence, would now show a universal presentation in terms of velocity defect law. However, Clauser (6) has shown that the pressure distribution could be adjusted to give similar boundary layer profiles in terms of velocity defect law. Such flows are called equilibrium boundary layers and the pressure parameter used for systematising data is $G = \frac{\delta^*}{\tau_0} \frac{\partial \bar{p}}{\partial x}$; hence

$$(U_o - \bar{u})/u_* = f\left(\frac{y}{\delta}, G\right)$$

as illustrated is Fig. 8.6.

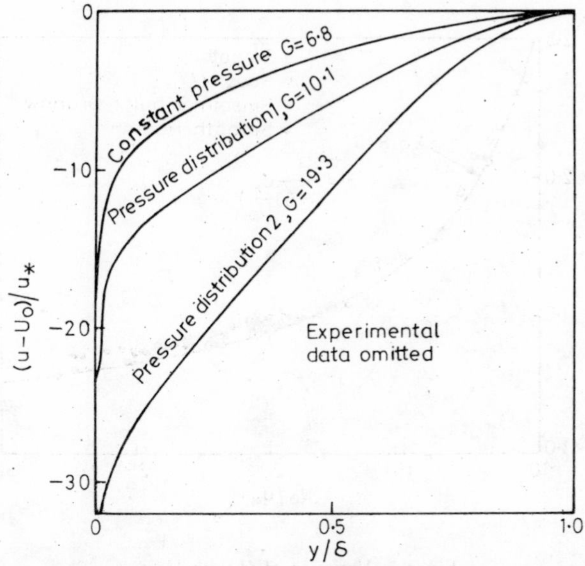

Fig. 8.6 Equilibrium boundary layer profiles on the basis of the velocity defect law, (after Clauser)

In the case of laminar boundary layer on a flat plate the ratio δ^*/θ assumes a constant value for given velocity distribution. If, for turbulent boundary layer on a flat plate, velocity distribution is assumed to follow the power law, viz.

Eq. 8.8, one can show that $\frac{\delta^*}{\theta} = \frac{2+n}{n}$, and since n decreases with increase in Reynolds number, n will change along the plate. Using velocity defect law it can be shown that (1)

$$\left.\begin{aligned} H &= \frac{\delta^*}{\theta} = \frac{1}{(1 - I u_*/U_o)} \\ \text{where } I\,(u_*/U_o) &= \int_0^\infty \left(\frac{U_o - \bar{u}}{u_*}\right)^2 d\left(\frac{y}{\delta_*}\frac{u_*}{U_o}\right) \end{aligned}\right\} \quad \ldots (8.19)$$

and I is expected to be aproximately constant for turbulent boundary layer on a flat plate. Figure 8.7 shows variation of H with U_o/u_* on which Eq. 8.19 with I = 6.10 has also been plotted. As the point of separation is approached, H value goes on increasing. In fact, it has been found that $\bar{u}/U_o$ depends on $\frac{y}{\theta}$ and H.

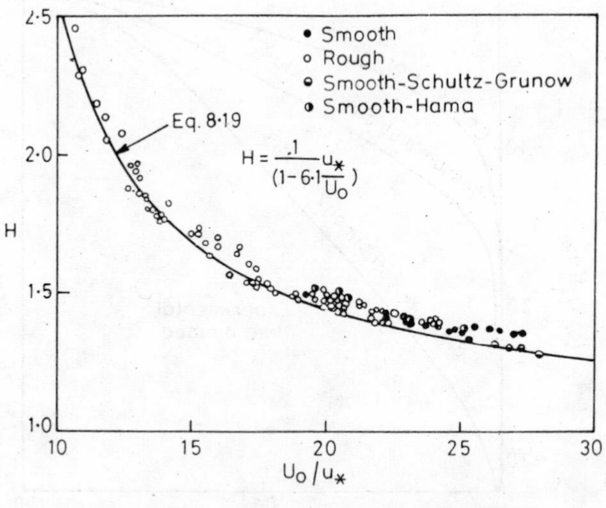

Fig. 8.7. Variation of H with U_o/u_*

In order to specify the velocity profile which is strongly affected by the pressure gradient Buri (9*) chose the quantity Γ defined as

$$\Gamma = \frac{\theta}{U_o}\left(\frac{U_o\theta}{\nu}\right)^{1/4}\frac{dU_o}{dx}$$

Here $\Gamma > 0$ corresponds to accelerated flows, $\Gamma < 0$ corresponds to decelerated flows, and according to Nikuradse, separation occurs when $\Gamma = -0.06$. Coles (8) carried out a detailed analysis of velocity distribution in boundary layers to obtain extension of the law of the wall. After examining carefully, practically all available experimental data on turbulent boundary layers in terms of the law of the wall and characteristic departure from it away from the wall, Coles concluded that the flow had a wake like structure modified by various degrees by the wall constraint. The wake like form could be reduced to a second universal law called "law of wake", a linear combination of the law of the wall and law of wake representing the complete profile for equilibrium and nonequilibrium flows. Following Coles' analysis velocity distribution over the entire boundary layer thickness can, therefore, be represented by

$$\bar{u}/U_o = f\left(\frac{u_* y}{\nu}\right) + g\left(\pi, \frac{y}{\delta}\right)$$

where π is a parameter independent of x and y for specific pressure distribution. In fact $g\left(\pi, \frac{y}{\delta}\right)$ gives departure from logarithmic law. In the most general case

$$\frac{\bar{u}}{U_o} = f\left(\frac{u_* y}{\nu}\right) + \frac{\pi(x)}{\kappa}\,\omega\left(\frac{y}{\delta}\right) \qquad \ldots (8.20)$$

where $\omega\left(\frac{y}{\delta}\right)$ is called law of the wake. If π does not depend on x, $g\left(\pi, \frac{y}{\delta}\right)$ is a function of $\frac{y}{\delta}$ alone. For general flows $\pi(x)$ is called a profile parameter, and $\omega(y/\delta)$ is a universal function such that $\omega(0) = 0$ and $\omega(1) = 2.0$ for all two dimensional flows. Further $\int_0^2 \left(\frac{y}{\delta}\right) d\omega = 1.0$. In order to determine variation of $\pi(x)$, the above equation can be written in its complete form as

$$\frac{\bar{u}}{u_*} = \frac{1}{\kappa}\, ln\left(\frac{u_* y}{\nu}\right) + C + \frac{\pi(x)}{\kappa}\,\omega\left(\frac{y}{\delta}\right) \qquad \ldots (8.21)$$

and when $\frac{y}{\delta} = 1$, $\bar{u} = U_o$. Hence,

$$\frac{U_o}{u_*} = \frac{1}{\kappa}\, ln\left(\frac{u_* \delta}{\nu}\right) + C + \frac{2\pi(x)}{\kappa}$$

It can be seen that $\pi(x)$ is related to local skin friction coefficient c_f, since $\frac{U_o}{u_*} = \sqrt{\frac{2}{c_f}}$. It can be shown that

$$\kappa\,\frac{\delta^*}{\delta}\,\frac{U_o}{u_*} = 1 + \pi \qquad \ldots (8.22)$$

For constant pressure flow when U_o is constant, $\pi = 0.55$. The universal wake function $\omega(\xi)$ is tabulated below.

Universal wake function

ξ	0	0.10	0.20	0.30	0.40	0.50	0.60	0.70	0.80	0.90	1.0
$\omega(\xi)$	0	0.084	0.168	0.396	0.685	0.994	1.307	1.600	1.840	1.980	2.000

It can be approximated by sine function, namely

$$\omega(\xi) = 2 \sin \xi$$

8.5 TURBULENCE CHARACTERISTICS IN THE BOUNDARY LAYER

Typical variations of $\sqrt{\overline{u'^2}}/U_o$, $\sqrt{\overline{v'^2}}/U_o$, $\sqrt{\overline{w'^2}}/U_o$ and $\overline{u'v'}/U_o^2$ with y/δ in boundary layer flow on a flat plate, as obtained by Klebanoff, are shown in Fig. 8.8.

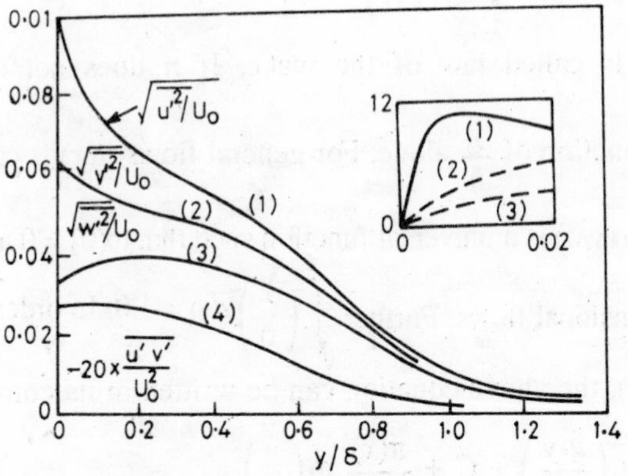

Fig. 8.8 Characteristics of turbulence in boundary layer on a flat plate

As can be seen from the inset of this figure, $\sqrt{\overline{u'^2}}/U_o$ reaches a maximum at about $y/\delta = 0.010$ and then decreases for larger values of y/δ. Further at any point, $\overline{u'^2} > \overline{v'^2} > \overline{w'^2}$; hence boundary layer turbulence is anisotropic. Also along the length, the turbulence characteristics change and, therefore, it is non-homogeneous. Further, it should be noted that the turbulent fluctuations do not vanish at y/δ equal to unity, but persist even for larger values of y/δ. Figure 8.9 shows variation of $\sqrt{\overline{u'^2}}/u_*$ with y/δ as obtained in some experiments in Roorkee (35) along with the mean curve of Lyles. Here also it can be noticed that turbulent fluctuations persist beyond y/δ equal to unity. It has been found that the boundary layer bounded by a free stream of vanishingly small turbulence usually has a sharp but very irregular boundary as shown schematically in Fig. 8.10. This phenomenon is common to all shear

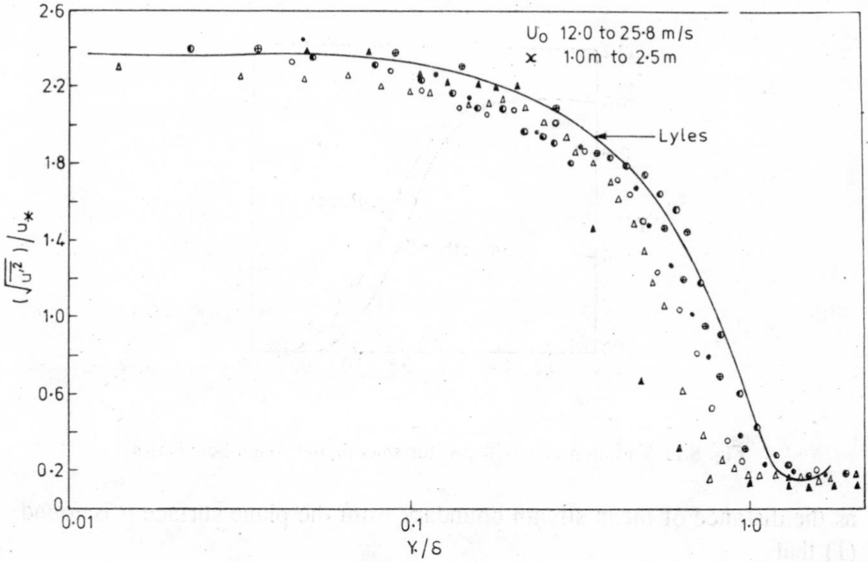

Fig. 8.9 Variation of $\sqrt{\overline{u'^2}}/u_*$ with y/δ in undisturbed boundary layer

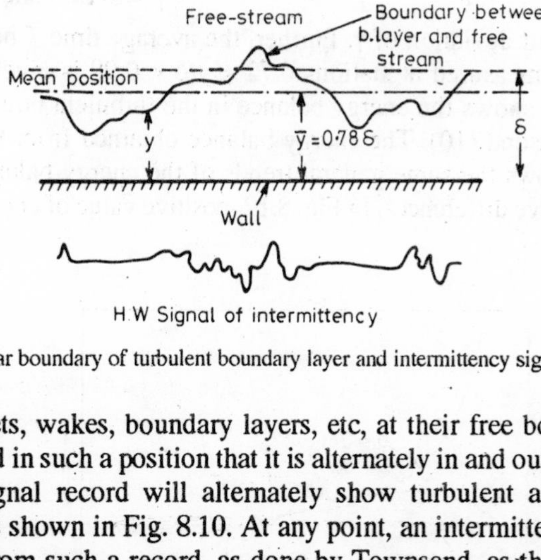

Fig. 8.10 Irregular boundary of turbulent boundary layer and intermittency signal in boundary layer

flows, e.g. jets, wakes, boundary layers, etc, at their free boundaries. If hot wire is placed in such a position that it is alternately in and out of the turbulent field, the signal record will alternately show turbulent and nonturbulent conditions as shown in Fig. 8.10. At any point, an intermittency factor γ can be defined from such a record, as done by Townsend, as the fraction of the time that the flow is turbulent. Figure 8.11 shows variation of γ with y/δ for turbulent boundary layers with smooth and rough surfaces. Immediately obvious is the fact that at the edge of the boundary layer, i.e. at y/δ equal to unity, γ is not zero but has definite magnitude. The effect of roughness on the variation of γ is also apparent. Since γ is not zero at y/δ equal to unity, and since δ is arbitrarily defined as that value of y at which $u = 0.99\ U_o$, it is some times suggested that the boundary layer thickness be defined as 1.2δ where $y = \delta$ when $u = 0.99\ U_o$, since at this value of y, γ is nearly zero. If $\bar{\gamma}$ is defined

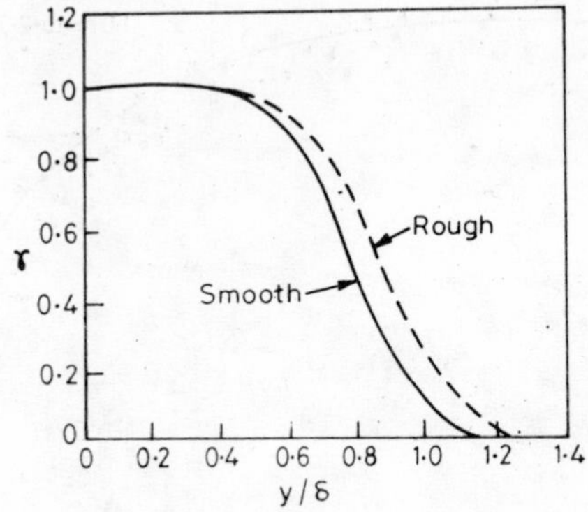

Fig. 8.11 Variation of γ with y/δ for smooth and rough boundaries

as the distance of mean stream boundary from the plane surface it is found (1) that

B.L. on smooth plate (Klebanoff's data) $\bar{\gamma} = 0.78\delta$ and $\sigma = 0.14\delta$

B.L. on rough plate (Corrsin and Kistler's data) $\bar{\gamma} = 0.82\delta$ and $\sigma = 0.15\delta$

where σ is standard deviation of γ. Further, the average time T between two turbulent regions measured at stations $0.72 < y/\delta < 0.98$ is given by $TU_o/\delta = 2.5$. Figure 8.12 shows the energy balance in the turbulent boundary layer as given by Townsend (10). The energy balance obtained from Klebanoff's measurements shows the same general trends of the energy balance, though there are quantitative differences. In Fig. 8.12, positive value of energy balance

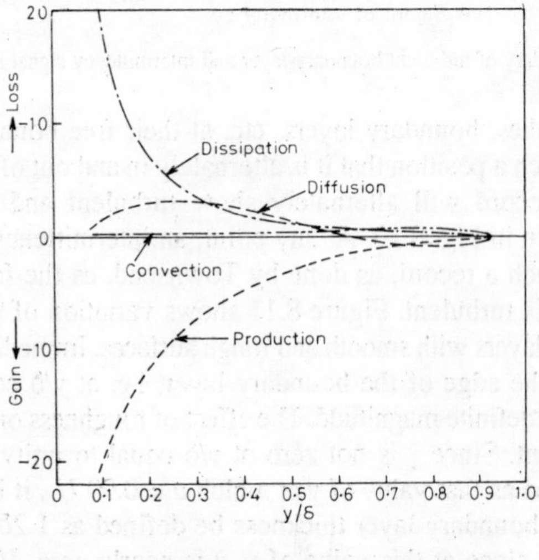

Fig. 8.12 Energy balance in the boundary layer along a smooth wall with zero pressure gradient

term means a negative contribution to turbulence kinetic energy of the volume element, i.e. a loss. It can be seen from this figure that the main contribution to the energy balance is given by the production and dissipation terms, if we exclude the region near outer edge of the boundary layer. Further, production and dissipation are nearly balanced in general, and this is particularly so near the wall. The contribution of convection by the mean motion is practically negligible, except near the outer edge of the boundary layer. This term being positive is interpreted as loss of energy. The difference between dissipation and production, plus the convection is counterbalanced by a gain by turbulence diffusion. In the wall region, the dissipation seems to outweigh the production, and gain occurs by turbulent diffusion. There is a transfer of energy from inner region of boundary layer to outer region.

8.6 LOCAL AND AVERAGE FRICTION COEFFICIENTS

The local skin friction coefficient is defined as

$$c_f = \tau_o \Big/ \frac{\rho U_o^2}{2}$$

where τ_o and U_o are the local boundary shear and local free stream velocity in the turbulent boundary layer, respectively. It can be shown that, in general, c_f should depend on $U_o\theta/\nu$, H and k/θ where k is the average height of surface roughness. For hydrodynamically smooth surfaces, Ludwieg and Tillman have obtained the following formula for c_f

$$c_f = (0.246)10^{-0.678H}\,(U_o\theta/\nu)^{-0.268} \qquad \ldots (8.23)$$

which yields reliable result for $U_o\theta/\nu$ greater than 1000 and H less than 2.0. Figure 8.13 shows variation of c_f with $U_o\theta/\nu$ and H for hydrodynamically

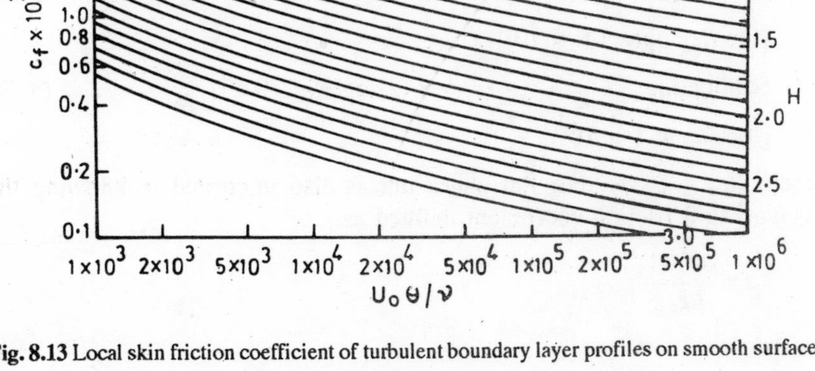

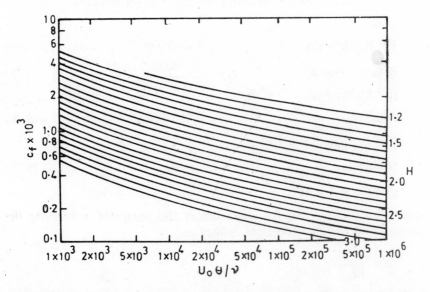

Fig. 8.13 Local skin friction coefficient of turbulent boundary layer profiles on smooth surfaces

smooth plates obtained from theoretical analysis using universal law of the wall and one parameter velocity defect law. For completely rough surface c_f will be a function of k/θ and H as shown in Fig. 8.14. Figures 8.13 and 8.14 are based on the equations which have been derived in reference 1. For boundary layer on a flat plate, it is desirable to have equation for c_f in terms of U_o,ν and x so that one need not calculate θ in order to obtain c_f. The following equations are found to be valid for hydrodynamically smooth plates.

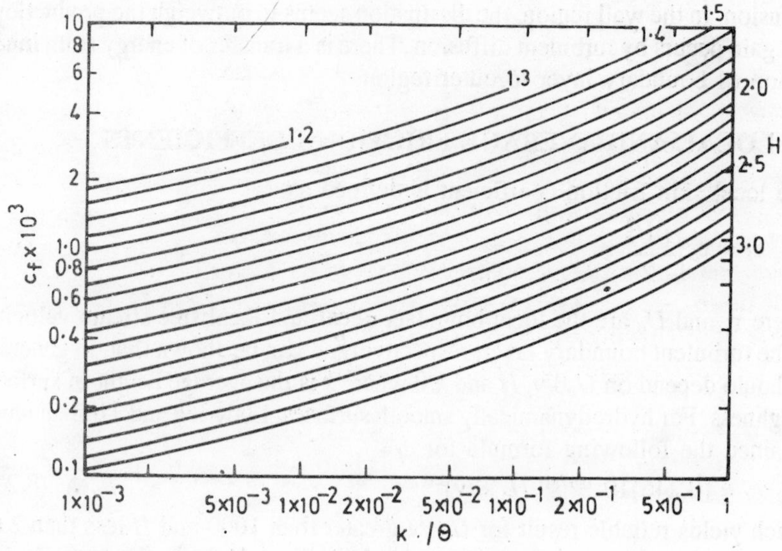

Fig. 8.14 Local skin friction coefficient of turbulent boundary layer profiles on completely rough sufaces with sand grain type roughness

For Re_x less than 10^7: $$c_f = 0.059/Re_x^{1/5} \qquad \ldots (8.24)$$

Schulz-Grunow:
(for higher Re_x) $$c_f = \frac{0.37}{(\log Re_x)^{2.584}} \qquad \ldots (8.25)$$

Nikuradse:
(for Re_x upto 1.8×10^7) $$c_f = 0.02296/Re_x^{0.139} \qquad \ldots (8.26)$$

Schlichting:
(for $Re_x < 1 \times 10^9$) $$c_f = (2 \log Re_x - 0.65)^{-2.3} \qquad \ldots (8.27)$$

where $Re_x = U_o x/\nu$. For flat plates one is also interested in knowing the average skin friction coefficient defined as

$$C_f = \frac{F_D}{BL\,\rho\,U_o^2/2}$$

where B is the plate width, L is its length and F_D is the total frictional force on the plate. When part of the plate is covered with laminar boundary layer and Re_L lies between 5×10^5 and 10^7, C_f is given by

$$C_f = \frac{0.074}{Re_L^{1/5}} - \frac{1700}{Re_L} \qquad \ldots (8.28)$$

The term $1700/Re_L$ corrects C_f for the fact that part of boundary layer on the front portion of plate is laminar, and the coefficient 1700 corresponds to critical Re_x of 5×10^5. For Re_L upto 10^9 one can use the formula

$$C_f = \frac{0.455}{(\log Re_L)^{2.58}} - \frac{1700}{Re_L} \qquad \ldots (8.29)$$

It may be mentioned that for high Reynolds numbers Re_L, the correction $1700/Re_L$ in Eqs. 8.28 and 8.29 becomes negligibly small compared to the first term and, hence, can be neglected.

In the case of practical applications connected with flat plates such as ships, turbine blades, aircraft wings, etc., the wall is not many times hydrodynamically smooth. In such cases the average skin friction coefficient C_f can be taken to the dependent on $U_o L/\nu$ and k/δ which corresponds to k/R or k/D in case of pipe flow. However, in a pipe k/R or k/D remains constant, whereas in the case of boundary layer flow k/δ would decrease in the downstream direction. As a result, in the front portion of the turbulent boundary layer on the flat plate, the boundary would act as rough, the next portion will be in transition followed by a smooth boundary. Prandtl and Schlichting (9) used the results of Nikuradse for pipe flow and converted them to the corresponding case for flat plate. The results can be expressed as $C_f = f\left\{\frac{U_o L}{\nu}, \frac{L}{k}\right\}$, (see Fig. 8.15).

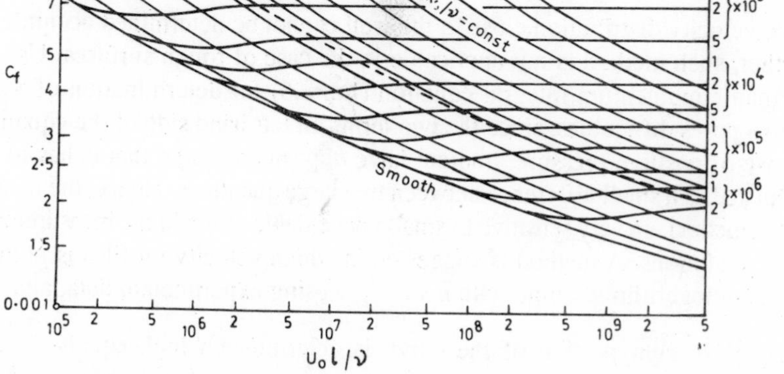

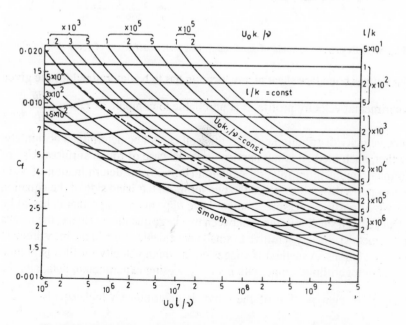

Fig. 8.15. Variation of C_f with $U_o L/\nu$ and L/k for sand roughened plate

The local friction coefficient c_f can also be expressed in terms of $U_o x/\nu$ and x/k as shown in Fig. 8.16 For completely rough surface, c_f and C_f can be expressed by the equations

$$c_f = \left(2.87 + 1.58 \log \frac{x}{k}\right)^{-2.5} \qquad \text{. . . (8.30)}$$

$$C_f = \left(1.89 + 1.62 \log \frac{L}{k}\right)^{-2.5} \qquad \text{. . . (8.31)}$$

which are valid for $10^2 < \frac{L}{k} < 10^6$.

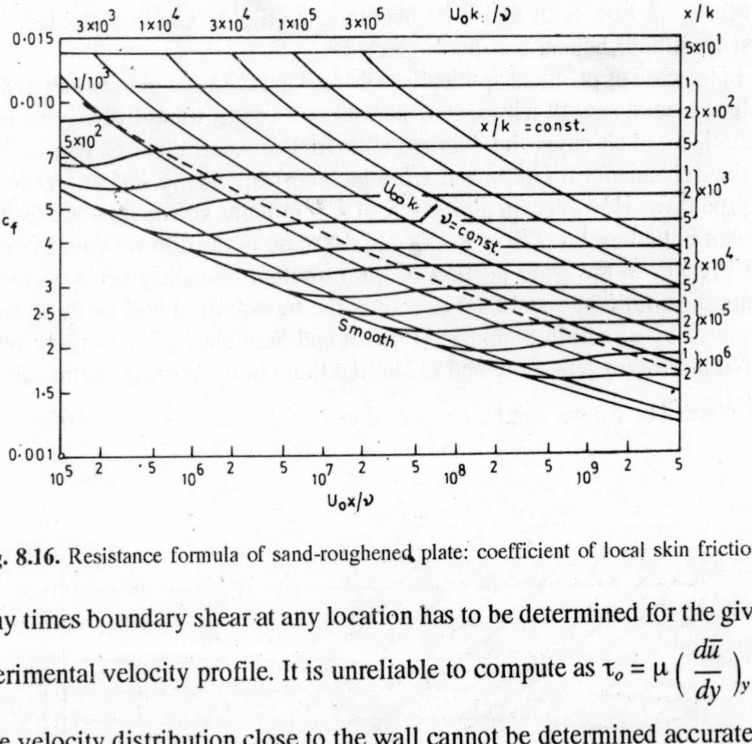

Fig. 8.16. Resistance formula of sand-roughened plate: coefficient of local skin friction

Many times boundary shear at any location has to be determined for the given experimental velocity profile. It is unreliable to compute as $\tau_o = \mu \left(\frac{d\bar{u}}{dy}\right)_{y=0}$, since velocity distribution close to the wall cannot be determined accurately. Further, such method is not applicable in the case of rough surfaces. Use of Karman's momentum integral equation (Eq. 8.6) for determination of τ_o or c_f also poses difficulties, since the two terms on left hand side of the equation in case of positive pressure gradient have opposite signs so that c_f has to be calculated as a small difference between two large quantities. Hence, the results of this method are very sensitive to small unavoidable errors in the measurement and calculations. A method is suggested in which velocity profiles is plotted on a semilogarithmic graph with $\bar{u}$ vs. log y using experimental data, and the slope of straight portion of the curve is determined which equals $\frac{2.3 u_*}{\kappa}$.

Assumption of Karman constant $\kappa = 0.39$ or 0.40 yields estimated value of shear velocity u_*. This method is based on the validity of law of the wall near the boundary, irrespective of the pressure gradient. Clauser, and Bradshaw (1*) have suggested slightly different technique for determination of τ_o or c_f from known velocity profile, but both make use of the law of the wall.

Clauser Expressed u_* as

$$u_* = U_0 \sqrt{\frac{c_f}{2}}, \quad \frac{\bar{u}}{u_*} = \frac{\bar{u}}{U_0}\sqrt{\frac{2}{c_f}} \quad \text{and} \quad \frac{u_* y}{\nu} = \frac{U_0 y}{\nu}\sqrt{\frac{c_f}{2}}$$

The law of the wall for boundary layer flow can now be expressed as

$$\frac{\bar{u}}{U_0}\sqrt{\frac{2}{c_f}} = 5.75 \log \frac{U_0 y}{\nu}\sqrt{\frac{c_f}{2}} + 5.2$$

Hence, observed velocity distribution data can be plotted as $\bar{u}/U_o$ vs. $\dfrac{U_o y}{\nu}$ on semi-log scale and the above equation can be plotted for various constant c_f values. By observing with which curve the experimental data match closely, one can determine c_f and hence τ_o.

8.7 SEPARATION (1)

Separation of the flow occurs when streamlines depart from the boundary. Separation occurs either on the sharp edges or on a smoothly curved or plane surfaces. In the former case the point of separation is given by the location of sharp edge, whereas in the latter case it is defined by vanishing shear at the wall and, therefore, depends entirely on the boundary layer development upstream of the point of separation. When flow separates, the boundary layer concept fails for the following reasons: (i) streamlines are no longer parallel to the surface: (ii) pressure is no longer constant across the boundary layer which is quite thick, and (iii) in incompressible flow the pressure distribution upstream of separation is influenced by the flow downstream of separation such that there is interaction between the boundary layer and external flow.

Some of the important characteristics of separation can be briefly mentioned. Usually separation is connected with an instability of the flow, as a result of which flow, many times, exhibits asymmetry and oscillating apperance. Further, near the region of separation, the law of the wall is not applicable even though it is found to be applicable in boundary layer development with positive pressure gradient. Some investigators have, on the basis of experimental results, suggested that separation occurs when $H = \dfrac{\delta^*}{\theta}$ lies between 1.8 to 2.7. However, this is a very vague criterion since, even in constant pressure layer on rough surfaces, H values upto 2.5 have been observed (see Fig. 8.7). Therefore, vanishing of c_f is a better criterion. Mean position of the point of separation can be found by extrapolating c_f vs x curve.

8.8 BOUNDARY LAYER CALCULATIONS

Let us consider calculation of turbulent boundary layer on a flat plate with zero incidence when pressure gradient is zero, i.e. $\frac{dU_o}{dx}$ is zero. The momentum equation with the omission of turbulence terms is

$$\frac{d\theta}{dx} = \frac{c_f}{2}$$

which can be expressed as

$$x = \int \frac{2}{c_f} \, d\theta$$

However, it may be remembered that in general

$$c_f = f\,(U_o, \theta, \nu, k)$$

and c_f cannot be explicitly expressed as a function of θ. Substitution of c_f in the above equation, and use of the boundary condition $\theta = \theta_o$ when $x = x_o$ can, in principle, give a method of calculation of variation of θ with x on a flat plate.

Calculation of boundary layer development with positive or negative pressure gradient is more complicated. Karman's momentum integral equation now involves $\theta(x)$, H, c_f and dU_o/dx or $d\bar{p}/dx$. As a first approximation c_f or τ_o at any value of x can be determined from equation for flat plate boundary layer development with zero pressure gradient on a smooth plate, namely

$$c_f = \frac{\tau_o}{\frac{\rho\, U_o^2}{2}} = \frac{0.0256}{(U_o\theta/\nu)^{0.25}} \quad \text{according to Prandtl}$$

or

$$c_f = \frac{0.0130}{(U_o\,\theta/\nu)^{1/6}} \quad \text{according to Falkner} \qquad \ldots (8.32)$$

or

$$c_f = \frac{0.0576}{[\log\,(4.075\;U_o\theta/\nu)]^2} \quad \text{according to Square and Yang}$$

In order to determine H, Von Doenhoff and Tetervin (9*) assumed that dH/dx would depend on τ_o and $d\bar{p}/dx$. From experimental data, they found that

$$\theta\,\frac{dH}{dx} = e^{4.68\,(H-2.975)}\left[\frac{\theta}{q}\,\frac{dq}{dx}\,\frac{2q}{\tau_o} - 2.035\,(H - 1.286)\right] \qquad \ldots (8.33)$$

where $q = \rho U_0^2/2$. Given $\frac{dq}{dx}$, momentum equation along with the above equations can be solved by step by step procedure for variation of θ and H with x if initial values of H_o, θ_o, $\frac{d\theta}{dx}$ and $\frac{dH}{dx}$ are known. It may be mentioned that there are a number of methods proposed for boundary layer calculations.

8.9 ATMOSPHERIC BOUNDARY LAYER

Engineers are often interested in wind effects on structures such as tall buildings, chimneys, bridges, and T.V. towers; in the prediction and abatement of atmospheric pollution; in wind induced discomforts in and around buildings; in the generation of.wind waves and the bumping of aeroplanes; and in the lifting of rockets and similar other phenomena. These phenomena involve, at most, a few kilometers of distance and heights of the order of 300-400 m. This layer in the vicinity of earth's surface is commonly known as the atmospheric boundary layer. In this layer, atmospheric motions corresponding to micro-scale (10^{-3} m to 10^1 m) and small scale (10^1 m to 10^4 m) are of primary significance in relation to the above mentioned problems involving mass transport and wind forces. However, these motions are affected by the meso-scale motions (10^4 m to 10^5 m) caused by topography, nonuniform heating of earth's surface and earth's rotation.

All types of winds are caused by the variable heating of earth's surface due to the sun. Solar radiation, which is far more intense at the equator than at the poles, tends to cause differential heating of the earth's surface which, in turn, gives rise to pressure gradients. Other factors, such as earth's rotation, cloud cover, precipitation, large topographical reliefs, surface roughness and their spatial variation, also affect winds. Elaborate numerical models based on fundamental fluid mechanics principles are being developed and tested to predict the future state of atmosphere for known initial state (12). However, they have not been fully developed and tested, and, hence, cannot be used by engineers, at present. Instead, engineers have to depend on the analysis of wind data to get design information. Winds, in general, can be classified into boundary layer winds, cyclones, anticyclones and tornadoes. We will deal only with the boundary layer winds near the earth's surface where wind velocity changes from zero at the surface of the earth to a near constant value at higher elevation. This is often known as planetary boundary layer or atmospheric boundary layer. Tornadoes, cyclones, severe thunder storms, and separation on the downstream side of hills, mountains, etc. are responsible for greatest perturbations in the structure of atmospheric boundary layers. Typical wind speed and direction profile in the planetary boundary layer measured at Digha, India in July 1979 under Monex experiment (13) is shown in Fig. 8.17. Characteristics of atmospheric layers have been described in detail by Plate (14) and present state of knowledge is summarised by Monin (15).

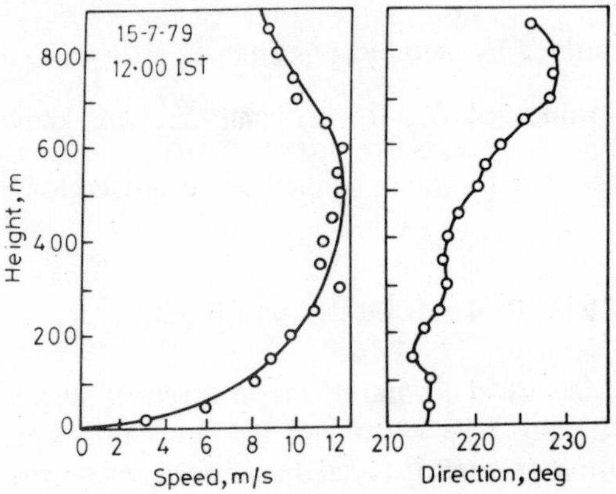

Fig. 8.17 Planetary boundary layer at Digha, West Bengal, India

The atmospheric motion is controlled by the following forces: (i) gravity (ii) force due to pressure gradient, and (iii) Coriolis force. The pressure gradient in the vertical direction is balanced by the gravity force. The force due to the pressure gradient in the horizontal direction is responsible for generation of wind. It may be mentioned that the former is over 10,000 times greater than the latter. Coriolis force (named after the French scientist) arises when the co-ordinate system of interest is rotating and the motion of the object (here air) occurs relative to the moving co-ordinate system.

Geostrophic Wind

Far away from the earth's surface, the wind velocity is governed by horizontal pressure gradient and Coriolis force. The resulting velocity is known as the geostrophic wind velocity, and denoted by U_g; if the isobars are curved, the centrifugal force will act in addition to pressure gradient and Coriolis forces. The resulting wind velocity is then called gradient wind velocity. The geostrophic or gradient velocity U_g is given by

$$U_g = \frac{\partial \bar{p}/\partial n}{2\rho\omega \sin \lambda_1} = \frac{\partial \bar{p}/\partial n}{\rho f} \qquad \ldots (8.34)$$

where ω is the angular velocity of the earth, λ_1 is the latitude of the place, and $f = 2\omega \sin \lambda_1 = 1.458 \times 10^{-4} \sin \lambda_1\ s^{-1}$. This velocity occurs at an elevation Y_g above the earth's surface; Y_g is of the order of 300-800 m. Within the vertical distance Y_g velocity will change from 0 to U_g due to the boundary friction.

Velocity Distribution

Thermal stratification in the planetary boundary layer is important in low velocity flows. Thermally stratified boundary layer flow is of great concern to environmental engineers concerned with dispersion and diffusion in the atmosphere. However, in connection with forces on buildings etc., since one is interested in maximum force which occurs at high velocities, the thermal stratification is destroyed in such a case. Such a boundary layer without thermal stratification is known as neutral boundary layer which is discussed here onwards.

Near the boundary, one can assume that the shear is constant and equal to the boundary shear τ_o. If mixing length l is assumed to be proportional to the distance from the boundary, i.e. $l \sim y$, one gets the logarithmic velocity distribution law in this surface layer as

$$\frac{\bar{u}}{u_*} = \frac{1}{\kappa} \ln \frac{y}{y'} \qquad \ldots (8.35)$$

in which $u_* = \sqrt{\tau_o/\rho}$ and Karman constant κ is 0.40; y' is the characteristic height of surface roughness such that at $y = y'$, $\bar{u} = 0$. As discussed earlier, this equation is valid for about 15–20 per cent thickness of the atmospheric boundary layer beyond which it deviates from Eq. 8.35. The length scale y' depends on the geometry, concentration and height of roughness on the earth's surface. For rough boundary made up of closely packed uniform sand particles of size κ, $y' = \kappa/30$. When y is not much larger than the height of roughness, Eq. 8.35 is modified to

$$\frac{\bar{u}}{u_*} = \frac{1}{\kappa} \ln \left(\frac{y - Y_1}{y'}\right) \qquad \ldots (8.36)$$

where Y_1 is called the zero plane displacement. The zero plane displacement Y_1 in the above equation is introduced to account for origin shift that is expected to occur for rough surfaces. Values of y' for crops vary from 5.0 mm to 65 mm, and are related to crop height h_c by the relation.

$$y' = 0.15\, h_c$$

As an approximation one can take

$$Y_1 = 0.63\, h_c$$

For cities Y_1 is obtained from the formula

$$Y_1 = H - 2.5\, y'$$

in which H is the general roof top level. The logarithmic law is valid upto elevation $y = y_m$ where y_m is given by

$$y_m = b u_*/f$$

and $b = 0.015$ to 0.030. Because of these limitations of logarithmic law, planetary boundary layer is often modelled using the power law

$$\frac{\bar{u}}{\bar{u}_h} = \left(\frac{y}{h}\right)^n \qquad \ldots (8.37)$$

where $\bar{u}_h$ is the velocity at elevation h. Alternately, it is written in terms of geostrophic wind velocity U_g and Y_g, the height at which it occurs, viz.,

$$\frac{\bar{u}}{U_g} = \left(\frac{y}{Y_g}\right)^n \quad \ldots (8.37a)$$

It is found (14) that Y_g and n are primarily functions of ground roughness. Values of Y_g, y' and n as a function of terrain, as given by Davenport (16) are shown in Fig. 8.18. These values can be used for a given terrain, so that for known $\bar{u}_h$ and h, one can determine velocity distribution in the vertical. Use of Fig. 8.18 and Eq. 8.37 (a) shows that, for the same geostrophic wind velocity, the velocity at a given elevation is lower if the surface is rougher.

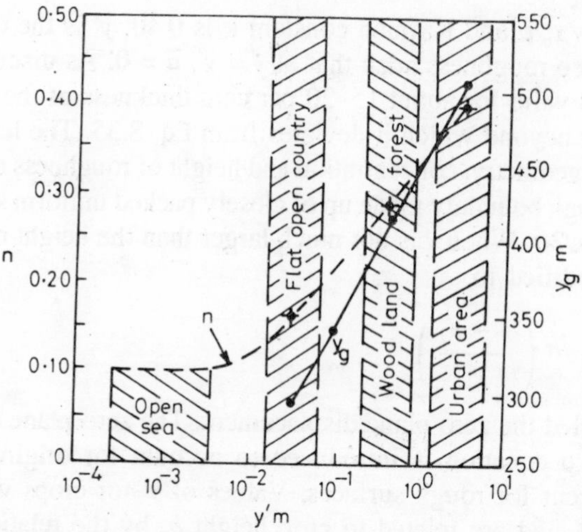

Fig. 8.18 Wind profile parameters for strong winds (16)

Turbulence Characteristics of Atmospheric Boundary Layer

From the consideration of dynamics of the system, the atmospheric boundary layer can be divided into two parts. There is a layer near the earth's surface, known as the dynamic sublayer, in which the effect of stratification due to humidity and temperature gradients can be neglected. This resembles the wall region of boundary layer in homogeneous fluid. All the dynamic parameters in the dynamic sublayer are well determined by u_*, ν and mean height of surface roughness. The thickness of the dynamic sublayer is of the order of $C_p \rho u_*^3/\beta(H_T + bH_L)$ where C_p is the specific heat at constant pressure, H_L is the vertical latent heat flux, H_T is the vertical turbulent heat flux, H_T/H_L being Bowen's ratio, $\beta = 1/T_o$ where T_o is the standard average value of temperature in the layer, and $b = \left(\frac{R_v}{R_d} - 1\right)\frac{C_p T_o}{L}$ where $R_v = 0.467$ J/g deg. and $R_d = 0.287$ J/g deg. are the gas constants for water vapour and dry air, respectively, and L is the latent heat. It may be mentioned that under neutral stratification $(H_T + bH_L)$ is almost zero and the whole surface layer is dynamic.

If the boundary acts as hydrodynamically smooth, adjacent to the boundary will be a thin sublayer within the dynamic sublayer in which the flow is laminar. This is known as the laminar sublayer. This sublayer is destroyed if the boundary acts rough, which is the usual case due to the presence of a variety of roughness elements on the earth's surface.

Above the dynamic sublayer is the surface layer in which the action of Coriolis force can be neglected. Above the dynamic sublayer, heat and humidity cannot be regarded as passive substances. In the surface layer with $y >> \left(\frac{\nu}{u_*}\right)$, the laws governing the statistical parameters of hydrodynamic fields are determined by the components of turbulence of not too small a scale. In this layer, a more appropriate length scale is

$$L_1 = \rho\ C_p\ u_*^3/\kappa\beta\ (H_T + bH_L)$$

known as Monin–Obukhov length scale.

In turbulent boundary layer portion in which logarithmic velocity distribution law prevails, $\sqrt{\overline{u'^2}}/u_*$, $\sqrt{\overline{v'^2}}/u_*$ and $\sqrt{\overline{w'^2}}/u_*$ assume (15) approximately constant values of 2.3, 1.7 and 0.90, respectively. Outside this region they are functions of y/L_1. The u' and w' correlation coefficient is negative, and under neutral stratification, it is close to -0.50.

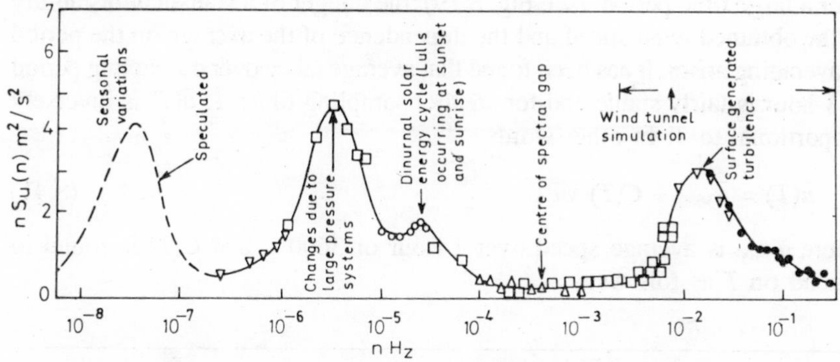

Fig. 8.19 Energy spectra of longitudinal velocity by Van der Hoven (17)

One can also consider the contribution of kinetic energy of eddies of given frequency n by the use of spectrum function. $S_u(n)$ can represent the kinetic energy contribution due to turbulent fluctuations in x direction with eddies of frequency n. Similarly, one can define $S_v(n)$ and $S_w(n)$ for turbulent fluctuations in y and z directions. Figure 8.19 shows variation of $nS_u(n)$ with n as obtained for atmospheric turbulence by Van der Hoven at 100 m above the ground level at Brook Haven laboratory in USA (17). On the figure are shown the processes responsible for generation of different ranges of the spectra. The spectral gap in the range of 10^{-3} to 10^{-4}Hz is attributed to the absence of any physical process capable of generating fluctuations in this range of frequencies. It has been found that

$$\frac{nS_u(y,n)}{u_*^2} = \frac{4.0\ x^2}{(1 + x^2)^{4/3}} \quad \ldots (8.38)$$

where $x = 1200\ n/\bar{u}_{10}$, $\bar{u}_{10}$ being the wind speed in m/s at 10 m above the ground. This equation gives spectral function for frequencies greater than spectral gap for the microscale fluctuations in the wind developed by thermal and mechanical effects. Spectra of vertical fluctuations upto 50 m are given by

$$\frac{nS_v(y,n)}{u_*^2} = \frac{3.36\ f_1}{(1 + 10\ f_1)^{5/3}} \quad \ldots (8.39)$$

while spectra for lateral fluctuations are given by

$$\frac{nS_w(y,n)}{u_*^2} = \frac{15\ f_1}{(1 + 9.5 f_1)^{5/3}} \quad \ldots (8.40)$$

where $f_1 = \dfrac{ny}{\bar{u}(y)}$ is sometimes known as *Monin's coordinate.*

Averaging of Wind Speed

Since at a given elevation the wind speed varies with time, in almost all the cases averaging in time is used. Since the turbulence spectrum is continuous over a large time period, (see Fig. 8.19), the danger of a statistical instability of the obtained wind speed and the dependence of the average on the period of averaging arises. It has been found that average taken over a sampling period of 1 hour is fairly stable and for smaller sampling times T, $\bar{u}(T)$ is inversely proportional to T. In other words

$$\bar{u}(T) = \bar{u}_{3600} + C(T)\ \sqrt{\overline{u'^2}} \quad \ldots . (8.41)$$

where $\bar{u}_{3600}$ is average speed over 1 hour or 3600 s, and $C(T)$ is found to depend on T as follows:

T s	1	10	20	30	50	100	200	300	600	1000	3600
$C(T)$	3.0	2.32	2.00	1.73	1.35	1.02	0.70	0.54	0.36	0.16	0.0

Wind Speed of Given Return Period

From the point of view of safety of a structure subjected to wind force, the wind velocity to be used in the analysis should not be maximum observed so far, but one which may occur once in T years (say 10 years or 50 years) in a probabilistic sense. This is determined by first finding the maximum U_g observed each year for a number of years, and forming a series arranged in an ascending order of mangitude. Such a series of extremal values is found to follow Gumbel's distribution.

$$F(x) = e^{-e^{-X}} \qquad \ldots (8.42)$$

where $F(x)$ is the probability that value of the variable will be equal to or less than x; $F(x) = m/(N+1)$ where N is the total observations in the series and m is the rank of the quantity in the ascending sereis. Here X is known as the reduced variable and is related to x as

$$X = a\,(x - x_{mo})$$

where x_{mo} and a are given as

$$x_{mo} = \bar{x} - 0.454\,\sigma$$

and $$a = 1/0.786\,\sigma$$

σ being the standard deviation of x and $\bar{x}$ is mean of x values. The return period will be $T = 1/[1 - F(x)]$. Fig. 8.20 shows the probability distribution of maximum yearly wind speeds at Cardington during 1932–1954. The data seem to follow Gumbel's distribution remarkably well.

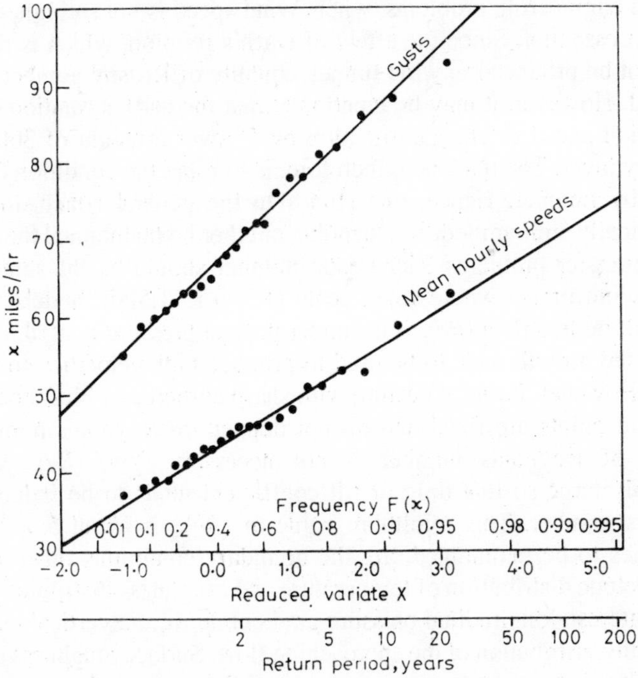

Fig. 8.20 Probability distribution of wind speeds

Simulation of Atmospheric Boundary Layer in Wind Tunnels

The nature of wind to which a structure is exposed is strongly conditioned by the geometry of upwind and surrounding structures, such as hillocks, large trees, cluster of buildings in the cities, etc. Since the wind conditions and its characteristics cannot be predicted correctly, model studies are carried out in wind tunnels, where the natural wind is simulated. Cermak (18) has shown that for dynamic similarity Reynolds number $Re = \dfrac{U_o L_o \rho_o}{\mu_o}$, Bulk Richardson number $R_i = \left(\dfrac{\Delta T_o}{T_o}\right)\left(\dfrac{L_o g}{U_o^2}\right)$, and Rossby number $R_b = U_o/L_o\omega$ must be maintained the same in model and prototype. Here ΔT_o is the difference of temperature between the wall and the fluid and T_o is the wall temperature; U_o and L_o are the characteristic velocity and length, respectively. The conditions of thermal similarity obtained from energy equation give equality of

Prandtl number $Pr = \dfrac{\nu_o}{(K_o/C_{po}\,\rho_o)}$

Eckert number $Ec = U_o^2/C_{po}\,\Delta T_o$

Here K_o is the thermal conductivity of fluid of mass density ρ_o, and C_{po} is specific heat at constant pressure. When the same fluid is used in model and prototype, the Prandtl number will automatically remain the same in both cases. Eckert number equality is essential at speeds approaching sonic speed. Hence, for wind engineering problems, where wind speed is not so high, equality of Ec is not essential. Since the effect of earth's rotation, which is reflected in ω, cannot be produced in wind tunnel, equality of Rossby number cannot be achieved. However, it may be mentioned that the earth's rotation causes the mean wind speed to change direction by 5° over a height of 300 m of the boundary layer. The tendency, therefore, is to relax the condition of equality of Rossby number. Hence, one comes to the general conclusion that for geometrically similar models, Reynolds number be maintained the same, and in heat transfer problems Richardson number should be the same.

In a wind tunnel where small scale (1:300 to 1:500) models are tested, $(Re)_m$ will be less than $(Re)_p$ if air under normal pressure is used. Otherwise, compressed air will have to be used to produce high velocities so that $(Re)_m = (Re)_p$. However, if one is dealing with sharp cornered or bluff bodies where separation points are fixed and do not depend on Reynolds number, strict equality of Reynolds number is not necessary. Yet, $(Re)_m$ should be reasonably large so that drag or lift coefficient tends to be independent of Reynolds number. This minimum value of $(Re)_m$ is about 5×10^5. Other similarities to be maintained are: the boundary conditions must be similar, which include distribution of temperature and roughness distribution over the area of interest, longitudinal pressure gradient $\partial\bar{p}/\partial x$ and vertical temperature and velocity distribution of the approaching flow. Surface roughness conditions are usually met if adequate care is taken. Surface roughness is scaled in accordance with the prototype roughness. However, when this results in an aerodynamically smooth surface, the roughness height in the model needs to be increased by an increment equal to viscous zone thickness $10\nu/u_*$. In summary, the conditions to be satisfied are

(i) geometrically similar model and kinematic similarity;
(ii) $(Re)_m$ to be greater than 5×10^5;
(iii) bulk Richardson number to be the same; and
(iv) boundary condition similarity.

Experimental Set-up

The experiments regarding wind engineering or atmospheric pollution are conducted in a meteorological wind tunnel of adequate cross-section, say 2 m to 4 m wide, 2 m deep and 15 m to 30 m long test section as shown in Fig. 8.21. The tunnel and method of testing are described by Cermak et al. (18, 19, 20).

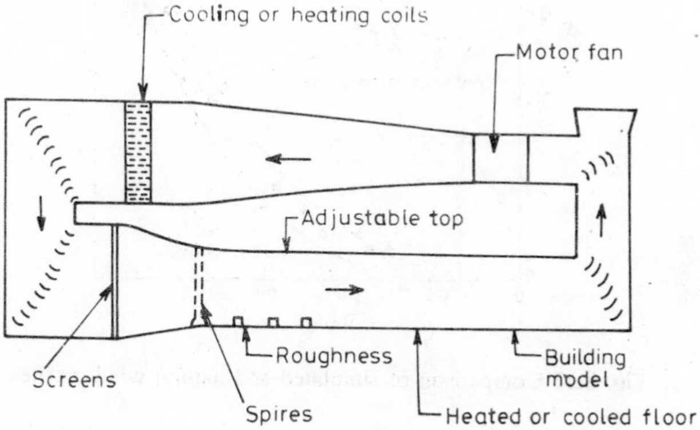

Fig. 8.21 Typical wind tunnel for wind engineering and pollution studies (19)

The lower boundary of the wind tunnel is heated if necessary and ambient air temperature can also be modified. This is done by providing heating and cooling arrangements. For studies related to forces on structures such heating or cooling is generally not required. The top ceiling of the tunnel is adjustable so that $\partial\bar{p}/\partial x$ can be adjusted to desired value. The boundary layer thickness of the order of 1.0 m at wind speeds of 40 m/s or so needs to be achieved. To get such a thick boundary layer, one would require large length of wind tunnel since boundary layer thickness increases as $L^{0.80}$. Hence, length requirement is reduced by using certain devices at the entrance to the tunnel. These include vortex generators or spires, curved screen or a screen with longitudinal bars whose spacing increases with increasing values of y, or sparsely placed gravel or other roughness at the entrance to the wind tunnel. The surface roughness downstream from the stimulator must be equivalent to the roughness which without the stimulator would have produced an equilibrium boundary layer with the same characteristics as those generated by stimulator. With these roughness and stimulators, velocity distribution should be plotted according to power law, and if the exponent n is the same for field data and wind tunnel data, one can assume that similarity of mean flow conditions is achieved. Normally even tall buildings are submerged in atmospheric boundary layer. Hence, boundary layer in wind tunnel has to be sufficiently thick so that model of a tall building is submerged in wind tunnel

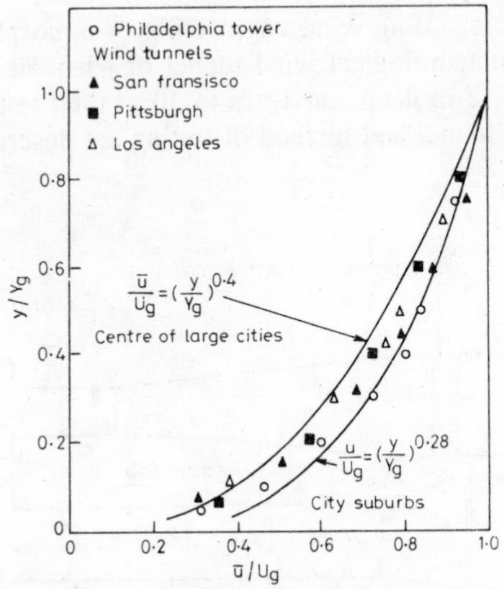

Fig. 8.22 Comparison of simulated and natural wind profiles

boundary layer. For this a minimum of 10 m length is required upstream of test section.

For a faithful modelling in wind engineering, atmospheric turbulence must be correctly modelled in the wind tunnel. Two aspects must be kept in mind which help in modelling turbulence. Firstly, turbulence being statistical in nature, for similarity of turbulence one would require equivalence of statistical functions, which is impossible. Further, since in wind tunnel it is not possible to simulate all forces that cause turbulence in atmosphere, all scales of turbulence over the entire range of wave number cannot possibly be simulated. One must usually identify those properties of turbulence which are important in transporting momentum, heat or mass. In certain cases, simulation of turbulent energy in a particular narrow eddy range would be required. For this, equivalence of normalised turbulence intensities and energy spectra in the prescribed wave number range may be considered as sufficient criteria for simulation. Turbulence intensities are comparatively easy to match. Matching of turbulence spectra requires thick boundary layer and high wind speeds. Figures 8.22 and 8.19 show simulation of mean velocity distribution and longitudinal turbulence spectra. It needs to be mentioned that very inadequate data are availabe about the turbulence intensities and spectra in winds over large cities. Figure 8.19 indicates that within model scale range of 1:100 to 1:1000, simulation is achieved down to scales associated from maximum turbulent energy to the end of spectral gap. This simulation is considered adequate. In such simulation, it is extremely important that all major upwind buildings be represented in the model since the wakes produced by them will add energy to the spectrum at wave lengths comparable to the dimensions of the wake.

Similarity of Motion Over Topographical Features

This aspect has been examined in detail by Cermak et al. (20) and Nemoto (21). They have found that to obtain similarity of flow patterns and turbulence structure in topographic modelling, Euler number, Froude or Richardson number, and Reynolds number need to the matched, in addition to geometric similarity. Matching of Euler number only requires geometric similarity of gross wind directions and distribution of vortices and eddies. It is well known that Reynolds number and Froude number cannot be maintained the same at the same time if model fluid is also air. However, it may be remembered that Froude number is a governing parameter when some special flow phenomena occurring in light winds and strong stratification, such as mountatin lee-waves, are of interest. When flow patterns in strong winds are of interest, Froude number can be ignored. As discussed earlier, exact *Re* number equality cannot normally be maintained. If topographical features are fairly sharp, mean flow patterns are independent of Reynolds number provided *Re* number exceeds a lower limit which depends on the topographical features. Smooth surfaces of topographical features should be roughned. In a few cases one may find it necessary to exaggerate the vertical scale.

FLOW PAST NONPOROUS PLATES AND POROUS FENCES

Flow past a fence or a two dimensional obstruction set normally on a plane boundary is of considerable interest to hydraulic engineers for a variety of reasons. Firstly, this illustrates the essential characteristics of flow past spurs in rivers, two dimensional roughnesses in channel flow, fences and flow past bluff buildings. Secondly, this case can be used to study the basic characteristics of flow separation and to illustrate the effect of boundary proximity on the separation zone. Needless to say, separation plays an important role in Hydraulic Engineering. In order to reduce continuous loss of energy, separation has to be avoided in many structures, such as open channel expansive transitions, by shaping the boundary properly. In other structures, such as energy dissipators, occurrence of separation is essential to cause maximum dissipation of energy. Thirdly, the flow past obstructions placed on the boundary can be utilised to illustrate dependence of drag coefficient C_D on the submergence ratio h/δ. Here $C_D = F_D/(h\rho U_o^2/2)$, h is the height of the plate, δ the boundary layer thickness of undistrubed flow at the section where plate is fixed, and F_D is the drag force on the plate per unit of its length. Flow past porous fence is considered in relation to its drag force and turbulence.

8.10 FLOW PAST NONPOROUS PLATES

Flow Past Plates in Midstream

Flow of a real fluid past a two dimensional plate held perpendicular to flow in the midstream has been studied extensively in the past. Typical pressure distribution on the front and the rear of the plate at Reynolds number $2U_oh/\nu$ greater than 10^3 is shown in nondimensional form in Fig. 8.23. The basic characteristics of flow, which are usually discussed in elementary fluid mechanics are enumerated below:

(i) At low values of *Re*, C_D decreases with increase in *Re*. Since the separation points are fixed, and the friction drag is negligible, C_D becomes constant at *Re* exceeding 10^3 and assumes a value of 1.90.

(ii) The separation zone is characterised by alternate shedding of vortices, a phenomenon common to two dimensional bodies such as cylinders, etc. This is known as Karman vortex trail. Alternate shedding of vortices leads to lateral oscillations. The frequency *f* of the vortex shedding is given by $\frac{2fh}{U_o} = 0.13$. In fact, for two dimensional bodies Strouhal number $S = fD/U_o$ where *D* is the characteristic dimension 2*h*, is related to C_D by the relation $S = 0.21/C_D^{0.75}$.

Uniform Flow Past Normal Plate on the Boundary

The role played by the separation pattern in the determination of drag can be seen by providing a long splitter plate as shown in Fig. 8.23. Such a long splitter plate stabilizes the vortices on both the sides converting them into stationary standing eddies and reducing the Strouhal number to near zero value.

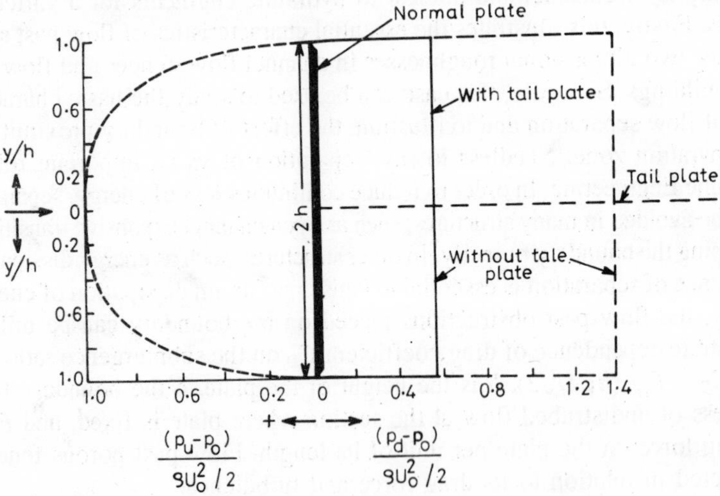

Fig. 8.23 Pressure distribution on a plate held normal to flow

Tail plate essentially changes the pressure on the the downstream side $(P_d - P_o)/\frac{\rho U_o^2}{2}$ from = −0.58 to −1.38. Hence, as shown by Arie and Rouse (22) the drag coefficient of the plate drops down from 1.9 to 1.38. This case essentially corresponds to a flow past a flat plate held perpendicular to the boundary except for the fact that approach flow is of constant velocity, unlike in the case of boundary layer flow.

The entire flow can now be divided into primary flow, secondary flow consisting of standing eddy, and the transition zone between these two zones as suggested by Rouse (23), (see Fig. 8.24). The transition zone is the zone of intense turbulence created by the high shear zone. The generation, convection, diffusion and dissipation of turbulence in the transition zone represent the most essential features of the separation phenomenon. The flow within the standing eddy is driven by the flow surrounding it; hence, the flow in the two zones is inter-related. The eddy shape is of primary importance because it governs the flow characteristics outside it. In fact, in a simplified analysis, one can assume the eddy profile to be the effective boundary and determine the pressure distribution within the main flow. In the uniform flow the eddy length is of the order of $17h$ (22). Further, the dimensionless pressure distribution $(p_d - p_o)\Big/\dfrac{\rho U_o^2}{2}$ on the downstream side of the plate has a value of -0.58, as compared to -1.38 in the case of flow past a normal plate held in midstream. Dimensionless pressure gradually increases in the downstream direction.

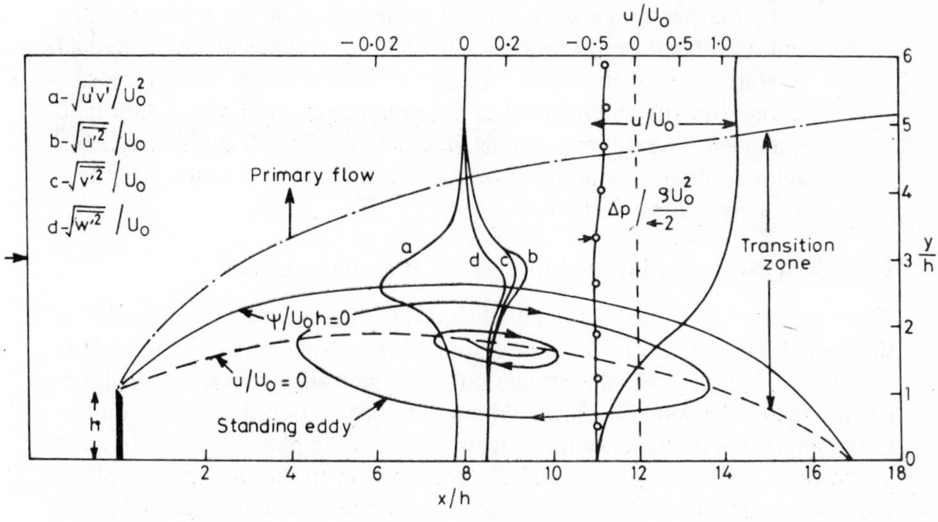

Fig. 8.24 Flow past a normal plate on the boundary (23)

The flow within the standing eddy slowly rotates in the direction shown, with the higher velocity at its periphery and smaller near its centre. As such, the flow within the eddy is rotational in its character. The turbulence generated at the interface of main flow and the standing eddy diffuses on both sides and creates a zone of transition in which the turbulence intensities are high and they decrease on either side of the separating line. Yet, the eddy zone is highly turbulent; hence, it can entrain sediment, etc. from the bed on the leeward side. The characteristics of the mean flow and turbulence within the standing eddy and outside are shown in Fig. 8.24. Critical study of such a flow has revealed (23) the following characteristics of the flow:

(i) Primary flow can be analysed as a state of irrotational motion around a solid body having the shape of standing eddy, augmented in the lateral dimension by the fraction of width of the diffusion zone, i.e. outside the broken line indicating the limit of diffusion zone.

(ii) Within the standing eddy the velocity decreases to zero at the neutral point and then reverses in direction on the lower side. The streamlines are shaped in such a manner that the continuity requirement is satisfied at every section.

(iii) The piezometric pressure or head at any section within the eddy is practically constant.

(iv) From equations of motion, it can be shown that in two dimensional flows $\bar{p} = p_o - \rho\overline{v'^2}/2$. Hence, in the zone of maximum turbulence along the edge of separating eddy, pressure is even lower than that in the eddy due to high value of $\rho\overline{v'^2}/2$, and it will fluctuate.

(v) The distribution of mean velocity in the field is determined usually by measuring the total head and the piezometric head separately and assuming the difference to represent the velocity head. This method of determination of mean velocity is satisfactory if the turbulence level is very low which is not the case in the zones of transition and standing eddy. Further, as has been pointed out in (iv), even the pressure varies and its value depends on $\rho\overline{v'^2}/2$. In addition, in the standing eddy and in zones of transition, the turbulence being intense, velocity vector at any time may be inclined to the axial direction in which pitot tube is aligned. These complications make it very difficult to obtain reliable velocity distribution and flow pattern in these two zones. This fact is many times overlooked.

Flow Past Normal Plate Submerged in Boundary Layer

The boundary layer gets disturbed when a flat plate is held perpendicular to the plane boundary layer flow. The disturbed flow conditions are shown (14) in Fig. 8.25. It can be seen that seven flow zones indicating different behaviours can be identified herein. The flow field in zone 1 is determined by the conditions in the undisturbed boundary layer upstream from the plate or fence. In zone 2, the flow field is displaced and distorted owing to the presence of plate, the lower boundary of this zone being given by the separation induced shear layer that starts at the top edge of the fence and forms the transition to a greatly retarded flow in zone 3. The flow in this zone is first accelerated and then decelerated. With solid fence, zone 6 of separation bubble or standing eddy is formed with the reattachment point at a distance L downstream from the fence. In zone 5, the flow velocity gradually increases until at some large distance the inner layer 5 blends with the outer flow and a new and thicker boundary layer is formed. This boundary layer adjusts gradually to the local boundary conditions at the ground. Hence, in zone 3 downstream of the standing eddy the velocity distribution is as shown. It takes a distance greater than 35h to 40h before typical boundary layer velocity profile is obtained. In zone 7 one gets potential flow condition. Zone 4 is the transition or blending region between zones 2 and 7.

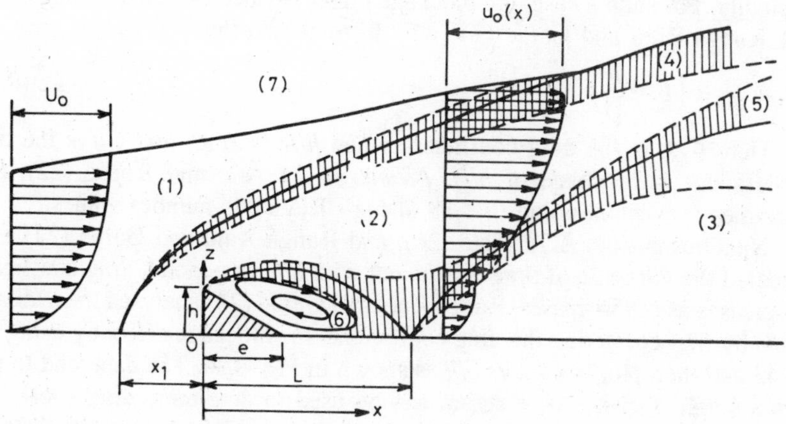

Fig. 8.25 The flow zones of a boundary layer disturbed by a shelterbelt on the boundary (14)

While dealing with flow past a fence or plate held perpendicular to the plane boundary in the boundary layer, one must decide about the characteristic velocity to be used in defining dimensionless parameters such as drag coefficient and Reynolds number. Four choices exist for characteristic velocity, namely, shear velocity at the section in undisturbed boundary layer, free stream velocity, velocity at the top edge of the plate and average velocity at the section. In this analysis the free stream velocity has been used.

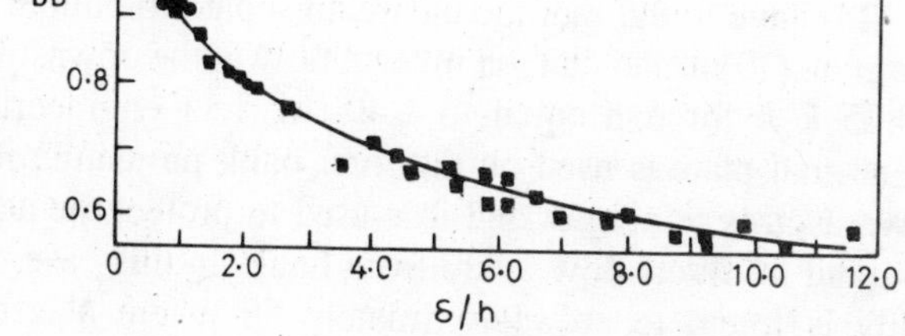

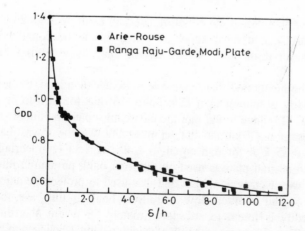

Fig. 8.26 Variation of C_{DO} with δ/h for flat plates held perpendicular to the boundary

The drag coefficient of the fence or plate in the boundary layer flow in the wind tunnel depends on the parameters *h/D* and δ/h where *D* is the depth of wind tunnel. The parameter *h/D* represents the blockage effect caused by the presence of the plate. If the plate is held in the mid-stream, it would correspond to the case if δ/h to be zero since the plate is subjected to a constant velocity. For such a case, the blockage effect has been studied among others by Ranga Raju and Garde (24) who have shown that

$$C_D/C_{DO} = \left(1 - \frac{h}{D}\right)^{-2.85} \quad .... (8.43)$$

Here C_{DO} is the drag coefficient when h/D is zero, and C_D is the drag coefficient at blockage of h/D. As shown by Arie and Rouse (22), C_{DO} assumes a constant value of 1.38 at high Reynolds number values.

Nagbhushanaiah (25), Plate (26), and Ranga Raju and Garde (24) have studied the variation of drag coefficient with δ/h. In general, drag coefficient decreases as δ/h increases. Ranga Raju and Garde (24) analysed the available data by first correcting the drág coefficients for blockage effect by using Eq. 8.43 and then plotting C_D vs δ/h as shown in Fig. 8.26. The data tend to plot on a single curve. This diagram can be used to determine drag coefficient of a long fence in the atmospheric boundary layer. The free stream velocity of flow and δ will have to be estimated. It may be mentioned that Ranga Raju et al. (27) have defined the drag coefficient C_* as $C_* = \dfrac{F_D}{h\rho u_*^2/2}$ where $u_* = \sqrt{\dfrac{\tau_o}{\rho}}$ is the shear velocity at the boundary in the undisturbed flow where the plate is kept. They have shown that in general

$$C_* = f\left(\frac{h}{y'},\ \frac{u_*}{U_o}\right)$$

where, y', is the parameter in velocity distribution law $\bar{u}/u_* = \dfrac{1}{\kappa}\ ln\ \dfrac{y}{y'}$; further, they have found the effect of u_*/U_o on C_* to be negligible. However, to compute C_*, one must know u_* which has to be computed from the velocity distribution data. Hence, earlier analysis involving Fig. 8.26 is preferred.

It may be mentioned that when δ/h is greater than zero, the length of the standing eddy is smaller than $17h$; Plate (26) has found it to be about $12h$. Pande et al. (28) have found that the upstream separation bubble is of length $1.22\,h$. Moss et al. (29) found that the reverse flow in the downstream standing eddy was $0.25\ U_o h$ for δ/h equal to 1.49. In river engineering, structure similar to a normal plate is used on the river bank projecting into the river. This is known as groyne or spur and it is used to protect the adjoining bank from erosion and to divert flow away from bank. In this case, the length of standing eddy is found to be approximately $5h$ where h is the length of projection. This difference in length of standing eddy is believed to be primarily due to Froude number effect.

Because of the presence of plate the boundary layer is disturbed. Downstream of the standing eddy the flow near the boundary is in the downstream direction and here the boundary layer starts redeveloping. Redeveloping boundary layer has been studied by many investigators (30).

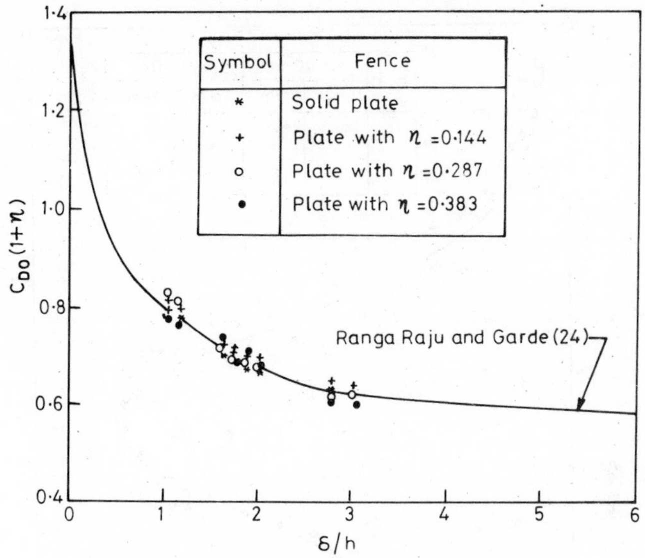

Fig. 8.27 Relationship for C_{DO} for porous fences

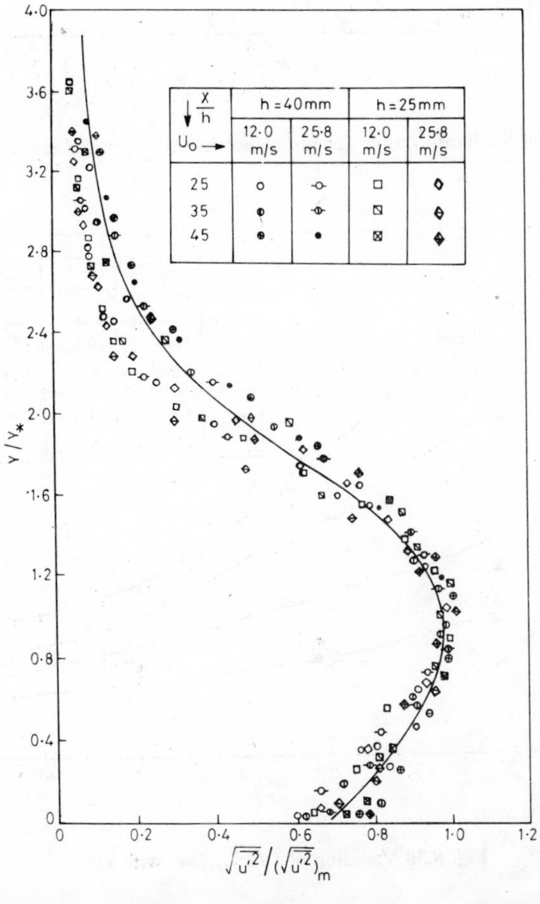

Fig. 8.28 Similarity profile for turbulence intensity for *x/H* greater than 16

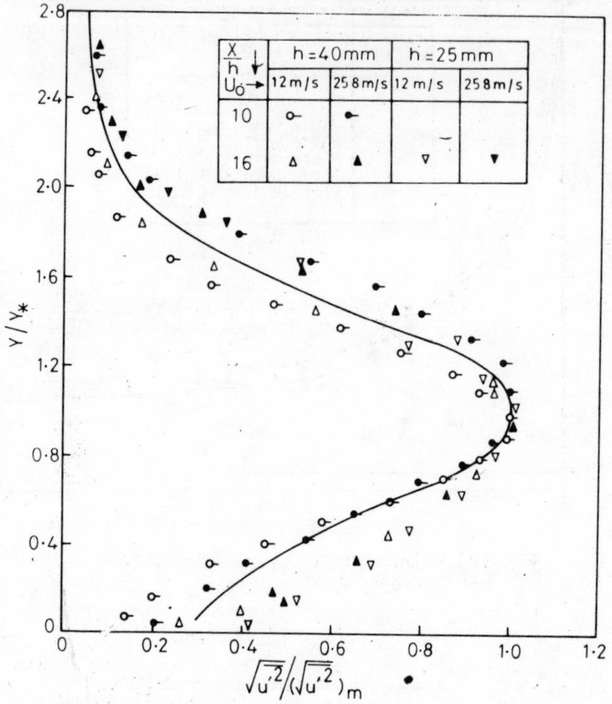

Fig. 8.29 Similarity profile for turbulence intensity for x/H less than 16

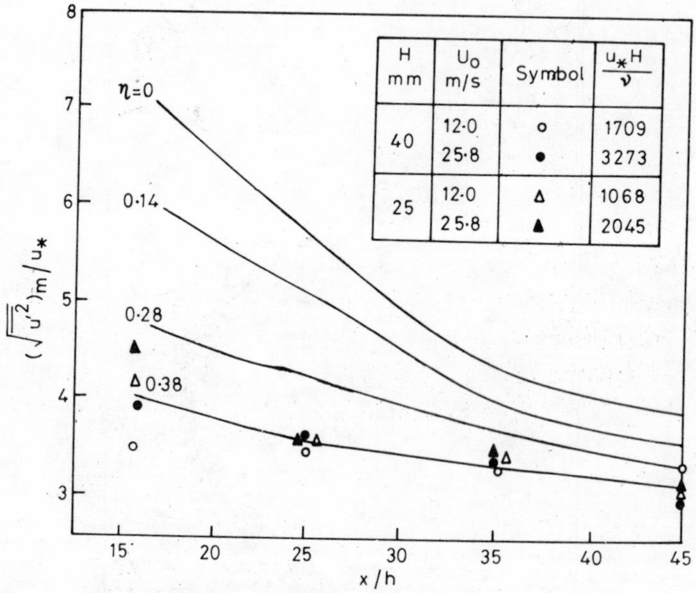

Fig. 8.30 Variation of $(\sqrt{\overline{u'^2}})_m/u_*$ with x/h

8.11 POROUS FENCES

Flow past porous fences has been studied by various investigators because porous fences find applications in the design of shelter belts, analysis of flow past tall buildings, protection of river banks by permeable spurs, and in the study of exchange processes like evaporation in redeveloping boundary layer flows. In the present context, attention is focussed on two aspects of flow past porous fences, namely, drag of porous fences and turbulence in the downstream flow. The porous fences are used to effect reduction of velocity in the downstream direction. Many investigators (30, 31, 32, 33, 34, 35), have conducted studies on porous fences with porosity ranging from zero to 0.70. These studies have led to the following characteristics of flow with porous fences. Porous fence is effective in reducing the velocity and shear on the downstream side of the fence. The shear on the downstream side varies as x^{-1} where x is the distance from the fence. As the porosity increases, there is a lesser reduction in the mean velocity and lower turbulence intensity level in the near wake region. The distribution of porosity in the vertical has a direct effect on the flow on the downstream side; a fence with porosity decreasing with height has a notable protective effect on the downstream side. Reverse is the case with a fence having its porosity increasing with height. If the porosity is greater than 0.30, the downstream standing eddy vanishes, and the flow downstream of the fence is only in the downstream direction. If the porosity exceeds 0.70, the porous fence is almost ineffective in reducing velocity in the downstream direction.

The drag coefficient of a porous fence with sharp edge, when corrected for blockage effect should depend on δ/h and the porosity η. Ranga Raju et al. (36) have studied this functional relationship using wind tunnel data collected by them with porosity η upto 0.383. They used Eq.8.43 for correcting for the blockage effect. Their results along with the curve proposed by Ranga Raju and Garde (24) for solid fence are shown in Fig. 8.27. The relation seems to be satisfactory but needs further verification with additional data. It may be noted that here C_D is defined with respect to U_o.

Variation of turbulence intensity $\sqrt{\overline{u'^2}}$ with height, and downstream of standing eddy indicates that a maximum value of $\sqrt{\overline{u'^2}}$ designated as $(\sqrt{\overline{u'^2}})_m$ occurs at height y_* above the boundary. However, very close to the fence two peaks occur. Leaving this region, plots can be prepared between $\sqrt{\overline{u'^2}}/(\sqrt{\overline{u'^2}})_m$ and y/y_*. For a given porosity, values of $\sqrt{\overline{u'^2}}/(\sqrt{\overline{u'^2}})_m$ and y_*/h vary with x/h. Further two unique curves are obtained for x/h values greater than 16 and less than 16 but greater than about 8. These are shown in Figs. 8.28 and 8.29. Upto a value of x/h = 35, y_*/h is related to η as

$$y_*/h = 2.4 - 2.75\,\eta \qquad \ldots (8.44)$$

Observed variation of $(\sqrt{\overline{u'^2}})_m/u_*$ with x/h and η is shown in Fig. 8.30. Using these figures, one can estimate the turbulence intensity at given values of y, x_*, h, and porosity η if u_* is known. The shear velocity can be estimated if one knows the velocity distribution in the undisturbed boundary layer. The velocity distribution in the redeveloping boundary layer downstream of solid and porous fences has been studied by Gupta and Ranga Raju (37).

DIFFUSERS

Diffusers are devices which are used for reducing the velocity of flow in a duct, thereby converting the velocity head into static pressure. As the fluid flows through the diffuser, the flow expands resulting in unavoidable energy loss. As the flow expands, there can be a transition in shape, e.g. from circular to square, etc., in addition to increase in the area of cross-section. Diffusers are used in piping systems, wind tunnels, venturimeters and other places. Sudden expansion is an extreme case of a diffuser. Various aspects of diffusers that are of interest to the hydraulic engineer are turbulence characteristics, energy balance, modelling of flow, pressure recovery, energy loss and means of making diffusers more efficient. These aspects are briefly discussed below.

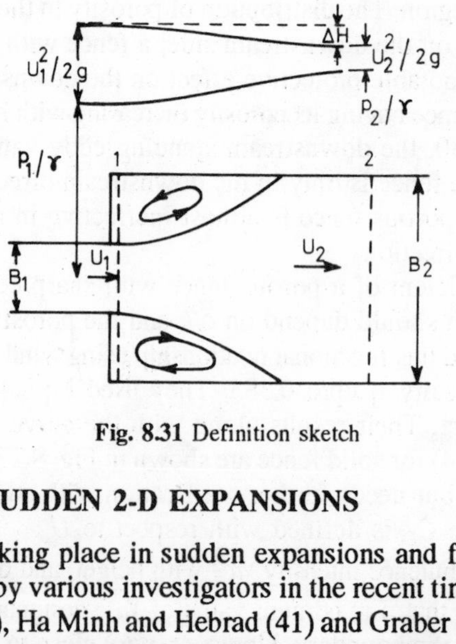

Fig. 8.31 Definition sketch

8.12 FLOW IN SUDDEN 2-D EXPANSIONS

The energy loss taking place in sudden expansions and flow characteristics have been studied by various investigators in the recent time, e.g. Chaturvedi (38), Mehta (39,40), Ha Minh and Hebrad (41) and Graber (42). In elementary fluid mechanics, an expression for head loss in sudden expansion between sections 1 and 2. (see Fig. 8.31) is derived on the basis of the following assumptions : (i) flow is steady, (ii) velocity distribution at both the sections is uniform; (iii) pressure distribution at both the sections is hydrostatic; (iv) shear force on the walls is negligible; and (v) at both sections, turbulence is negligible. Use of momentum and energy equations then yields

$$\left.\begin{aligned} \Delta p &= \frac{\rho}{2}(U_1 - U_2)^2 \\ \Delta H &= \frac{U_1^2}{2g}\left(1 - \frac{A_1}{A_2}\right)^2 \\ \Delta p &= r\,\Delta H \end{aligned}\right\} \qquad \ldots (8.45)$$

where Δp is the equivalent energy loss.

Here U_1 and U_2 are the average velocities at sections 1 and 2, A_1 and A_2 are the areas there and ΔH is the energy loss or head loss between the two sections. This is found to be true for two dimensional expansions as well as sudden expansions in pipes. Abbott and Kline (43), while studying the subsonic flow over single and double backward steps, found that for B_2/B_1 less than 1.50, the separation bubbles on both the sides are symmetric and of equal length; however, for larger ratios unsymmetric flow is obtained. Here, B_1 and B_2 are widths of channel before and after expansion. Mehta (39, 40) obtained symmetric flow for B_2/B_1 equal to 1.25, but unsymmetric flow for B_2/B_1 values of 2.0, 2.5 and 3.0. Similar results were obtained by Filleti and Kays, and Fray et al. as reported by Graber (42). These experimental studies indicate that symmetric flow with standing eddies of equal length is obtained if B_2/B_1 is less than 1.5. It has been also reported that as long as the flow is fully turbulent there is no effect of Reynolds number on asymmetric expansion. Further, it has been reported that long and short eddies can exchange the positions if a disturbance is introduced. Graber (42) has given an explanation and criterion for the onset of instability, according to which the cause of this behaviour is the static instability of the system occurring for certain expansion ratios. The static stability of the system, in the case of the fluid jet (coming out from the unexpanded portion) and downstream zone of pressure recovery, refers to its ability to return to a position from which it has been deflected, once the cause of the deflection has been removed. If an infinitesimal disturbance results in a spontaneous change in the position of the system, the original system was one of unstable equlibrium. If a temporary cause of any magnitude results in no permanent deflection, the jet is in a stable equilibrium state and is considered to be statically stable.

The instability is considered in terms of two forces acting on the system when the jet undergoes a small deflection from its symmetric position as shown in Fig. 8.32. These are :

F_m, the reaction force in y direction resulting from change in momentum; and
F_p, the pressure force resulting from the difference in pressure on two sides of the jet caused by its deflection. For the shown position $p_{xb} > p_{xa}$; hence F_p will act in downward direction.

Thus, F_m is the stabilising force, whereas F_p will be the destabilising force for a small deflection y_o of the jet.

For stability $$\left(\frac{\partial F_m}{\partial y_o}\right)_{\substack{x_o \\ y = y_o}} \geq \left(\frac{\partial F_p}{y_o}\right)_{\substack{x_o \\ y = y_o}}$$

Evaluating the axial pressure gradients from Mehta's experiments for symmetric flows, the term $(\partial F_p/\partial y_o)$ was evaluated, while $\partial F_m/\partial y_o$ was evaluated using momentum equation and an assumed function for variation of p along x as obtained by Mehta. From such as analysis Graber found that at $B_2/B_1 = 1.509$, instability would occur, and at this value $x_o = 6.2\ (B_2 - B_1)$. This value of B_2/B_1 agrees well with the maximum value of B_2/B_1 above which asymmetric flow develops.

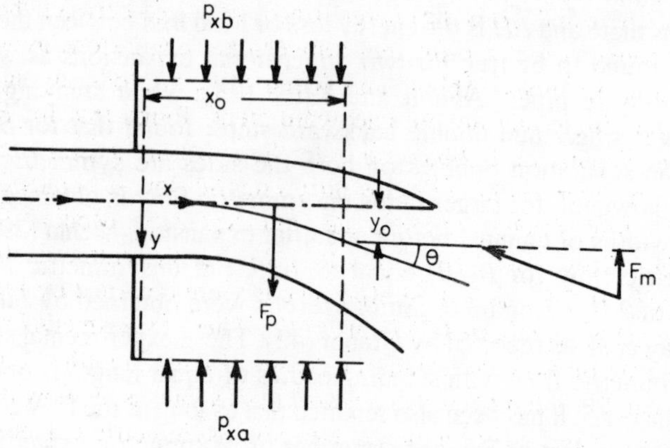

Fig. 8.32 Dynamics of a curved jet

Mehta (39,40), has conducted detailed investigations of flow in two dimensional conduit expansions with B_2/B_1 values of 1.25, 2.0, 2.5 and 3.0 using air as the fluid, and measured $\overline{u'^2}$, $\overline{v'^2}$, $\overline{u'v'}$, mean velocity distribution and the pressure distribution. The flow in expansion with expansion ratio 1.25 was symmetric, while that in other expansions was asymmetric. Figure 8.33 shows the mean curves for the lengths of short and small eddy as proposed by Abbott and Kline on which Mehta's data have been plotted along with subcritical open channel expansion data of Ashida, Garde et al. and Nashta and Garde. Here it can be seen that the lengths of short and small eddy can be satisfactorily predicted by Abbott and Kline's curves.

As a result of asymmetry of flow for large expansion ratios, the location of point of maximum velocity shifts from the centre line. This shift in the location of maximum mean velocity is higher for larger expansion ratios. The value of maximum mean velocity decreases very fast beyond the shorter eddy, and flow gets established in a length $2x/(B_2 - B_1)$ ranging from 28 to 18 for B_2/B_1 ranging from 1.25 to 3.00. This value is 28, 20 and 18 for B_2/B_1 value of 1.25 and 2.0 and 2.5 and above. This indicates that larger the expansion ratio, shorter is the magnitude of $2x/(B_2 - B_1)$ in which the flow gets established. Typical mean flow pattern in asymmetric condition is shown in Fig. 8.34.

Figure 8.35 shows typical distributions of $\sqrt{\overline{u'^2}}/U_1$ and $\sqrt{\overline{v'^2}}/U_1$ across the expanded portion for various values of $2x/(B_2 - B_1)$ for expansion ratio of 2.5. Maximum value of these parameters attained was about 0.20 in the region of high shear layer. One of the important feature of these fluctuations as observed by Mehta was the development of secondary peak values of turbulence intensities in the eddy pocket for measurements near the boundary for all expansion ratios except 1.25. This was attributed to a complex interaction between return flow in the eddy pocket with the turbulent boundary layer in

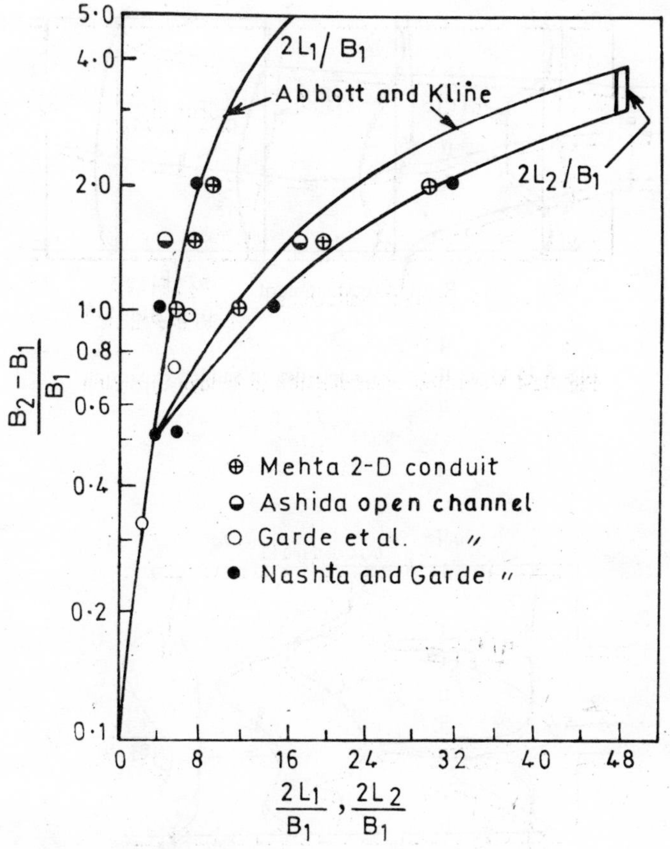

Fig. 8.33 Eddy lengths in 2-D expansions

the eddy and the shear layer. Experimental results indicated that the larger the expansion ratio, the earlier is the development of peak values of turbulence intensities and turbulent shear and greater is the rate of decay of turbulence. For larger expansion ratios, the boundary being farther away from the shear layer, turbulence can develop freely. Hence, for larger expansion ratios, greater is the rate of mixing and earlier is the establishment of uniformity after the reattachment points.

Mehta (39) also obtained numerical solution for velocity and pressure distribution using two equation, i.e. k–ε model as described in Chapter VI but using vorticity and stream function as two variables. His choice of length scale has already been shown in Fig. 6.3. The conformity between measured and calculated values of stream function ψ was reasonably good in the main flow, but $\psi = 0$ lines differed significantly for expansion ratio of 1.25. For larger expansion ratios the numerical solution differed significantly from measured flow pattern, since the former produced symmetrical flow.

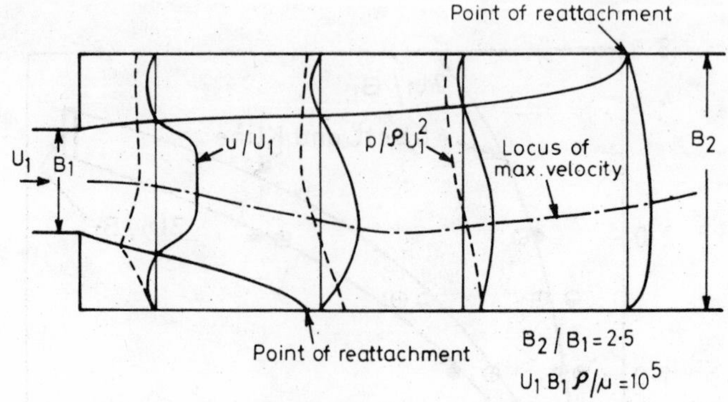

Fig. 8.34 Mean flow characteristics in sudden expansion

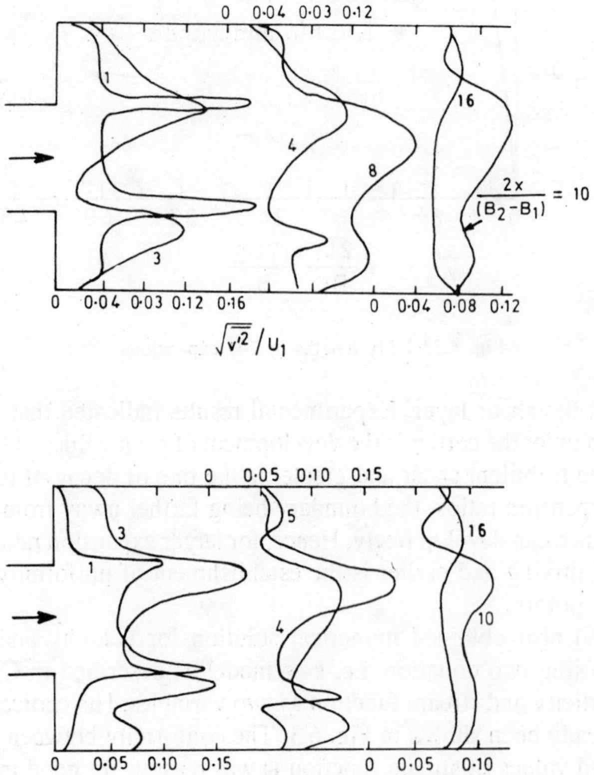

Fig. 8.35 Variation of $\sqrt{\overline{u'^2}}/U_1$ and $\sqrt{\overline{v'^2}}/U_1$ for sudden expansion

8.13 CONICAL DIFFUSERS

Kalinske (44) reported results of an extensive experimentation on diffusers in circular pipes with total expansion angles of 7.5°, 15°, 30° and 180° connecting 70 mm and 120.7 mm diameter pipes carrying water. The data collected included distribution of mean velocity and wall pressure, along with distribution of $\overline{u'^2}$ and $\overline{v'^2}$ along the diffuser length. Turbulence was measured by photographing the motion of 1 to 2 mm diameter droplets formed from a mixture of carbon tetrachloride and benzene in which powdered anthracite was suspended to give them milky appearance. Earlier studies by Nikuradse as well as those of Kalinske indicate that for expansions with total angle exceeding 5° to 7.5°, the flow separation occurs. Up to about 10° expansion angle, the energy loss is primarily due to friction and remaining due to form loss. Beyond this total expansion angle, friction loss along the boundary is negligible whereas form loss accounts for the major loss. In discussion on this paper, Bakhmeteff referred probably to one of the earliest references of Andres to cite experimental evidence to the effect that condition of incoming flow has much to do with the onset and degree of separation. He found that one of the best ways to counteract the pressure loss in a diffuser was to impose rotary motion on the expanding jet.

Chaturvedi (38) has studied characteristics of flow in axisymmetric expansions of total angle 30°, 60°, 90° and 180° using 108 mm and 216 mm dia pipes with air as the fluid. Independent head loss measurements were carried out in water pipe assembly using 50°, 70° and 180° expansion angles. Making use of detailed data concerning mean velocity, pressure, turbulent fluctuations and turbulent shear measurements, the dynamics of the flow was studied using momentum and energy equations.

One of the striking features of sudden expansions in pipes (expansion angle of 180°) is that the flow is always symmetrical, unlike in the case of two dimensional sudden expansions. This is because the separation zone is interconnected.

Figure 8.36 shows typical distribution of mean pressure, velocity and turbulent fluctuations in a 45° diffuser. Studies of such data for other diffuser angles revealed that

(i) mean velocity varies intensely both axially and radially in such a manner that conditions of near constancy of velocity across the section tend to be achieved rapidly;
(ii) the variation of pressure is not directly related to the variation of velocity, and even though there is appreciable variation in the axial direction, that in the radial direction is small;
(iii) pressure within the standing eddy is almost the same as outside it;
(iv) turbulent field is highly non-homogeneous; and
(v) as the boundary divergence increases from 30° to 60°, there is a marked change in flow pattern; however, flow patterns for 60°, 90°, and 180° are similar.

Chaturvedi used the integral forms of momentum and energy equations for turbulent flow and evaluated various terms in order to check the balance.

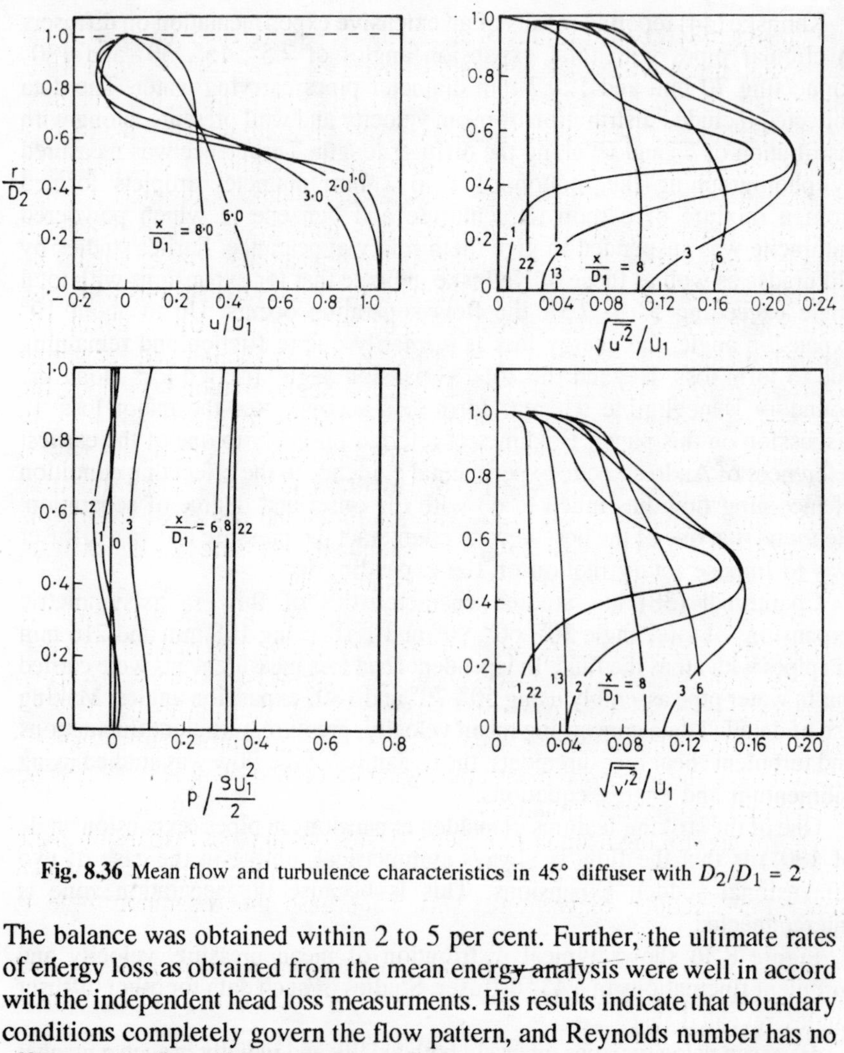

Fig. 8.36 Mean flow and turbulence characteristics in 45° diffuser with $D_2/D_1 = 2$

The balance was obtained within 2 to 5 per cent. Further, the ultimate rates of energy loss as obtained from the mean energy analysis were well in accord with the independent head loss measurments. His results indicate that boundary conditions completely govern the flow pattern, and Reynolds number has no effect on the flow if it is large.

Turbulence is produced at the maximum rate immediately at the downstream of expansion and its intensity is nearly the same for all diffuser angles; $\sqrt{\overline{u'^2}}/U_1 = 0.20$ and $\sqrt{\overline{v'^2}}/U_1 = 0.15$ are typical maximum values. The diffusion of turbulence is inhibited by boundary proximity upto diffuser angle of 60°, beyond which the effect of boundary proximity is negligible. The diffusion and convection of turbulence in the downstream direction make the velocity distribution more and more uniform. Because the boundary proximity has no effect on flow for angles exceeding 60°, Chaturvedi's experiments indicated that dimensionless energy loss $\Delta H / \frac{U_1^2}{2g}$ increased with diffuser angle upto 60° and beyond this value it remained nearly constant, and could be predicted by $\Delta H = \left((1 - (\frac{D_1}{D_2})^2\right)^2 \frac{U_1^2}{2g}$

Point of reattachment or standing eddy length varied between 3.3 D_2 to 4.6 D_2 for diffuser angles varying from 30° to 180°, and, hence, on the average it can be taken as 4 D_2.

Ha Minh and Hebrad (41) have experimentally studied the mean flow characteristics as well as turbulence characteristics in diffusers of D_2/D_1 ratio 2.0 and 2.67, and applied one and two equation models to predict the velocity distributions not only for his data but also for data collected by Chaturvedi. Figure 6.4 shows the comparison between observed velocity distribution in sudden expansion in a pipe, and computed distributions using one and two equations models. It can be seen that two equation model gives better results than one equation model. Similar comparisons were also made for distribution of turbulent shear $\overline{u'v'}$ and mean kinetic energy of turbulence, which were quite satisfactory.

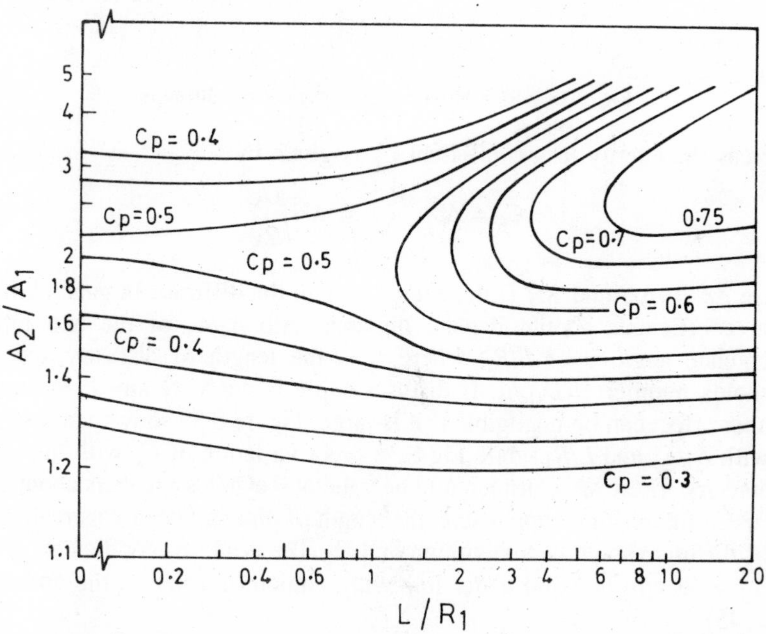

Fig. 8.37 Variation of C_P with A_2/A_1 and L/R_1 for conical diffusers

The diffuser performance is usually studied in terms of two parameters, namely, pressure recovery coefficient C_P and energy loss coefficient C_L. The pressure recovery coefficient is defined as

$$C_p = (p_2 - p_1)/ \frac{\rho U_1^2}{2}$$

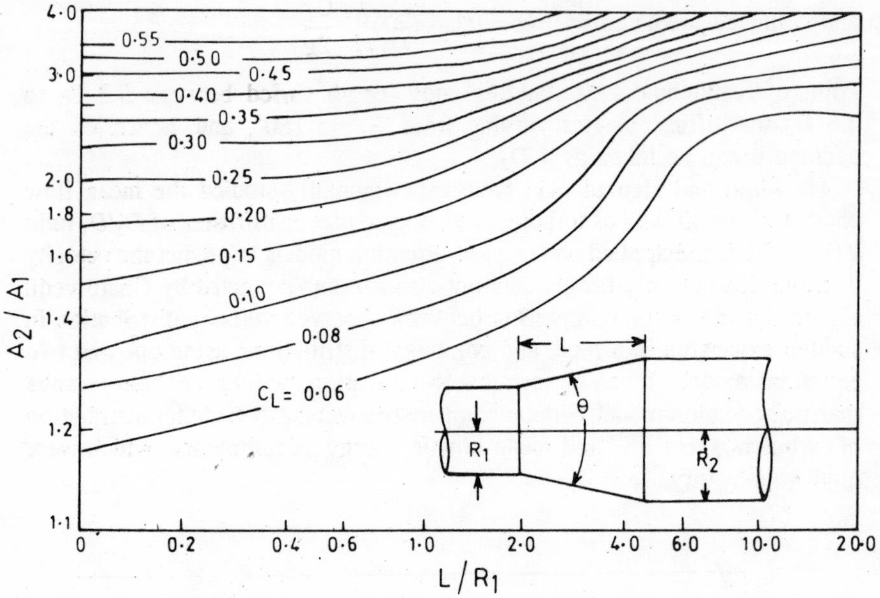

Fig. 8.38 Loss coefficients for axisymmetric diffusers

whereas the energy loss coefficient C_L is given by

$$C_L = \Delta p \Big/ \frac{\rho U_1^2}{2} = \Delta H \Big/ \frac{U_1^2}{2g}$$

where $\Delta p = \gamma \Delta H$ and ΔH is the energy loss in the diffuser. In general, both these coefficients should depend on area ratio A_2/A_1 of the two pipes, divergence angle θ or L/R_1, where L is the length of diffuser, and the Reynolds number $U_1 D_1 \rho/\mu$. If diffuser angle is not very small, Reynolds number effect can be negligible if it is large. Figure 8.37 shows variation of C_p with A_2/A_1 and L/R_1 while Fig 8.38 gives variation of C_L with the same parameters. These are considered to be valid at Reynolds numbers about 10^6. For these figures to be applicable, the length of pipe on the downstream side of the diffuser should be a minimum of 4 D_1. The head loss coefficients given in Fig. 8.38 will be valid under following nonuniformities on the upstream side (45):

1. Diffuser located one inlet diameter or more after a bend of centreline radius to pipe diameter ratio greater than unity;
2. Diffuser located three or more diameters after a component having a loss coefficient of less than inlet mean velocity head; and
3. For diffusers located four or more diameters after any component.

Graber (42) and Miller (45) have discussed various methods for improving the performance of two dimensional sudden expansions and diffusers in rectangular ducts, respectively. Methods suggested by Graber tend to make the flow symmetrical. The first method is to equalise the pressures in the two standing eddies; this can be done by either internally or externally connecting

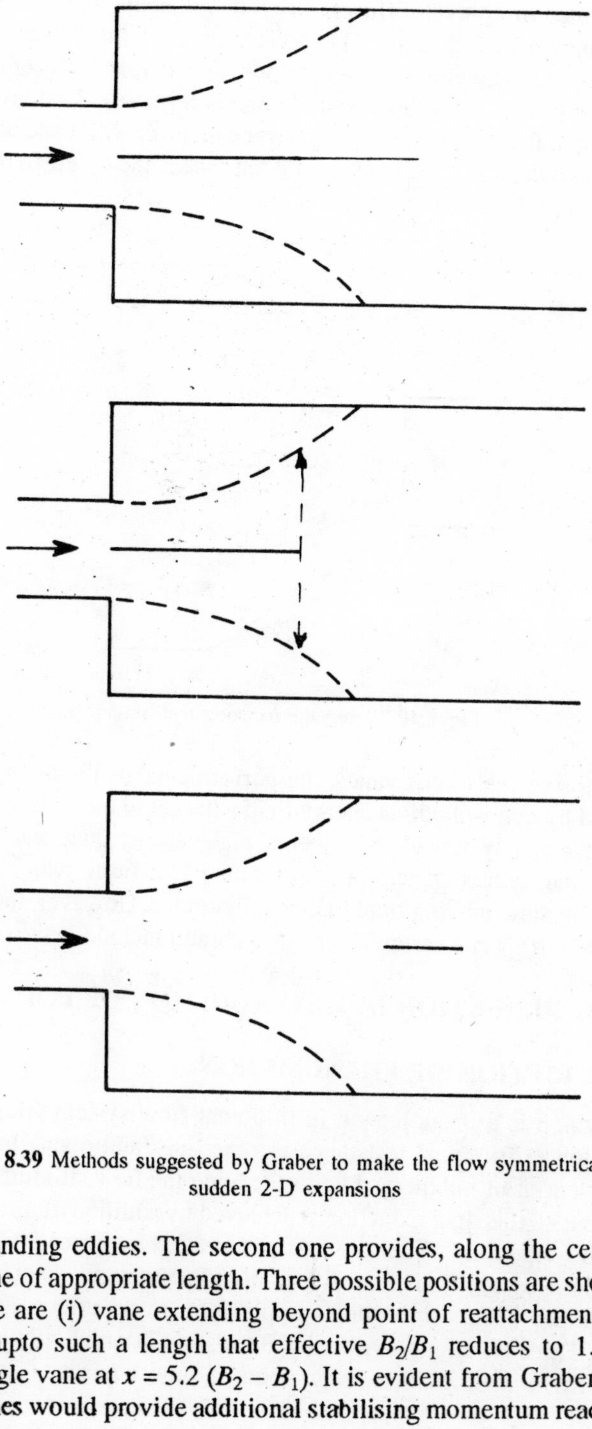

Fig. 8.39 Methods suggested by Graber to make the flow symmetrical in sudden 2-D expansions

the two standing eddies. The second one provides, along the centre line, a straight vane of appropriate length. Three possible positions are shown in Fig. 8.39. These are (i) vane extending beyond point of reattachment; (ii) vane extending upto such a length that effective B_2/B_1 reduces to 1.5 and (iii) placing single vane at $x = 5.2\ (B_2 - B_1)$. It is evident from Graber's analysis that the vanes would provide additional stabilising momentum reaction force.

Miller (45) has given dimensions of vanes and their location for improving the performance of conical diffusers in rectangular ducts. Basic three vane diffuser is shown in Fig. 8.40. The performance of three vane diffuser is satisfactory upto total diffuser angle upto 50°. A two vane diffuser is the same as that shown in Fig. 8.40 when central vane is removed. Two vane diffuser works well upto θ equal to 40°. A single vane diffuser with vane at the centre does not reduce the loss coefficient much but makes the downstream velocity distribution slightly more uniform.

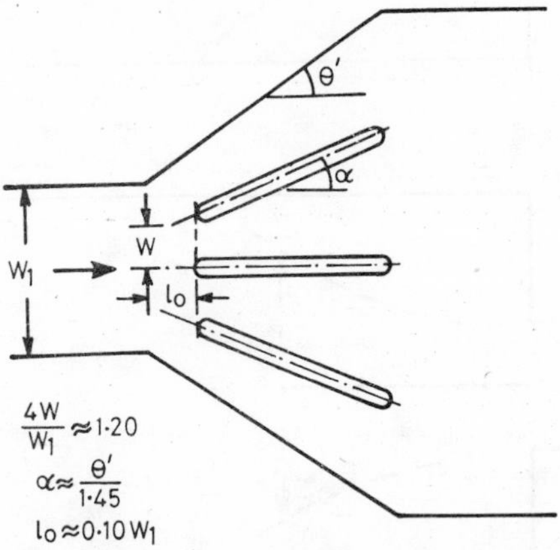

Fig. 8.40 Three vane rectangular diffuser

In addition to the use of vanes, the performance of the diffuser can also be improved by removal of low energy fluid adjacent to the walls by energising the flow in the vicinty of walls, by injecting high velocity fluid, and by providing roughness and vortex generators upstream of diffuser which would add momentum to slow moving fluid in the wall regions. However, these methods are used only in special cases. The most common method is the use of vanes.

DRAG REDUCTION DUE TO ADDITION OF POLYMERS

8.14 DESCRIPTION OF PHENOMENON

Drag reduction is a phenomenon in turbulent flow system which results in the reduction in frictional resistance in pipe lines and open channels, or in drag experienced by submerged bodies leaving the flow still turbulent. In the present context, this drag reduction is achieved by addition of a small quantity of polymer to the fluid. After the first observation by Toms (46) in 1948, on drag reduction by addition of small quantity of polymer to water, considerable experience has been gained in its application to a variety of engineering problems. Three excellent reviews by Berman (47), and Sellin et al. (48, 49) give the present state of knowledge about the mechanism of drag reduction and its applications. This section is primarily based on the information in these reviews. These reviews also give very extensive lists of references.

Polymers are a large assembly of identical subunits linked by covalent bonds to make a single large molecule. There are three types of polymers, namely, linear polymers, branched polymers, and copolymers; their structures are shown below :

Linear : $A' - A - A - A - A.......A''$

Branched : $A' - A - A''' - A - A.......A''$ (with a side chain from A''': $A - A - A'''$)

Copolymers : $A' - A - B - B - A - A - B.......A''$

Where A, A', A'', A''' and B are organic units e.g. (C_2H_4O), $(C_2\,H_5\,O)$, etc. Experiments have shown that polymers with molecular weight exceeding 10^6, highly soluble in water, and those with linear structure are most effective in drag reduction. Some of the polymers which find wide use as drag reducing agents include polyethlyene oxide (e.g. polyox), polyacrylamide—polyacrylic acid, copolymers (e.g. separan), carboxymethyl cellulose, and polystyrene, all synthetic, and other natural plant extracts, such as guar gum. Some of them are soluble in water while others are soluble in organic solvents. Polymers have also been used to reduce drag in the transport of crude oils and kerosene. For this purpose organosilicon polymers, such as polydimethylsil methylene and polydimethylsiltrimethylene, have been used in Russia for transport of kerosene and α-olefin polymer in kerosene solvent for oil extraction. In general, synthetic polymers, with higher molecular weights, are used at concentrations in the range of 5 to 100 ppmw (parts per million by weight) whereas natural polymers, which have smaller molecular weights, are used in the concentration range of 1000 to 5000 ppmw. Synthetic polymers degrade more easily while the latter are more stable. It has been found that inert fibres in suspension also produce drag reduction; hence, it seems clear that polymers act with some kind of 'mechanical effect', i.e. chemical composition of a polymer may be of secondary importance, with the general requirement of linear structure.

Figure 8.41 shows effect of addition of polymer on the Darcy-Weisbach friction factor f, defined as $f = \dfrac{h_f}{L}\dfrac{2gD}{U_m^2}$ for smooth pipes. Here U_m is the average velocity of flow in a pipe of diameter D and length L, and h_f is the head loss in this length. For a given concentration of known polymer, drag or friction reduction starts if flow is turbulent, and Reynolds number (computed using water viscosity) is above a certain minimum. This point on the Karman–Prandtl line on friction factor diagram is called drag reduction onset Reynolds number and is shown as A in Fig. 8.41. Use of water viscosity in the calculation

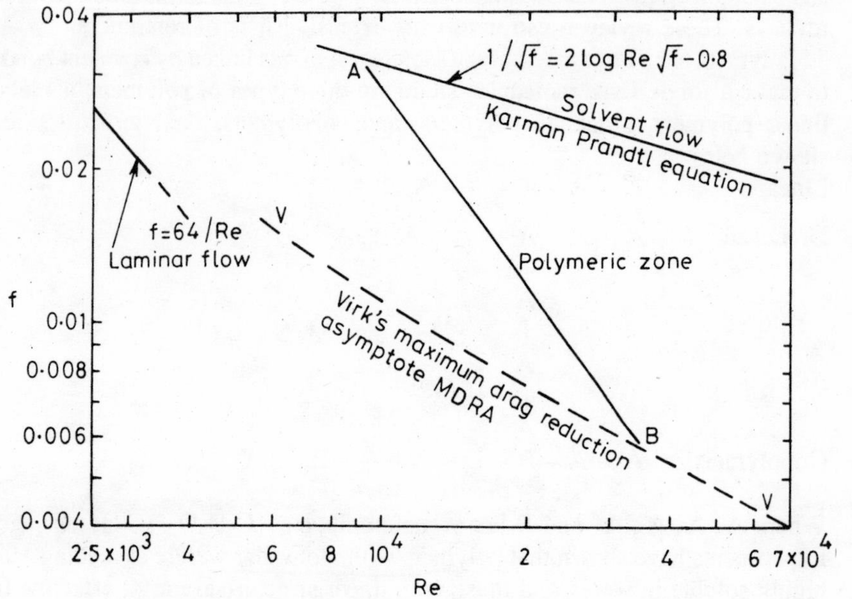

Fig. 8.41 Effect of polymer addition on *f-Re* relation for smooth pipes

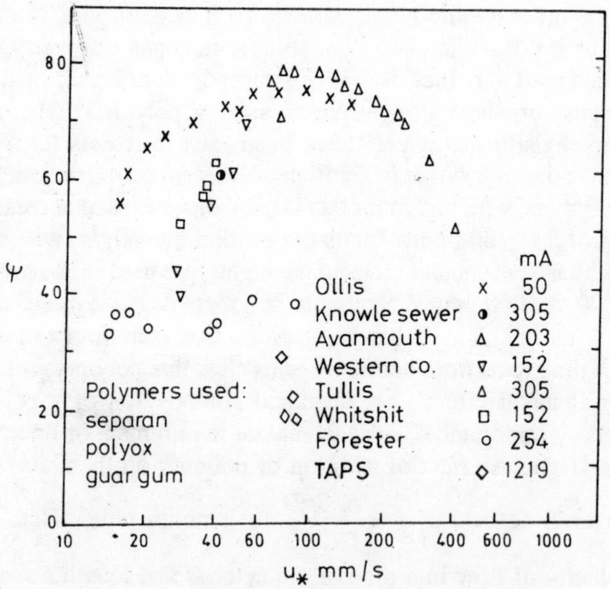

Fig. 8.42 Variation of ψ with u_* for large diameter pipes (48, 49)

of Reynolds number is justified, because addition of small amount of polymer causes negligible change in the viscosity of water. Higher the concentration, slightly smaller is the Reynolds number at which onset of drag reduction occurs.

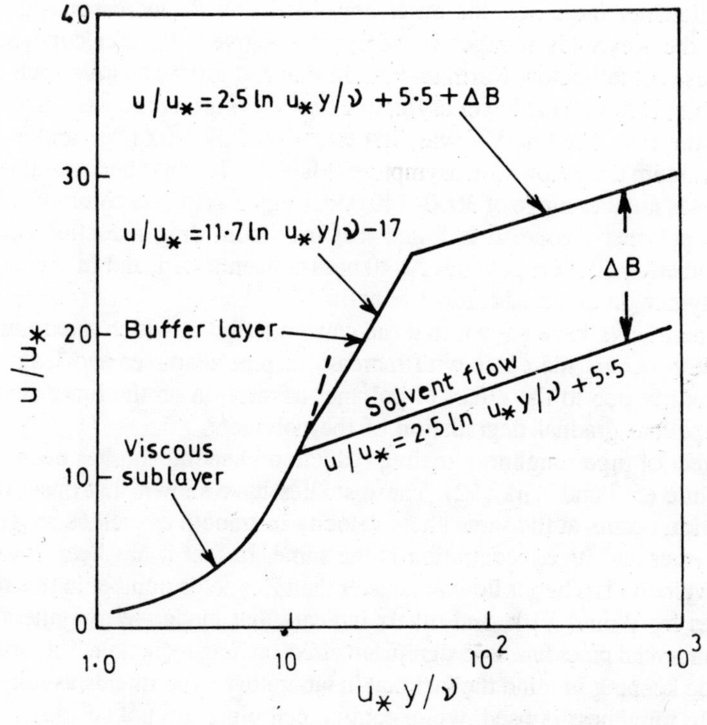

Fig. 8.43 Velocity distribution in pipes in the presence of polymers

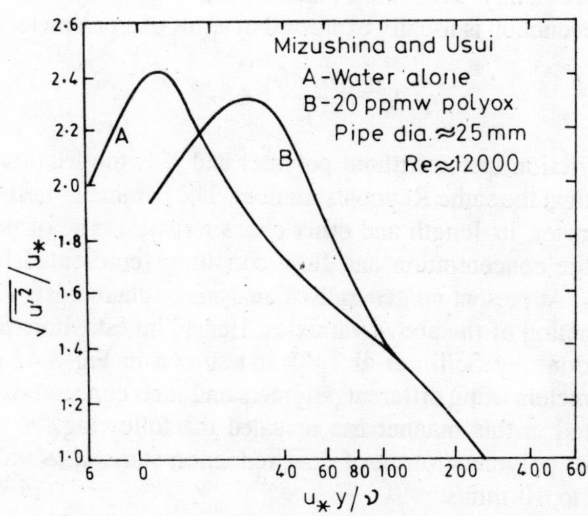

Fig. 8.44 Effect of polymer on turbulence intensity variation (47)

For very high concentration of the polymer, onset drag reduction occurs as soon as flow becomes turbulent. For a given polymer concentration, as the pipe diameter increases, the onset drag reduction *Re* increases slightly.

As the Reynolds number is increased for given polymer concentration, f values will fall below Karman–Prandtl line and follow a curve such as AB. The point B lies on the lower asymptote line VV. Further increase in *Re* would trace line BV. The line VV was first established by Virk (50), and is known as maximum drag reduction asymptote (MDRA). This has been obtained over Reynolds number range of 3000–140,000. Region A to B is commonly known as the polymeric zone. It is found that maximum drag reduction occurs in all synthetic polymers at about 20-40 ppm concentration, and further addition of polymer gives no additional benefits.

Experiments have shown that the concentration at which maximum drag reduction occurs increases with increase in pipe diameter and length. This is primarily due to the effect of polymer adsorption on the inner surface of the pipe and gradual degradation of the polymer.

Effect of pipe roughness in drag reduction phenomenon has been studied by White (51) and Virk (52). These studies have shown that onset of drag reduction occurs at the same shear velocity in smooth as well as rough pipes if polymer and its concentration is the same. In fact it has been found that shear velocity is a better flow parameter than Reynolds number in this respect. Studies by White, Virk, and others indicate that moderate roughness found in commercial pipes has little significant effect on drag reduction. This statement is made keeping in mind the fact that in laboratory experiments, usually larger relative roughness is used, while commercial pipes are not so rough. It may also be mentioned that similar drag reduction results have been obtained in noncircular conduits; but local effects have been sometimes observed due to changes in secondary circulation pattern (48).

The drag reduction is usually expressed in terms of a parameter ψ defined as

$$\psi = \left(\frac{f_s - f_P}{f_s}\right) \times 100$$

where f_s is friction factor without polymer and f_p is the friction factor with polymer, both at the same Reynolds number. The parameter ψ should depend on pipe diameter, its length and other characteristics, type of polymer used along with the concentration and flow condition represented by Reynolds number or u_*. At present no generalised analysis or chart exists for prediction of ψ as a function of the above variables. Hence, investigators prefer to plot ψ vs u_* as done by Sellin et al. (49) and shown in Fig 8.42 for pipes of different diameters using different polymers and their concentration. Analysis of data plotted in this manner has revealed the following:

(1) For most polymers, onset of drag reduction starts at u_* values ranging from 5 to 10 mm/s.

(2) For a given concentration of a specific polymer the drag reduction is a unique function of wall shear stress.

(3) For a given polymer and pipe size, ψ increases as u_* increases, until a maximum value of ψ is obtained at a certain value of u_* which depends on the type of polymer used. Beyond this value of u_*, ψ decreases gradually with increase in u_* which is due to degradation of the polymer. For most synthetic polymers, this takes place at u_* values exceeding 70–80 mm/s.

In the absence of a generalised criterion for prediction of ψ for known conditions, the question of scaling up of drag reduction data from laboratory experiments to a larger scale assumes importance. This has been studied by Savins and Seyer (53) and Granville (54). Savins and Seyer suggest that, using a given concentration of a given polymer, small scale experiments be conducted covering a range of u_* values expected to be observed in the prototype, and ψ vs u_* curve prepared. This curve can be used for predicting ψ for larger diameter pipe for known u_* and fixed concentration. Granville has shown that, for a given concentration of a known polymer in two pipes of diameters D_1 and D_2 having Reynolds numbers Re_1 and Re_2 and friction factors f_{p1} and f_{p2}, the following relation holds

$$R_{e2}\sqrt{f_{p2}}/R_{e1}\sqrt{f_{p1}} = D_2/D_1$$

Hence, knowing R_{e1}, f_{p1}, D_1, D_2 and R_{e2}, f_{p2} can be predicted. Knowing f_{p2}, ψ for the prototype can be determined. It may be mentioned that this relationship needs further confirmation.

8.15 VELOCITY DISTRIBUTION AND STRUCTURE OF TURBULENCE

Virk (55) has suggested a three layer model for describing the velocity distribution for turbulent flow in pipes in the presence of polymers. The three layers are the viscous sublayer, elastic intermediate layer (often called buffer layer) where the mixing length is derived from the maximum drag reduction asymptote, and the outer or the core region in which Karman constant of pure solvent, i.e. 0.4 is valid (see Fig 8.43). The buffer layer velocity distribution has been obtained as

$$\frac{\bar{u}}{u_*} = 11.7 \, ln \, \frac{u_* y}{\nu} - 17 \tag{8.46}$$

while in the core region it is given by

$$\frac{\bar{u}}{u_*} = 2.5 \, ln \, \frac{u_* y}{\nu} + 5.5 + \Delta B \tag{8.47}$$

where ΔB is the displacement from Newtonian profile. If the core region dominates, integrating the above equation over the pipe radius and using the subscript s to denote pure solvent, one gets

$$\frac{1}{\sqrt{f_p}} = 2 \log Re\sqrt{f_s} - 0.80 + 8^{-1/2} \Delta B \tag{8.48}$$

But $$2 \log Re\sqrt{f_s} - 0.8 = \frac{1}{\sqrt{f_s}}$$

Hence, $$\frac{1}{\sqrt{f_p}} - \frac{1}{\sqrt{f_s}} = 8^{-1/2}\,\Delta B \quad (8.49)$$

Available experimental data give maximum observed ΔB in the range of 10 to 26 at Re in the range of 44,000 to 65,000 and at concentration of 10 ppmw. Several studies, both experimental as well as theoretical, indicate the dependence of ΔB on polymer concentration, its unit weight and u_{*p}/u_{*s}.

Based on theoretical and experimental studies, the following broad description of polymer action can be given. Rigid structures are not effective drag reducers. Polymer molecules are effective only when they are elongated. Polymer molecules become extended in the core of the flow where near irrotational flow exists. Here coiled molecules get stretched. In the laminar sublayer, the rotation and elongation rates are nearly equal and, hence, little stretching takes place here. Of course, the onset of stretching must occur in the buffer layer just outside the laminar sublayer as shown in many experiments. Extension of molecules in the turbulent flow is very complex and is related to the statistics of vorticity, strain rate and time scales of turbulence. The strain rate would be responsible for initiating molecular stretching and alignment, while the persistence time or time scale would control the amount of stretching.

Limited data on the variation of longitudinal turbulence intensity indicate that in the presence of polymer,

(i) close to the wall, the turbulence intensity is lower for polymer solution compared to pure solvent;

(ii) as the buffer region is increased, the peak in turbulence intensity is shifted to higher $u_* y/\nu$ values; and

(iii) little change is observed at the centre of the pipe. Figure 8.44 shows typical results of variation of turbulence intensity as obtained by Mizushina and Usui (47*). According to Lumley (47*), the turbulent production and dissipation are maximum in the buffer region. The addition of polymer expands the buffer layer and shifts the location of greatest production and dissipation closer to the pipe centre as the drag reduction is increased. Danohue (47*) found that the bursting period T_B did not change when polymer was added to a solvent keeping u_* the same. This suggests that either $T_B u_*^2/\nu$ = constant, or $T_B u_*/D$ is constant. In the spanwise direction, the mean spacing between 'coherent structures' is increased in the presence of polymers. Thus, the indication is that the three dimensional structure of the flow is altered in the presence of polymers. Berman and Cooper (55) found that a periodic disturbance in pure solvent rapidly decayed downstream into random fluctuations while dilute polymer solution retained the fundamental frequency much farther downstream. In drag reducing flow small scales are found to be suppressed while large ones are enhanced.

8.16 DEGRADATION OF POLYMERS

The term degradation is used to describe loss of drag reduction effectiveness and other polymer properties. It is found that simple shear flows cause the degradation of dilute polyethylene oxide solutions. It seems that this is due to the fact that elongation and alignment of polymer molecules in extreme cases could lead to cission and distangling of molecular agglomerates. Higher the molecular weight of the polymer, more succeptible it is for early degradation and vice versa. Passing of polymer solution through pump causes it to degrade because of intense shear and high turbulence intensity. In a single pass through pump, dilute polymer solution (concentration less than 5 ppmw) may lose 10 per cent of its effectiveness, and, hence, to compensate for this degradation, somewhat higher comcentration needs to be used. Experiments have indicated that in pipes negligible degradation would occur if shear velocity is less than 50 mm/s, while above 70–80 mm/s it would be significant. It may also be mentioned that part of the effectiveness of a polymer is reduced because of its adsorption on the pipe surface.

In industrial applications, e.g. sewers and pipelines, mean velocities range from 0.5 to 2.0 m/s, and corresponding shear velocities range from 20 to 80 mm/s. In this range, degradation of polymer is not a serious problem.

8.17 APPLICATIONS (49)

The applications of polymers in engineering problems as drag reducing agents are numerous. Some have been on prototype scale, while a few are at laboratory scales. The economic benefit of polymer use in these processes lies in the significant saving of energy or in increase in capacity of system for carrying water. These applications include : sewers, open channels, hydropower penstocks, culverts, fluid machinery, fire fighting, marine applications, oil pipelines and oil well fracturing.

In Milwaukee and Dallas, USA, 1067 mm and 610 mm diameter sewers were tested using polyethylene oxide with 10 to 60 ppmw concentration. With this, 30 to 120 per cent increase in velocity has been observed. Addition of polymer can be resorted to if the sewer is occasionally overloaded. In short penstocks with length/diameter ratio less than 100, polymers have been used effectively to reduce the head loss. In sprinkler irrigation 100 ppmw of polyox FRA increased waterspread from 42 m^2 to 129 m^2 of crop area. It was also found that polymer also improved the soil structure, thereby increasing infiltration and percolation.

Polymers have been used for hydraulic fracturing operations in oil wells. The reduction of friction in small bore down pipe enabled the process to be carried out in wells which were considered too deep previously. Polymers have been used in pipelines carrying crude oil, kerosene and other oils under turbulent flow condition. In investigations in Russia, it has been found that certain orgnosilicon polymers, such as polydimethylsilmethylene and polydimethylsiltrimethylene, are quite effective in kerosene. Field tests in 343 mm and 1194 mm diameter pipe lines of Trans-Alaska pipeline have shown significant reduction in friction. Similar studies on laboratory and pilot scale pipe lines carrying slurries have shown drag reduction to the tune of 10 to 15 per cent.

Some experiments conducted on discs rotating in a stationary fluid have shown drag reduction in the presence of polymers. Improvement in the pump efficiency, delivery pressure and flow output have been observed when polymers are added to the flow entering the centrifugal pumps. However, controversial results have been obtained on the performance of propellers in the presence of polymers and, hence, more work needs to be done.

Another interesting application of polymers is about ships and motion of submerged bodies. Experiments were conducted by Chanham et al. on a 42.7 m long ship HMS Highburton by injecting polymer from two slots on the front portion. Drag was found to reduce by 12.7 per cent when polyethylene oxide concentration in the boundary layer was estimated as 10 ppmw. Pressure distribution and turbulence around submerged bodies towed in fluid is found to change significantly, thereby reducing the drag. This has important bearing on the performance of submarines, torpedoes, etc. Addition of polymers also improves the performance of nozzles which can be fruitfully used in fire fighting applications.

It can, thus, be seen that basic experimental work has been done to prove the benefits resulting from addition of polymers in a variety of applications of relevance to hydraulic engineer. Information is also available on method of injection, amount needed etc. (48,49). Hence, concerted effort needs to be made to popularise their use to effect economy.

REFERENCES

1. Ferri, A.et al. (Ed.) *Progress in Aeronautical Sciences*, Vol. 2, Boundary Layer Problems. Pergamon Press, 1962.
2. Kline, S.J., W.C. Reynolds, F.A. Schraub and P.W. Runstadler. The Structure of Turbulent Boundary Layer. JFM, Vol. 30, 1967.
3. Kline, S.J. and P.W. Runstadler. Some Preliminary Results of Visual Studies on the Flow Model of the Wall Layers of the Turbulent Boundary Layer. Jour. Appl. Mechanics, Vol. 26, 1959.
4. Kovasznay, L.S.G. The Turbulent Boundary Layer. Annual Review of Fluid Mechanics, Vol. 2, 1970.
5. Einstein, H.A. and H.Li. The Viscous Sublayer Along a Smooth Boundary. JEM, Proc. ASCE, Vol. 82, Paper 945, April 1956.
6. Clauser, F.H. *The Turbulent Boundary Layer*. Advances in Applied Mechanics, Vol. 4, Academic Press, 1956.
7. Schubauer, G.B. and C.M. Tchen. *Turbulent Flow*. Princeton University Press, P.U.P. Paperbacks No. 9, 1961.
8. Coles, D. The Law of the Wake in Turbulent Boundary Layer, JFM, Vol.1, Pt. 2, 1956.
9. Schlichting, H. *Boundary Layer Theory*. McGraw Hill Book Co. Inc., 1st Edition, Chapters 21 and 22, 1955.
10. Townsend, A.A. *The Structure of Turbulent Shear Flow*. Cambridge University Press, 1956.
11. Clauser, F.H. Turbulent Boundary Layers in Adverse Pressure Gradients. Jour. Aero. Sci. Vol. 21, 1954.
12. Smagorinsky, J. General Circulation Experiments with the Primitive Equations. Monthly Weather Reviews, Vol. 91, No.3, 1963.

13. Sethuraman, S., P. Michael, W.A. Tuthill and J. McNeil. Atmospheric Boundary Layer Measurement During Summer Monex 79 at Digha, India, Brook Haven National Laboratory, USA, 1979.
14. Plate, E.J. *Aerodynamic Characteristics of Atmospheric Boundary Layers*. AEC Critical Review Series. Atomic Energy Commission, USA, 1971.
15. Monin, A.S. The Atmospheric Boundary Layer. Annual Review of Fluid Mechanics, Vol. 2, 1970.
16. Devenport, A.G. The Relationship of Wind Structures to Wind Loading. Symposium No. 16, Wind Effects on Buildings and Structures, London 1965.
17. Van der Hoven, I. Power Spectrum of Horizontal Wind Speed in the Frequency Range 0.0007 to 900 Cycles per Hour. Jour. of Meteorology, Vol. 14, 1957.
18. Cermak, J.E. Wind Tunnel Testing of Structures. ASCE-EMD Speciality Conference, Univ. of California (CEP 75-76 JEC 15), March 1976.
19. Cermak, J.E. Laboratory Simulation of the Atmospheric Boundary Layer. AJAA Journal, Vol. 9. No. 9, Sept. 1971.
20. Cermak, J.E., V.A. Sandborn, E.J. Plate, G.H. Binder, H.Chung, R.W. Meroney and S. Ito-Simulation of Atmospheric Motion by Wind Tunnel Flows. Fluid Mechanics Programme, CSU, CER-66-JEC-VAS-EJP-GHB-HC-RNM-SI, Colorado State University, USA, 1966.
21. Nemoto, S. Similarity Between Natural Load Wind in the Atmosphere and Model Wind in the Wind Tunnel. Papers, Meteorological Geophysics. Vol. 19, 1968.
22. Arie, M. and H. Rouse. Experiments on Two Dimensional Flow Over a Normal Wall. JFM, Vol. 1, Pt. 2, 1956.
23. Rouse, H. Energy Transformation in the Zone of Separation. Proc. of 9th Congress of IAHR, Dubrovnik, 1961.
24. Ranga Raju, K.G. and R.J. Garde. Resistance of an Inclined Flat Plate Placed on a Plane in Two Dimensional Flows. Trans ASME, Jour. of Basic Engineering, Series D., Vol. 92, 1970.
25. Naghbushanaiah H.S. Separation of Flow Downstream of a Plate Set Normal to a Plane Boundary. Ph.D. Thesis, Colorado State University, Fort Collins (USA), 1962.
26. Plate, E.J. The Drag on a Smooth Flat Plate with a Fence Immersed in its Turbulent Boundary Layer. Proc. of the 1964 ASME Fluids Engineering Conference, Philadelphia, USA.
27. Ranga Raju, K.G., J. Loesser and E.J. Plate. Velocity Profile and Fence Drag for a Turbulent Boundary Layer Along Smooth and Rough Plates. JFM, Vol. 76, 1976.
28. Pande, P.K., R. Prakash and M.L. Agarwal. Flow Past Fences in Turbulent Boundary Layer, JHD., Proc.ASCE, Hy-1, Vol.106, 1980.
29. Moss, W.D., Baker, S. and L.J.S. Bradbury. Measurement of Velocity and Reynolds Stress in Some Region of Recirculating Flow. First International Symposium, Pennsylvania University, USA, Apr. 1977.
30. Gupta, V.P. Experimental Study of Disturbed Turbulent Boundary Layer Flow. Ph.D. Thesis, University of Roorkee, India, 1980.
31. Raine, J.K. and D.C. Stevenson. Wind Protection by Model Fences in a Simulated Atmospheric Boundary Layer. Jour. Ind. Aerodynamics, Vol.2, 1977.
32. Grandmer, J. Wind Shelters. Jour. Wind Engg. and Ind. Aerodynamics, Vol. 4, 1979.
33. Counihan, J., J.C.R. Hunt and P.S. Jackson. Wakes Behind Two Dimensional Surface Obstacles. JFM, Vol. 64, 1974.
34. Bradley, B.F. and P.J. Mulhearn. Development of Velocity and Shear Stress Distributions in the Wake of Porous Shelter Fence. Jour. Wind Engg. and Ind. Aerodynamics. Vol. 15, 1983.
35. Singh, S.K., Study of Turbulence in the Boundary Layer Flow Past Porous Fences. M.E. Thesis, Civil Engg. Dept. University of Roorkee, India, 1987.
36. Gupta, V.P. and K.G. Ranga Raju. Separated Flow in Lee of Solid and Porous Fences. JHE Proc. ASCE, Vol. 113, No. 10 Oct. 1987.
37. Gupta, V.P. and K.G. Ranga Raju. Mean Flow Characteristics of Redeveloping Flows. JHE Proc.ASCE. Vol. 113, No. 9, Sept. 1987.

38. Chaturvedi, M.C. Flow Characteristics of Axisymmetric Expansions JHD, Proc. ASCE. Vol.83, no. Hy-3, May 1963.
39. Mehta, P.R. Flow Characteristics in Two—Dimensional Expansions. JHD, Proc. ASCE, Vol. 105, No. Hy-5, May, 1979.
40. Mehta, P.R. Separated Flow Through Large Sudden Expansions JHD, Proc. ASCE. Vol. 107, No. HY-4, April 1981.
41. Ha Minh, H. and P. Hebrad. Etude Theorique et Experimentale d'un E' coulement De'colle' en Aval d' un E' largissement Brusque de Revolution. JHR, IAHR, Vol. 15, No. 1, 1977.
42. Graber, S.D. Asymmetric Flow in Symmetric Expansions. JHD, Proc. ASCE. Vol. 108, No. HY-10, Oct. 1982.
43. Abbott, D.E. and S.J. Kline. Experimental Investigation of Subsonic Turbulent Flow Over Single and Double Backward Facing Steps. Jour. of Basic Engg, Trans. ASME, Vol. 84, Series D. Sept. 1962.
44. Kalinske, A.A. Conversion of Kinetic Energy to Potential Energy in Flow Expansions. Trans. ASCE, Vol. 111, 1946.
45. Miller, D.S. *Internal Flow Systems*. Vol. 5 in the BHRA Fluid Engineering Series, U.K. 1978.
46. Toms, B.A. Some Observations on the Flow of Linear Polymer Solutions through Straight Tubes at Large Reynolds Numbers. Proc. of 1st Int. Congress on Rheology Vol. 2 1948.
47. Berman, N.S. Drag Reduction by Polymers. Annual Review of Fluid Mechanics. Vol. 10, 1978
48. Sellin, R.H.J., J.W. Hoyt and O. Serivener. The Effect of Drag Reducing Additives on Fluid Flow and Their Industrial Applications: Part 1. Basic Aspects, JHR, IAHR, Vol. 20, No. 1, 1982.
49. Sellin, R.H.J., J.W. Hoyt, J. Pollert and O. Serivener. The Effect of Drag Reducing Additives on Fluid Flows and Their Industrial Applications: Part 2. Present Applications and Future Proposals. JHR. IAHR, Vol. 20, No.3, 1982.
50. Virk, P.S. An Elastic Sublayer Model for Drag Reduction by Dilute Solutions of Linear Macromolecules. JFM, Vol.45, No. 3, 1971.
51. White, A. Some Observations on the Flow Characteristics of Certain Dilute Macromolecular Solutions. Viscous Drag Reduction (Ed. Wells, C.S), Plenum Press, 1969.
52. Virk, P.S. Drag Reduction in Rough Pipes, JFM Vol. 45, No. 2, 1971.
53. Savins, J.C. and F. A. Seyer. Drag Reduction Scale-up Criteria. Physics of Fluids, Vol. 20, No. 10, 1977.
54. Granville, P.S. A Method for Predicting Additive Drag Reduction from Small Diameter Pipe Flows. 3rd Int. Conference on Drag Reduction, Univ. of Bristol, 1984.
55. Berman, N.S. and E.E. Cooper. Stability Studies in Pipe Flows Using Water and Dilute Polymer Solutions, Jour. AICHE, Vol. 18, 1972.

CHAPTER IX

Description of Turbulent Flows-II: Free Surface Flows

9.1 INTRODUCTION

Several investigations have been carried out to study the characteristics of turbulent flow in open channels. As mentioned in Chapter VII, initially turbulence in water was measured using the method of diffusion of dye or other soluble material. Later total head tube, current meters, hot film anemometers, hydrogen bubble method and, finally, laser doppler method have been used. The knowledge of distribution of turbulence intensities, micro and macro scales, and spectrum of turbulence is important in the understanding of certain basic phenomenona, such as energy dissipation, diffusion and dispersion of solid and liquid pollutants, settlement of solids in the flow, initiation and transport of bed material in the streams, suspension of sediment, scour around hydraulic structures and fluctuating forces on objects submerged in flow, such as baffle blocks, aprons, canal linings, etc.

In order to illustrate the complexities of the flow and information obtained so far, following typical flows are considered:

(i) Turbulence in rigid bed channels;
(ii) Diffusion and dispersion; and
(iii) Hydraulic jump.

In general, it has been found that the small scale fluctuations in the velocities and pressure in open channel flow follow normal or Gaussian distribution.

TURBULENCE IN RIGID BED OPEN CHANNELS

9.2 TURBULENCE CHARACTERISTICS

Turbulence Intensity Variation in Open Channels

The turbulent fluctuations in the velocity $\sqrt{\overline{u'^2}}$, $\sqrt{\overline{v'^2}}$ and $\sqrt{\overline{w'^2}}$ have been usually normalised by dividing them by local time averaged velocity $\bar{u}$, shear velocity u_* or average velocity in the vertical, and plotted against y/y_o or u_*y/ν; here y is the distance from the bed and y_o is the depth of flow. Figure 9.1 shows variation of $\sqrt{\overline{u'^2}}/\bar{u}$ with y/y_o for hydrodynamically smooth boundaries as obtained by Blinco and Partheniades (1). Two characteristics are obvious from this figure. First, for a given y/y_o, as Reynolds number increases, $\sqrt{\overline{u'^2}}/\bar{u}$ decreases; however. it may be emphasised that Reynolds number effect is asymptotic, i.e. as Reynolds number becomes large, the relative reduction in $\sqrt{\overline{u'^2}}/\bar{u}$ reduces appreciably. Further, Reynolds number effect is

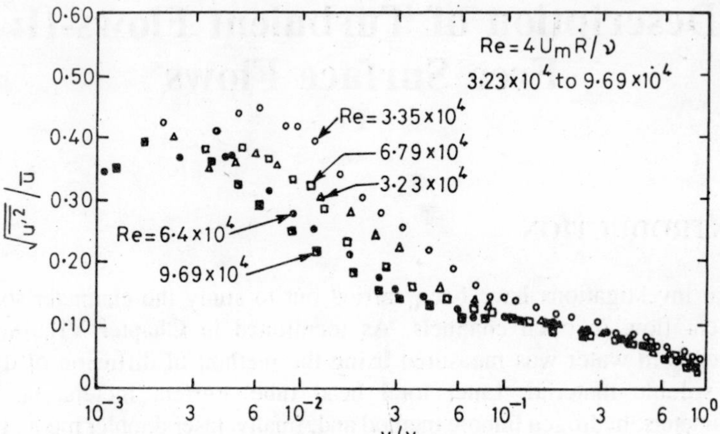

Fig. 9.1 Variation of $\sqrt{\overline{u'^2}}/\bar{u}$ with y/y_o for smooth channels (1) .

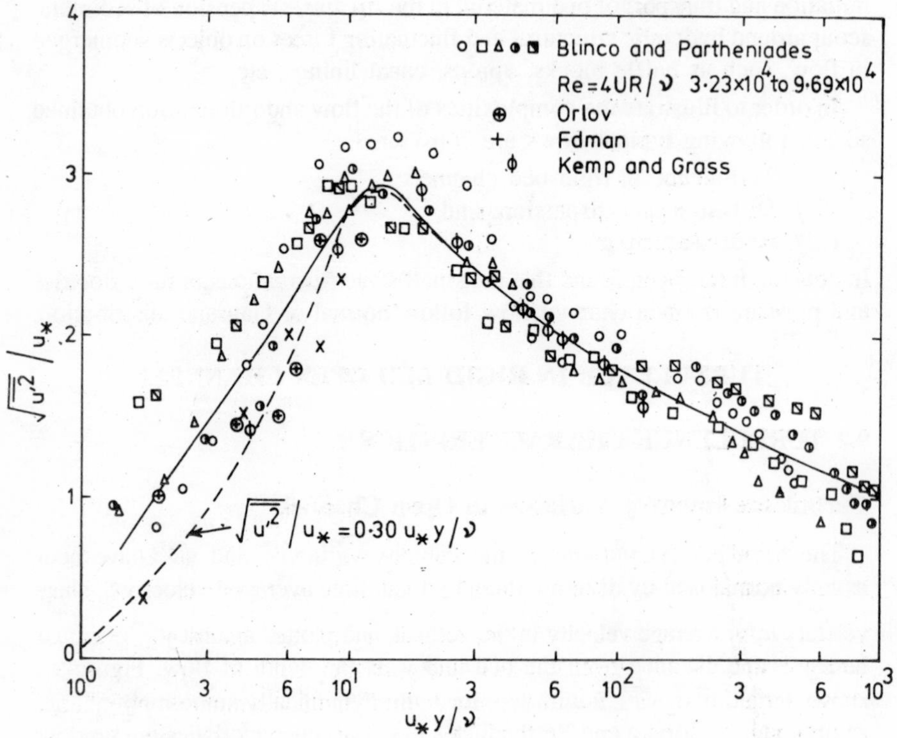

Fig. 9.2 Variation of $\sqrt{\overline{u'^2}}/u_*$ with u_*y/ν for smooth channels

more pronounced near the bed upto y/y_o equal to 0.10. It may be mentioned that the choice of u_* for normalisation of the fluctuations is more appropriate.

Near the boundary the turbulence intensity would increase from zero at the wall to a maximum, and then reduce as shown in Fig. 9.1. This is evident from Fig. 9.2, in which $\sqrt{\overline{u'^2}}/u_*$ is plotted against u_*y/ν for smooth boundary data. In addition to the data of Blinco and Partheniades, data of Orlov, Fidman (2*), and Kemp and Grass (3) have also been plotted. The data collected by Nalluri and Novak (4) in smooth circular channels, and those collected by Richardson and McQuivey (5) also fall in the same range. It may be noticed that at the edge of laminar sublayer, when u_*y/ν is between 10 and 15, the turbulence intensity is maximum and reaches a value of $\sqrt{\overline{u'^2}}/u_* = 3.0$. For u_*y/ν less than 10 to 15, Ljatkher (2) has proposed a linear relation

$$\sqrt{\overline{u'^2}}/u_* = 0.30\ u_*y/\nu \qquad \ldots (9.1)$$

However, Fig. 9.2 suggests a different relationship shown by full line. Figure 9.3 shows variation of $\sqrt{\overline{v'^2}}/u_*$ with u_*y/ν near smooth boundary as obtained by Ljatkher. Plotted on this figure are also data of Kemp and Grass (3). Ljatkher has proposed a relation

$$(\sqrt{\overline{v'^2}}/u_*)^{0.50} = 0.07\ u_*y/\nu \qquad \ldots (9.2)$$

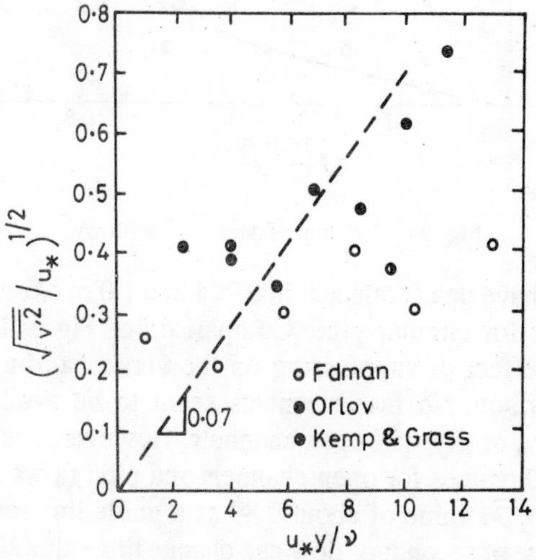

Fig. 9.3 Variation of $(\sqrt{\overline{v'^2}}/u_*)^{0.50}$ with u_*y/ν near smooth boundary

However, it can be seen that a lower coefficient, say 0.06 suits the data better. Figure 9.4 shows variation of $\sqrt{\overline{v'^2}/\overline{u'^2}}$ with y/y_o for open channel flows, both smooth as well as rough. It is interesting to note that for $0.15 < y/y_o < 0.80$, $\sqrt{\overline{v'^2}/\overline{u'^2}}$ assumes a near constant value of 0.60. These values of $\sqrt{\overline{v'^2}/\overline{u'^2}}$ may be compared with values obtained by Reichardt, and Laufer for enclosed flows as tabulated below:

$\frac{y}{R}$ or $2y/H$	0.10	0.20	0.30	0.40	0.60	0.70	0.80	0.90
Laufer $\sqrt{\overline{v'^2}}/\sqrt{\overline{u'^2}}$	0.55	0.58	0.60	0.65	0.67	0.80	0.82	0.96
Reichardt	0.48	0.56	0.62	0.65	0.66	0.69	0.76	0.86

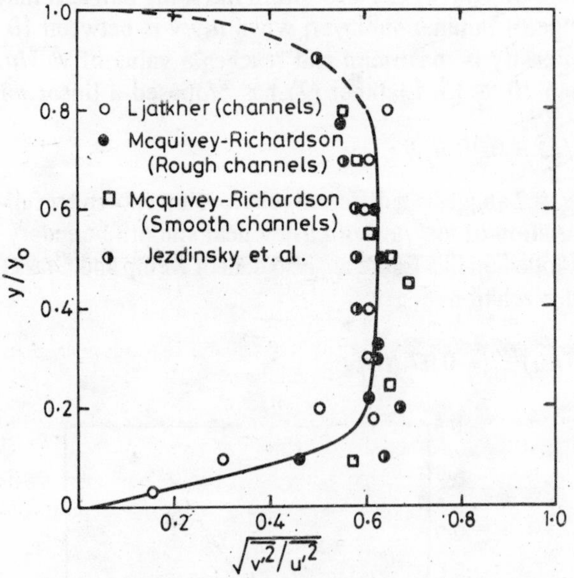

Fig. 9.4 Variation of $\sqrt{\overline{u'^2}}\ /\ \overline{u'^2}$ with y/y_o

Reichardt's data have been collected in 0.244 m × 1.0 m closed channel, while Laufer's data are for circular pipe. Comparison of Fig. 9.4 with the above table shows the effect of suppressing on the vertical turbulent fluctuations near the free surface. No measurements seem to be available for $\sqrt{\overline{w'^2}}/u_*$ variation with y/y_0 or $u_* y/\nu$ in open channels. However, considering that for $0.15 < y/y_0 < 0.80$, values for open channels and pipe flows are nearly same, one can use $\sqrt{\overline{w'^2}/\overline{u'^2}}$ value of about 0.80 as a guide line for open channels. However, presence of secondary flow can change this value significantly. This has been shown by the recent investigations of Tominaga et al. (17). Experiments have been conducted by Rao (6), McQuivey and Richardson (7), Sushil Kumar et al. (7a) and other to study variation of turbulence intensity in hydrodynamically rough channels. One can expect that for channels with roughness on the bed

$$\sqrt{\overline{u'^2}}/\bar{u} \text{ or } \sqrt{\overline{v'^2}}/\bar{u} = f\left(\frac{y}{y_0}, \text{Re}, \frac{k_s}{y_0}\right)$$

Here k_s is the equivalent roughness. For large Reynolds number values, effect of Re on the turbulence intensity will be very small and can be neglected.

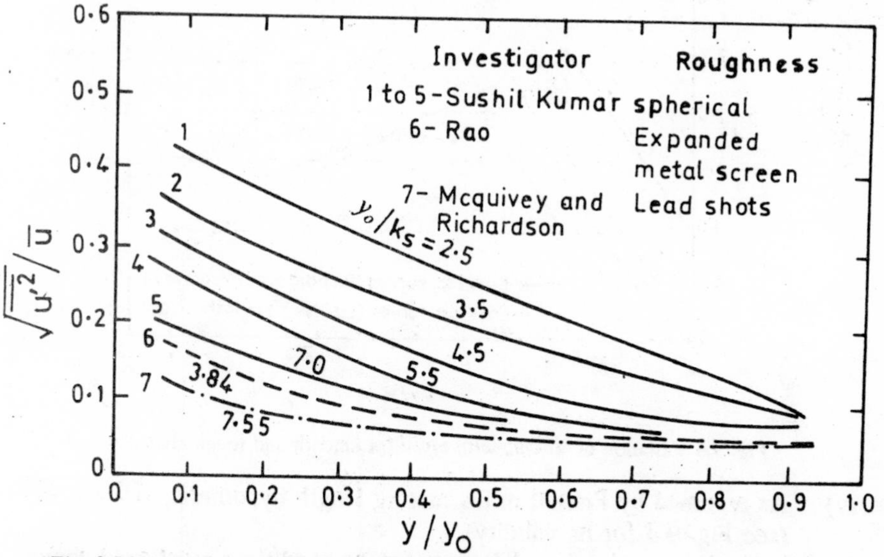

Fig. 9.5 Variation of turbulence intensity with relative depth in open channels with roughness on the bed

Figure 9.5 shows variation of $\sqrt{\overline{u'^2}}/\bar{u}$ with y/y_o for various roughnesses as obtained for the data collected by Sushil Kumar, Rao, and McQuivey and Richardson. It can be seen that, for given value of y/y_o, the intensity increases as k_s/y_o increases. It may further be noted that away from the bed, i.e. for larger y/y_o values, effect of roughness on $\sqrt{\overline{u'^2}}/\bar{u}$ is much smaller. Blinco and Partheniades have plotted $\sqrt{\overline{u'^2}}/u_*$ with $u_* y/\nu$ for smooth and rough channel data collected by them (see Fig. 9.6). In the same figure are shown envelop curves based on Richardson and McQuivey's data for smooth as well as rough channels. Also in the same figure is shown band of scatter for Laufer's data for air flow in a duct which shows similarity in variation. This figure indicates the band in which smooth and rough channel data fall. It can be seen that for small values of $u_* y/\nu$ i.e. closer to the boundary, $\sqrt{\overline{u'^2}}/u_*$ is greater for rough boundary than for smooth boundary. Farther away from the wall, $\sqrt{\overline{u'^2}}/u_*$ falls within the same band for smooth and rough boundaries. However, it cannot be used to accurately predict turbulent intensity for rough channels. Further investigation is needed in this regard.

Li et al. (8) have presented a semi analytical model for predicting $\sqrt{\overline{u'^2}}$ and $\sqrt{\overline{v'^2}}$, distribution in open channels.

The model is based on the following assumptions :

(i) Flow is steady and uniform in the conventional sense.

(ii) Turbulence velocity fluctuations in the longitudinal and lateral directions follow Gaussian distribution.

(iii) Fluctuating velocity in the streamwise direction is the result of momentum transfer only in y direction.

(vi) Mean velocity in y direction is zero across the entire depth.

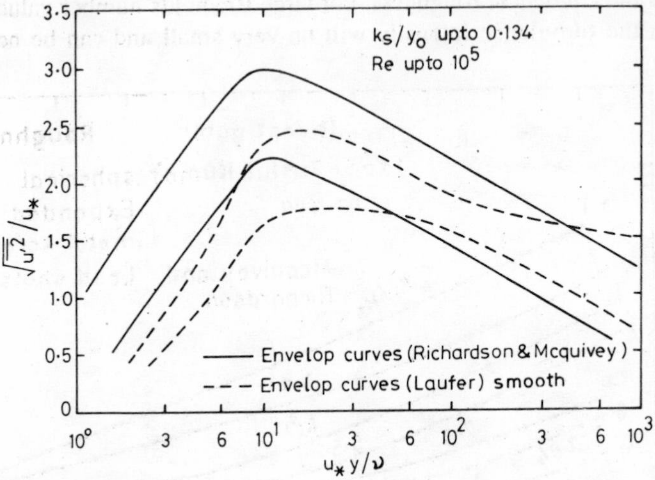

Fig. 9.6 Variation of $\sqrt{\overline{u'^2}}/u_*$ with $u/y/\nu$ for smooth and rough channels

(v) As assumed by Prandtl in his mixing length hypothesis, $\sqrt{\overline{v'^2}} = c\sqrt{\overline{u'^2}}$ (see Fig. 9.4 for its validity).

From Reynolds equation in x direction, for the conditions mentioned above, one gets,

$$\overline{u'v'} = \nu\frac{\partial \bar{u}}{\partial y} - u_*^2\left(1 - \frac{y}{y_o}\right) \quad \ldots (9.3)$$

and $\overline{u'v'}$ can be expressed as

$$\overline{u'v'} = R_1\sqrt{\overline{u'^2}}\sqrt{\overline{v'^2}} \quad \ldots (9.4)$$

where R_1 is the correlation coefficient. Substituting $\frac{\partial \bar{u}}{\partial y} = \frac{u_*}{\kappa y}$ where κ is the Karman constant, one gets

$$R_1 c\overline{u'^2} = \frac{u_*\nu}{\kappa y} - u_*^2\left(1 - \frac{y}{y_o}\right) \quad \ldots (9.5)$$

From the experimental results of Reichardt, McQuivey and Richardson, and Laufer it was found that

$$\left.\begin{aligned} &R_1 = R_{1c} \text{ for } y/y_o < y_* \\ \text{and} \quad &R_1 = R_{1c}K\left(1 - \frac{y}{y_o}\right) \text{ for } y_* < \frac{y}{y_o} < 1.0 \end{aligned}\right\} \quad \ldots (9.6)$$

in which $K = 1/(1 - y_*)$.

Analysis of data was used to determine the optimum values of y_*, R_{1c} and c. It was found that $y_* = 0.80$ for smooth boundary and $y_* = 0.60$ for rough boundary, whereas $c = 0.55$ and $R_{1c} = -0.70$ for both smooth and rough boundaries. Hence, for known kinematic viscosity ν, y_o, y_*, u_*, $\kappa = 0.40$ and k_s, one can determine $\sqrt{\overline{u'^2}}$ and $\sqrt{\overline{v'^2}}$ for any value of y as follows :

(i) Determine whether the boundary is smooth or rough, and hence, choose correct values of y_*, c and R_{1c}

(ii) For known y/y_o determine R_1 using Eq. (9.6) and determine $\sqrt{\overline{u'^2}}$ using Eq. (9.5)

(iii) Determine $\sqrt{\overline{v'^2}}$ using the relation $\sqrt{\overline{v'^2}} = c\,\sqrt{\overline{u'^2}}$

It was found that this model predicted $\sqrt{\overline{u'^2}}$ and $\sqrt{\overline{v'^2}}$ fairly well for y/y_o greater than 0.15. But this was done using Richardson and McQuivey's data only, and needs further verification using additional data. Figure 9.4 indicates that the assumption of constancy of $\sqrt{\overline{v'^2}/\overline{u'^2}}$ is valid only for $0.15 < y/y_o < 0.80$.

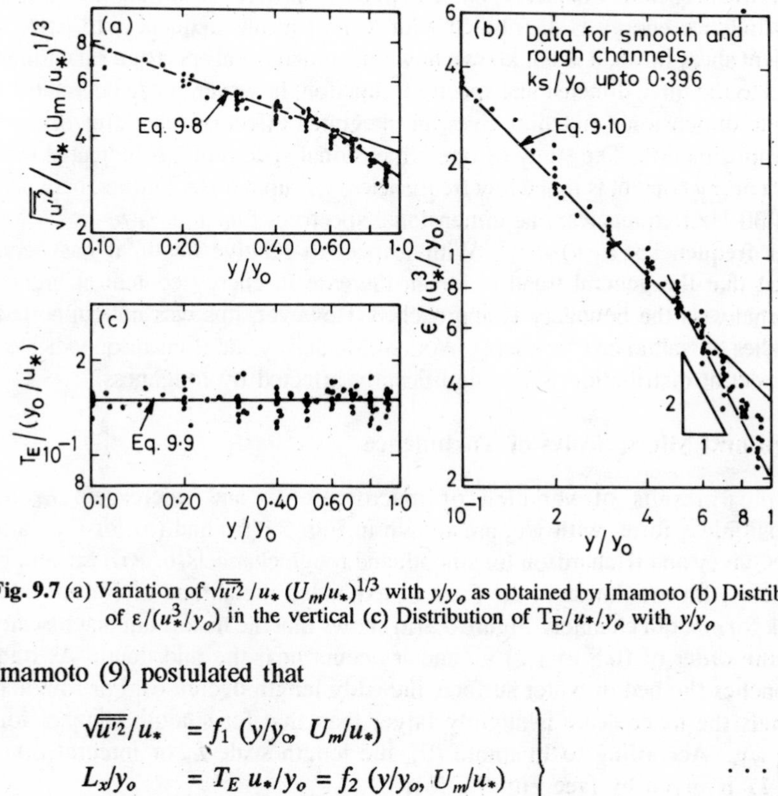

Fig. 9.7 (a) Variation of $\sqrt{\overline{u'^2}}/u_*\,(U_m/u_*)^{1/3}$ with y/y_o as obtained by Imamoto (b) Distribution of $\varepsilon/(u_*^3/y_o)$ in the vertical (c) Distribution of $T_E/u_*/y_o$ with y/y_o

Imamoto (9) postulated that

$$\left.\begin{aligned} \sqrt{\overline{u'^2}}/u_* &= f_1\,(y/y_o,\ U_m/u_*) \\ L_x/y_o &= T_E\,u_*/y_o = f_2\,(y/y_o,\ U_m/u_*) \\ \varepsilon/(u_*^3/y) &= f_3\,(y/y_o,\ U_m/u_*) \end{aligned}\right\} \qquad \ldots (9.7)$$

Here ε is the rate of energy dissipation per unit mass. It is evident that effect of relative roughness k_s/y_o is indirectly taken into account by inclusion of Chezy's coefficient in dimensionless form, viz., U_m/u_*.

Figure 9.7 (a) shows variation of $\sqrt{\overline{u'^2}}/u_*\left(\dfrac{U_m}{u_*}\right)^{1/3}$ with y/y_o. The data indicate that for $0.1 < y/y_o < 0.6$

$$\sqrt{\overline{v'^2}}/u_*\left(\frac{U_m}{u_*}\right)^{1/3} = 0.36\,(y/y_o)^{-1/3} \qquad \ldots (9.8)$$

and for $0.6 < y/y_o < 0.90$

$$\sqrt{\overline{u'^2}}/u_* \left(\frac{U_m}{u_*}\right)^{\frac{1}{3}} = 0.30\ (y/y_o)^{-2/3}$$

These relations need to be verified with additional data.

Turbulence Spectra

One dimensional spectrum of turbulent fluctuations has been studied by Ishihara and Yokosi (10), Nalluri and Novak (4), and McQuivey and Richardson (6). As mentioned in Chapter V, energy spectrum provides information on how the turbulence energy is distributed with respect to the frequency. In case of turbulent shear flows, it is not known how one dimensional spectrum function is related to the three dimensional spectrum function; however, it can be assumed that one dimensional function gives an integrated effect of three dimensional spectrum function. The study of one dimensional spectrum has indicated that major energy content is in the low frequencies, viz. upto 5 Hz. Further, between 5 and 30 Hz frequencies, one dimensional spectrum function $F_1(n) \sim n^{-5/3}$; for higher frequencies $F_1(n) \sim n^{-1/7}$. With respect to relative depth, it has been noticed that the general trend is for an increase in energy content at higher frequencies as the boundary is approached. However, this was not supported by studies of Nalluri and Novak (4). Work of McQuivey and Richardson indicates that spectral distribution is not significantly affected by roughness.

Macro and Micro Scales of Turbulence

Typical results of variation of macroscale L_x and microscale λ_x in dimensionless form, with y/y_o are shown in Fig. 9.8 (a) and (b) for the data of McQuivey and Richardson for smooth and rough channels, of Rao for rough channels with roughness in the form of expanded metal, and of Nalluri and Novak for smooth channels. Figure 9.8(a) shows that the maximum macroscale is of the order of (0.8 to 1.2) y_o, and it occurs near the mid depth. As one approaches the bed or water surface, the eddy length decreases. For rougher channels the macroscale is slightly larger than that for smooth channel for same y/y_o. According to Imamoto (9), the length scale L_x or integral time scale T_E is given by [see Fig. 9.7 (b)]

$$\frac{L_x}{y_o(U_m/u_*)} = \frac{T_E}{(y_o/u_*)} = 0.15 \qquad \ldots (9.9)$$

for $0.10 < y/y_o < 0.90$ as shown in Fig. 9.7 (b). From this equation it can be inferred that, if dimensionless Chezy's coefficient were to change from 20 to 10, the macro-scale L_x will change from 1.5 y_o to 3.0 y_o. Another startling fact regarding variation of L_x/y_o can be seen from the experimental results of Komura and Kubota (11) presented in Fig. 9.9. As B/y_o increases from 1.95 to 9.5, maximum L_x increases from y_o to 3.5 y_o. Here B is the channel width. Thus, roughness and width/depth ratio seem to influence L_x/y_o

to a significant extent. Since L_x/y_o is larger in shallow wide channels, one can expect greater mixing in shallow wide channels than in deep narrow channels.

Figure 9.8(b) shows variation of micro scale with y/y_o. It is seen that, for limited data available. λ_x/y_o attains a maximum value of about 0.07 to 0.14 at about middepth. Energy dissipation rate ε is related to $(\sqrt{\overline{u'^2}})^3/\lambda_f$. Imamoto (9) has found that $\varepsilon/(u_*^3/y_o)$ variation is given by:

$$\left.\begin{array}{l} \text{For } 0.10 < y/y_o < 0.60 \\ \qquad \varepsilon/(u_*^3/y_o) = 0.35\ (y/y_o)^{-1.0} \\ \text{For } 0.60 < y/y_o < 0.90 \\ \qquad \varepsilon/(u_*^3/y_o) = 0.20\ (y/y_o)^{-2.0} \end{array}\right\} \qquad \ldots (9.10)$$

See Fig 9.7 (c).

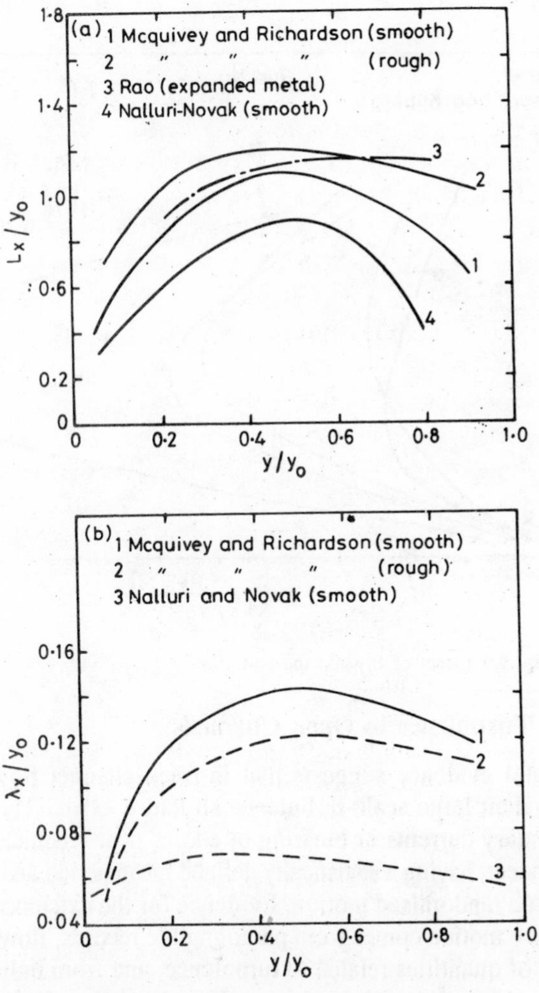

Fig. 9.8 Variation of Lx/y_o and λ_x/y_o with y/y_o

Pressure Fluctuations on Channel Bed

Aki (12) studied pressure fluctuations on the bottom of the channel and its relation to the bed shear. He showed that, as proposed by Kraichnan, $\sqrt{\overline{p_w'^2}}$ is related to the shear by the equation

$$\sqrt{\overline{p_w'^2}} = \beta\,\tau_o = \beta c_f \frac{\rho U_o^2}{2} \qquad \text{. . . (9.11)}$$

where p_w' is pressure fluctuation on the wall and β was found to vary between 3 to 5. In turbulent boundary layer β is related to d/δ_* where d is the diameter of presure transducer and δ_* is the displacement thickness. The coefficient β was found to reduce from 4.0 to 1.5 as d/δ_* increased from 0.2 to 5.0. In the case of turbulent flow over the chute spillway, it was found that $\sqrt{\overline{p_w'^2}}$ was about 60 per cent of pressure due to flow depth. He also determined experimentally the integral macro scales L_x and L_y of pressure fluctuations; they were found to be $L_x = 20\delta_*$ and $L_y = L_x/7 \approx 3\delta_*$.

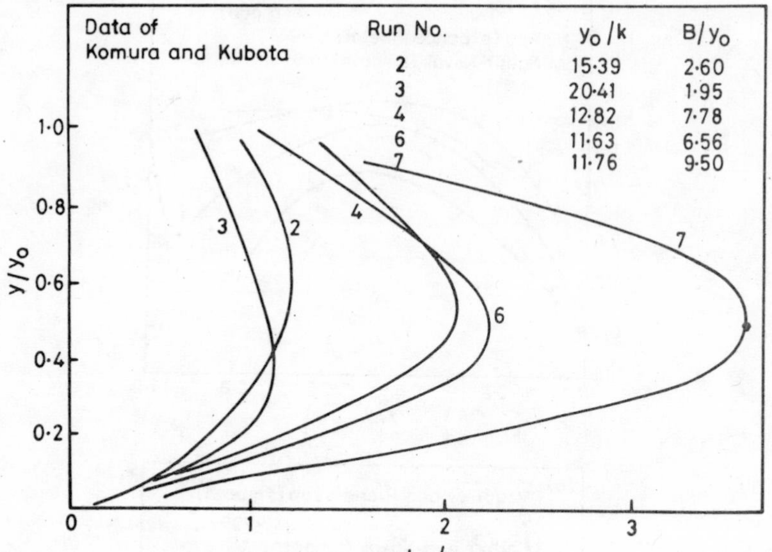

Fig. 9.9 Effect of B/y_o on the variation of L_x/y_o with y/y_o (33)

Large Scale Turbulence in Open Channels

Experimental evidence suggests that in open channel flow, intermittent or time dependent large scale turbulence structure exists. This can take the form of secondary currents or bursting of eddies near channel bottom which are time dependent, having a statistically defined mean period and which interact with small scale randomised motion. Evidence for the existence of such large scale turbulence motion comes from photographic records, flow visualization, measurement of quantities related to turbulence, and from indirect inference. Hayashi et al. (13) refer to photographs of the Kiso river in Japan, taken by

Kinoshita, showing presence of coherent structure in the form of longitudinal vortices. McQuivey (14) refers to dye studies on the Mississippi, Missouri, and Alabama rivers in USA, and conveyance channels, indicating existence of large scale motions that assume importance in understanding of turbulence mechanism, mixing and transport processes.

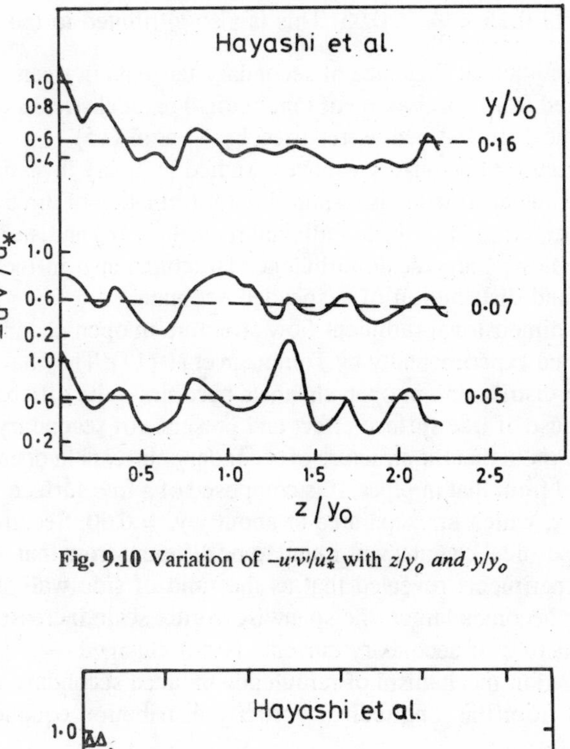

Fig. 9.10 Variation of $-\overline{u'v'}/u_*^2$ with z/y_o *and* y/y_o

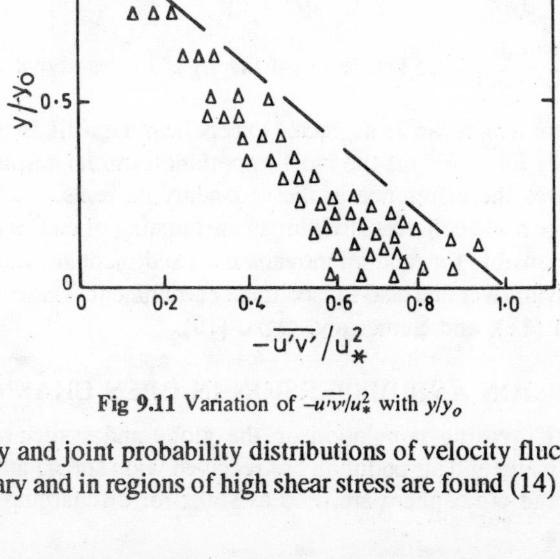

Fig 9.11 Variation of $-\overline{u'v'}/u_*^2$ with y/y_o

Probability and joint probability distributions of velocity fluctuations near rough boundary and in regions of high shear stress are found (14) to be skewed with large positive kurtosis (from 24 to 12) indicating presence of intermittency or large scale time dependent structures. Variation of turbulent shear (i.e. Reynolds stress) as well as $\sqrt{\overline{u'^2}}/u_*$ and $\sqrt{\overline{w'^2}}/u_*$ across the channel at any given value of y/y_o show a certain periodicity, see Fig. 9.10, which is attributed to the presence of secondary currents. In addition, as shown in Fig. 9.11, the variation of turbulent shear in any vertical deviates from the theoretical one in the range of $0.15 < \frac{y}{y_o} < 0.95$. This is also attributed to the existence of large scale turbulence. Presence of secondary currents in open channels can also be inferred from observation of longitudinal parallel streaks of deposition of fine sediment on the bed as reported by Vanoni (15).

Secondary current patterns have been studied by many investigators. Such large scale turbulent structure is responsible for formation of three dimensional sand waves occurring in the bed of alluvial rivers (13, 16) and, hence, it affects resistance to flow. Large scale turbulence structures also influence velocity distribution and distribution of suspended sediment.

The three dimensional turbulent flow structure in open channels has been recently studied experimentally by Tominaga et al. (17). They have found that $\sqrt{\overline{v'^2}}$ and $\sqrt{\overline{w'^2}}$ distribution in open channels becomes quite different from that in pipes because of free surface effect and presence of secondary currents. It was found by them that the structure of secondary currents in open channels is quite different from that in pipes. It is composed of a free surface vortex and a bottom vortex, which are separated at about $y/y_o = 0.60$. Secondary current pattern in trapezoidal channel was found to be different from that in recangular channels. Experiments revealed that as the ratio of side wall shear to the bottom shear becomes larger, the spanwise vortex scale increases; however, the basic structure of secondary currents is not changed.

The production mechanism of turbulence induced secondary currents can be explained from the longitudinal vorticity distribution equation

$$\bar{v}\frac{\partial \xi}{\partial y} + \bar{w}\frac{\partial \xi}{\partial z} = \frac{\partial^2 (\overline{u'^2} - \overline{w'^2})}{\partial y \partial z} + \left(\frac{\partial^2}{\partial z^2} - \frac{\partial^2}{\partial y^2}\right)\overline{u'w'} + \nu\left(\frac{\partial^2 \xi}{\partial y^2} + \frac{\partial^2 \xi}{\partial z^2}\right) \quad (9.12)$$

where $\xi = \left(\frac{\partial \bar{w}}{\partial y} - \frac{\partial \bar{v}}{\partial z}\right)$. The last term on the right in the above equation is the viscous term which can be neglected except near the wall. In this equation, distribution of $(\overline{u'^2} - \overline{w'^2})$ is the most important term. Its variation for each condition causes the difference in the secondary currents.

Another large scale coherent structure is the bursting of eddies near the bed. These are responsible for random movement of sediment at isolated spots on the bed of alluvial river and also for sediment entrainment. These are discussed by Sutherland (18), and Sumer and Ogzu (19).

DIFFUSION AND DISPERSION IN OPEN CHANNELS

With ever increasing population on the globe and associated industrial development, water and air pollution has received wide spread attention, since water bodies and atmosphere are used as sinks for discharging sewage and industrial wastes. In this connection, one has to distinguish between inert pollutants and organic pollutants. Inert pollutants dó not degrade due to bacterial or chemical action, whereas organic and certain chemical pollutants can undergo degradation or changes due to bacterial action or chemical reaction. Only inert pollutants are considered here. Further, unless it is otherwise stated, pollutant does not cause change in the density of fluid.

When the pollutant is introduced in a steady uniform flow in an open channel in the form of a point or line source, it spreads in the vertical, lateral and longitudinal directions due to turbulent motion. It has been emphasised earlier that the turbulent flow has a great diffusivity as compared to the laminar flow; hence, the spread is relatively rapid. This process is known as diffusion and it is dependent on motion of individual particles. It is, therefore, a Lagrangian space time process of mixing in the three directions. Once the mixing is complete in the vertical and lateral directions, it disperses in the longitudinal direction as a cloud and the average concentration continues to decrease in the downstream direction. Longitudinal dispersion is a result of molecular and turbulent diffusion and convection caused by the velocity variation in the entire cross section. It may be mentioned that molecular diffusion is negligibly small and in the dispersion process the contribution of turbulent diffusion in longitudinal direction is only a few per cent of the total.

9.3 DIFFUSION ÉQUATION

Mass transport in a fluid can take place due to diffusion in a stationary fluid and advective diffusion due to the fact that fluid is moving. It is assumed that the two processes are additive. German physiologist Adolf Fick in 1855, propounded the law of molecular diffusion according to which, as in the case of Fourier's law of heat flow, mass of solute crossing a unit area per unit time q_s in any given direction is proportional to the negative gradient of concentration of solute in that direction. In other words,

$$q_s = -D_m \frac{\partial C}{\partial x}$$

Here D_m is the constant of proportionality, and is known as the molecular diffusion coefficient having dimensions of L^2/T, and C is the concetration of a the solute. For diffusion in three directions, the above equation can be written as

$$q_s = -D_m \nabla C$$

where q_s is now mass vector flux with components in x, y, z directions.

In case of one dimensional diffusion, one can consider inflow and outflow rates of solute through faces of area $\delta y \delta z$ distance δx apart. Then the net rate of inflow through these two faces is

$$q_s \delta_y \delta_z - \left(q_s + \frac{\partial q_s}{\partial x}\delta x\right)\delta y \delta z, \text{ i.e. } -\frac{\partial q_s}{\partial x}\delta x \delta y \delta z$$

Therefore, mass within the parallellopiped $\delta x \delta y \delta z$ will increase at the rate $\frac{\partial C}{\partial t}\delta x \delta y \delta z$. Equating these two, one gets

$$\frac{\partial C}{\partial t} = -\frac{\partial q_s}{\partial x}$$

which when combined with Fick's law yields

$$\frac{\partial C}{\partial t} = D_m \frac{\partial^2 C}{\partial x^2} \qquad \ldots (9.13)$$

This is the classical heat conduction or diffusion equation. Corresponding equation in three dimensions is

$$\frac{\partial C}{\partial t} = D_m\left(\frac{\partial^2 C}{\partial x^2} + \frac{\partial^2 C}{\partial y^2} + \frac{\partial^2 C}{\partial z^2}\right) \qquad \ldots (9.13a)$$

Advective transport through a plane perpendicular to x direction and with area $\delta y \delta z$ is $(uC)\,\delta y \delta z$ where u is flow velocity in x direction. Total transport due to advection and molecular diffusion through this plane is

$$\left\{uC + \left(-D_m \frac{\partial C}{\partial x}\right)\right\} \delta y \delta z$$

Total outflow through similar face distance δx apart (see Fig. 9.12) is

$$\left\{uC + \frac{\partial}{\partial x}(uC)\,\delta x + \left\{-D_m \frac{\partial C}{\partial x} - \frac{\partial}{\partial x}\left(D_m \frac{\partial C}{\partial x}\right)\right\}\delta x\right]\delta y \delta z$$

Hence, net inflow rate into the paralleopiped by flow in x direction is

$$-\frac{\partial}{\partial x}(uC)\,\delta x \delta y \delta z + \frac{\partial}{\partial x}\left(D_m \frac{\partial C}{\partial x}\right)\delta x \delta y \delta z$$

Determining similar quantities in other two directions, and equating the result to the net rate of solute increase in the parallelopiped, one gets

$$\frac{\partial C}{\partial t} + u\frac{\partial C}{\partial x} + v\frac{\partial C}{\partial y} + w\frac{\partial C}{\partial z} = \frac{\partial}{\partial x}\left(D_m \frac{\partial C}{\partial x}\right) + \frac{\partial}{\partial y}\left(D_m \frac{\partial C}{\partial y}\right) + \frac{\partial}{\partial z}\left(D_m \frac{\partial C}{\partial z}\right) + S \qquad \ldots (9.14)$$

Here S is the source or sink contribution per unit volume and is usually zero. This can be written in tensor form as

$$\frac{\partial C}{\partial t} + u_i \frac{\partial C}{\partial x_i} = \frac{\partial}{\partial x_i}\left(D_m \frac{\partial C}{\partial x_i}\right) + S \qquad \ldots (9.14a)$$

Equation (9.14) is valid for laminar flow. For turbulent flow, one can substitute $C = \overline{C} + C, u = \overline{u} + u', v = \overline{v} + v'$ and $w = \overline{w} + w'$ in Eq. (9.14) and use Reynolds rules of averages to get

$$\frac{\partial \overline{C}}{\partial t} + \overline{u}\frac{\partial \overline{C}}{\partial x} + \overline{v}\frac{\partial \overline{C}}{\partial y} + \overline{w}\frac{\partial \overline{C}}{\partial z} = \frac{\partial}{\partial x}\left(D_m \frac{\partial \overline{C}}{\partial x} - \overline{u'C'}\right) + \frac{\partial}{\partial y}\left(D_m \frac{\partial \overline{C}}{\partial y} - \overline{v'C'}\right)$$

$$+ \frac{\partial}{\partial z}\left(D_m \frac{\partial \overline{C}}{\partial z} - \overline{w'C'}\right) + S \qquad \ldots (9.15)$$

or in tensor notation

$$\frac{\partial \overline{C}}{\partial t} + \overline{u}_i \frac{\partial \overline{C}}{\partial x_i} = \frac{\partial}{\partial x_i}\left(D_m \frac{\partial \overline{C}}{\partial x_i} - \overline{u'_i C'}\right) + S \qquad \ldots (9.15a)$$

The term $\overline{u'_i C'}$ is the transport of solute or pollutant in the ith direction due to turbulence. Using analogy with Fick's law of diffusion or Boussineq's hypothesis, $\overline{u'C'}$, $\overline{v'C'}$, and $\overline{w'C'}$ can be expressed as $\overline{u'C'} = -\varepsilon_x \frac{\partial \overline{C}}{\partial x}$, $\overline{v'C'} = -\varepsilon_y \frac{\partial \overline{C}}{\partial y}$ and $\overline{w'C'} = -\varepsilon_z \frac{\partial \overline{C}}{\partial z}$. Further, the turbulent diffusion coefficients $\varepsilon_x, \varepsilon_y$ and ε_z being much larger than molecular diffusion coefficient D_m, the latter can be neglected in preference to the former. Hence, Eq. 9.15 reduces to

$$\frac{\partial \overline{C}}{\partial t} + \overline{u}\frac{\partial \overline{C}}{\partial x} + \overline{v}\frac{\partial \overline{C}}{\partial y} + \overline{w}\frac{\partial \overline{C}}{\partial z} = \frac{\partial}{\partial x}\left(\varepsilon_x \frac{\partial \overline{C}}{\partial x}\right) + \frac{\partial}{\partial y}\left(\varepsilon_y \frac{\partial \overline{C}}{\partial y}\right) + \frac{\partial}{\partial z}\left(\varepsilon_z \frac{\partial \overline{C}}{\partial z}\right) + S \qquad \ldots (9.16)$$

This is the basic diffusion-dispersion equation for turbulent flow. The diffusion coefficients ε_x, ε_y, ε_z have the dimensions of L^2/T and they must be determined. It may be mentioned that since turbulent diffusion takes place due to presence of eddies and turbulent fluctuations, from dimensional considerations the diffusion coefficient can be expressed either as $\left(\frac{L^2}{T^2}\right) T$ or $\left(\frac{L}{T} L\right)$ i.e. either as (characteristic velocity)^2 × characteristic time, or (characteristic velocity) × characteristic length.

Diffusion by Continuous Movements

Even an elementary theory of turbulent diffusion dealing with transport of material particles from one point to another suggests use of the Lagrangian description. However, eventually diffusion of any physical property must be accompanied by molecular process which is more conveniently described by Eulerian approach. This makes the complete description of turbulent diffusion very complicated. Taylor (20) has given the theory of turbulent diffusion in statistically steady homogeneous isotropic field. Since this theory explains the dependence of diffusion coefficients on the characteristics of turbulence, it is given below. It may be kept in mind that u', v', w' in this discussion are Lagrangian velocity components of a particle.

The distance Y_1 travelled by the particle in time t is given by

$$Y_1 = \int_0^t u'(\tau)\, d\tau \qquad \ldots (9.17)$$

Hence, the variance of Y_1, i.e. $\overline{Y_1^2}$ will be given by

$$\frac{1}{2}\frac{d}{dt}\left(\overline{Y_1^2}\right) = \overline{Y_1 \frac{dY_1}{dt}} = \overline{Y_1 u'} \qquad \ldots (9.18)$$

Hence, substituting Eq. 9.17 in Eq. 9.18, one gets

$$\frac{1}{2}\frac{d}{dt}\left(\overline{Y_1^2}\right) = \overline{u' \int_0^t u'(\tau)\, d\tau} \qquad \ldots (9.19)$$

One can make use of the Lagrangian correletion coefficient with respect to time

$$R_L(\tau) = \frac{\overline{u'(t)u'(t-\tau)}}{\overline{u'^2}}$$

then

$$\overline{u'^2}\int_o^t R_L(\tau)d\tau = \overline{u'\int_o^t u'(\tau)\,d\tau} = \frac{1}{2}\cdot\frac{d}{dt}\,(\overline{Y_1^2})$$

Hence,

$$\frac{1}{2}\,\frac{d}{dt}\,(\overline{Y_1^2}) = \overline{u'^2}\int_o^t R_L(\tau)\,d\tau \qquad \ldots (9.20)$$

The Lagrangian time correlation coefficient $R_L(\tau)$ varies with τ in such a manner that $R_L(\tau) \sim 1.0$ for very small time t. Hence, integration of Eq (9.20) for small t will give

$$\overline{Y_1^2} = \overline{u'^2}\,t^2 \quad \text{or} \quad \sqrt{\overline{Y_1^2}} = \sqrt{\overline{u'^2}}\;t \qquad \ldots (9.21)$$

Similarly, for large times

$$\int_o^t R_L(\tau)d\tau = T_L$$

where T_L is the Lagrangian integral time scale. Hence,

$$\frac{1}{2}\,\frac{d\overline{Y_1^2}}{dt} = \overline{u'^2}\,T_L$$

or, $$\overline{Y_1^2} = 2\overline{u'^2}\,T_L \cdot t = 2D_L t \qquad \ldots (9.22)$$

where $D_L = \overline{u'^2}T_L$ is the Lagrangian diffusion coefficient. Since Lagrangian turbulence characteristics are difficult to measure or obtain, it is desirable to know the relationship between Lagrangian and Eulerian turbulence characteristics, the later being easier to measure. This has been done by Hay and Pasquil (21) according to which

$$T_{Lx} = \beta T_E \qquad (9.23)$$

where, $T_{Lx} = \int_0^\infty \dfrac{\overline{u'(t)u'(t-\tau)}}{\overline{u'^2}}\,d\tau$

and u' here is Lagrangian velocity fluctuations, while

$$T_E = \int_0^\infty \frac{\overline{u'(t)u'(t-\tau)}}{\overline{u'^2}}\,d\tau$$

and u' refers to Eulerian velocity fluctuations at a point. For turbulence that is carried downstream at a constant velocity U, β is given by

$$\beta = 0.40\;U/\sqrt{\overline{u'^2}} \qquad \ldots (9.24)$$

and $$T_E = L_x/U \qquad \ldots (9.25)$$

where L_x is the Eulerian integral macroscale of turbulence. Hence, with Eulerian description, the diffusion coefficient ε_x in x direction is given by

$$\varepsilon_x = \overline{u'^2}\;T_{Lx} = \overline{u'^2}\;\beta\;\frac{L_x}{U} \quad \text{or} \quad \varepsilon_x = 0.40\;\sqrt{\overline{u'^2}}\;L_x$$

In a similar manner, transverse diffusion coefficient ε_z can be expressed as

$$\varepsilon_z = 0.40\;\sqrt{\overline{w'^2}}\;L_z$$

Miller and Richardson (22), Engelund (23), Fisher (24), and others have related

ε_z to u_* and hydraulic radius in open channels and their data indicate that

$$\varepsilon_z = (0.22 \text{ to } 0.24)\ u_*\ R$$

and it is unaffected by channel roughness. The longitudinal diffusion coefficient ε_x has been found to be approximately three times ε_z. In natural streams with bends, meanders and irregular channel sections ε_z observed is larger and values as high as $0.70 u_* y_o$, where y_o is the depth of flow, have been obtained. For channels, Engelund has expressed ε_x as

$$\varepsilon_x = 0.045\ u_* y_o$$

9.4 SEDIMENT SUSPENSION

When diffusion equation is derived for suspended sediment in the case of water, one must take into account the fact that the sediment will settle at a velocity ω_o (see Fig. 9.12). As a result, the term $v C \delta x \delta z$ in Fig. 9.12 should be replaced by $(v - \omega_o)\ \overline{C}\ \delta x \delta z$ and, correspondingly, the outflow term to $(v - \omega_o)\overline{C}\ \delta x\, \delta z + \frac{\partial}{\partial y}(v - \omega_o)\ \overline{C}\ \delta x \delta y \delta z$. With these changes and subsequent simplifications using $S = 0$, one gets

$$\frac{\partial \overline{C}}{\partial t} + \overline{u}\frac{\partial \overline{C}}{\partial x} + \overline{v}\frac{\partial \overline{C}}{\partial y} + \overline{w}\frac{\partial \overline{C}}{\partial z} = \frac{\partial}{\partial x}\left(\varepsilon_x \frac{\partial \overline{C}}{\partial x}\right) + \frac{\partial}{\partial y}\left(\varepsilon_y \frac{\partial \overline{C}}{\partial y}\right)$$

$$+ \frac{\partial}{\partial z}\left(\varepsilon_z \frac{\partial \overline{C}}{\partial z}\right) + \omega_o \frac{\partial \overline{C}}{\partial y} \qquad \ldots (9.26)$$

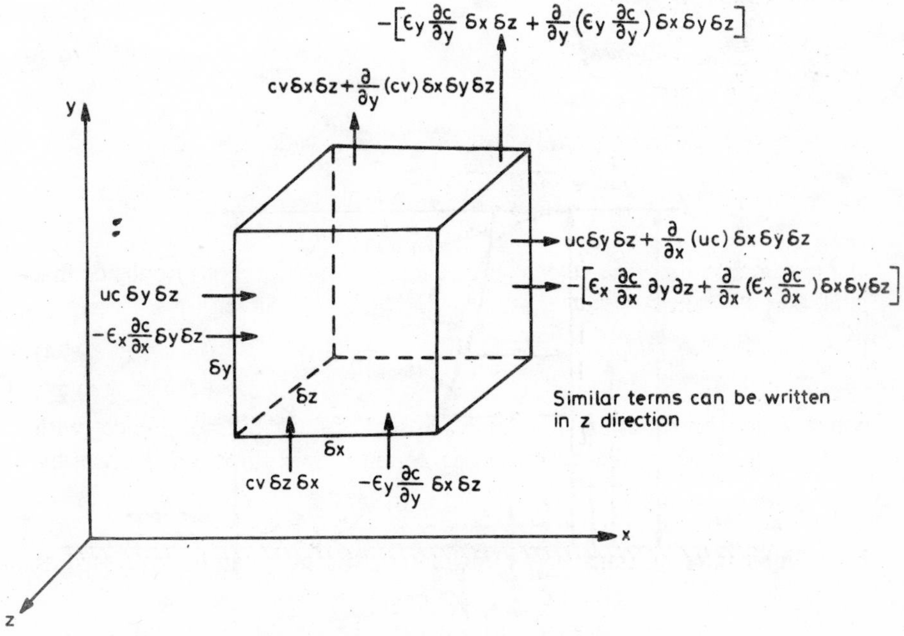

Fig. 9.12 Mass balance in diffusion process

It is instructive to reduce this equation for the case of steady two dimensional uniform flow in an open channel. Since flow is steady, $\frac{\partial}{\partial t} = 0$; for two dimensional flow $\frac{\partial}{\partial z} = 0$. Further, the flow being uniform $\frac{\partial}{\partial x}$. Continuity equation gives $\bar{v} = \bar{w} = 0$. Hence, Eq. 9.26 reduces to

$$\frac{\partial}{\partial y}\left(\varepsilon_y \frac{\partial \bar{C}}{\partial y}\right) + \omega_o \frac{\partial \bar{C}}{\partial y} = 0$$

which on integration yields

$$\varepsilon_y \frac{\partial \bar{C}}{\partial y} + \omega_o \bar{C} = C_1 (x, z, t) \qquad \ldots (9.27)$$

where C_1 is constant of integration. For steady two dimensional uniform flow, C_1 can be either constant or zero. Since there is no sediment transfer across the free surface $C_1 = 0$. Hence,

$$\varepsilon_s \frac{\partial C}{\partial y} + \omega_o C = 0 \qquad \ldots (9.28)$$

Note that ε_y is replaced by ε_s to denote it as the sediment transfer coefficient, and bar over C has been removed for convenience. Integration of Eq. (9.28) can be done if suitable assumption about variation of ε_s with y is made. In general, the diffusion coefficient for momentum, heat or sediment need not be the same. If ε_s is assumed to be constant with respect to y, Eq. 9.28 can be integrated to yield

$$\frac{C}{C_a} = e^{-\omega_o (y-a)/\varepsilon_s} \qquad \ldots (9.29)$$

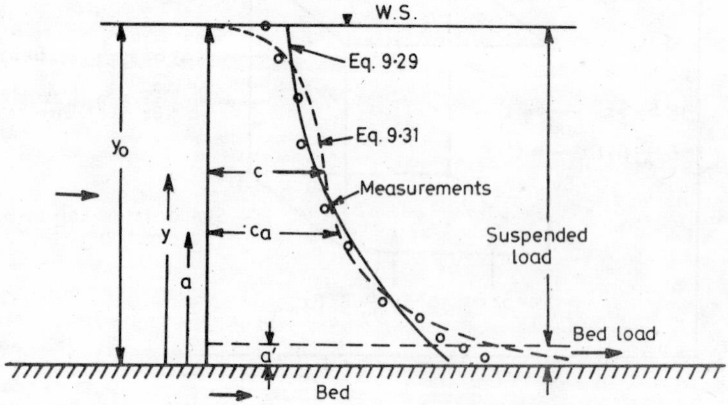

Fig. 9.13 Distribution of suspended load in open channels

where the boundary condition $C = C_a$ at $y = a$ has been used (see Fig. 9.13). This equation was first obtained by Schmidt (25). This equation yields finite concentrations at the boundary and at the free surface (see Fig 9.13). However. assumption of constancy of ε_s over the depth is unrealistic in view of the fact that momentum transfer coefficient ε_m varies with y as (36)

$$\varepsilon_m = u_* \kappa y \left(1 - \frac{y}{y_o}\right) \qquad \ldots (9.30)$$

Here κ is Karman constant with a value of 0.40. If it is assumed that $\varepsilon_s = \varepsilon_m$, substitution of Eq. 9.30 in Eq. 9.28, and subsequent integration yields

$$\frac{C}{C_a} = \left(\frac{y_o - y}{y}\,\frac{a}{y_o - a}\right)^{\omega_0/u_* \kappa} \qquad (9.31)$$

This equation was first published by Rouse in 1937, but independently derived earlier by Ippen, and Hayami. This is also plotted on Fig. 9.13. It may be noted that according to this distribution, $C = 0$ at $y = y_o$ and $C \to \infty$ as $y \to 0$, both these conditions being unrealistic. However, in the major portion of the depth, Eq. 9.31 gives distribution which is realistic.

Even though it is not intented to discuss suspended sediment transport in detail, a few observations are in order. Firstly, one must know, or should be able to predict, C_a at a given elevation a, in order to compute distribution of C with y. Secondly, if velocity distribution law is known or velocity measurements in the vertical are available $\int_a^{y_o} Cudy$ will give the rate of suspended load transport. Here a is the lower limit of integration. This is usually taken as some multiple of bed material size, and it is assumed that bed load transport will take place between y = zero and a. Lastly, it may be mentioned that several attempts have been made to have more refined alternatives to Eq. 9.31. However, Eq. 9.31 is more often used. Other sediment distribution equations as well as limitations of Eq. 9.31 are discussed in reference 26.

Interaction between Sediment and Turbulence

One of the tacit assumptions in the suspended sediment distribution theory presented above is that this diffusion theory considers the suspended particles as passive substances which are carried by the transporting fluid, but which do not affect the turbulence in the fluid. Also, it is assumed that the flow structure does not affect the characteristic fall velocity of the sediment. This is not completely true. Two effects need to be mentioned in this regard.

The first is the effect of concentration of suspended particles and turbulence in the stream on the fall velocity ω_0 of the suspended particles. Investigations by Cunningham, Smoluchowsky and Burgers (27*), McNown and Lin (28) and others have shown that in the presence of suspended particles, the fall velocity of a particle is reduced. If ω is the fall velocity when sediment concentration in absolute volume is C_v, and ω_o is its value in quiescent fluid, then

$$\omega_o/\omega = 1 + 1.56\, C_v^{1/3} \qquad \ldots (9.32)$$

Maude and Whitmore (29) after analysis of a large volume of data showed that

$$\omega_o/\omega = (1 - C_v)^{-\beta_1} \qquad \ldots (9.33)$$

where β_1 is a function of particle Reynolds number $\omega d/\nu$ as indicated in the following table

$\omega d/\nu$	< 0.40	1.0	10	100	1000
β_1	4.65	4.40	3.70	2.80	2.35

For a given ω_o, d, ν and C_v, the above equation has to be solved by successive approximations. It may be mentioned that for given ω_o, d, v and C_v Eqs. 9.32 and 9.33 give different ω values; however both the equations indicate significant reduction in fall velocity.

The turbulence present in the flowing liquid also affects the fall velocity of the particle falling under gravitational action. This effect is more complex. Bechteler et al. (30) found that the fall velocity is reduced in the presence of turbulence. Field (31) also arrived at the same conclusion. Indirect computation of fall velocity of sediment particles in open channel by Jobson and Sayre (32) indicated that for 0.12 mm and 0.39 mm particles the fall velocity increased in the presence of turbulence, however, the effect was more pronounced in the case of finer particles. The role of turbulence characteristics on the change in fall velocity has been studied by Boillat and Graf (33) who conducted experiements in a tank in which homogeneous isotropic turbulence ($\overline{u'^2} \approx \overline{v'^2} \approx \overline{w'^2}$) was created and measured its characteristics along with fall velocities. They compared C_D vs. Re curves for spherical particle under different conditions of turbulence with curve in quiescent water. It was found that the data with different values of $\sqrt{\overline{u'^2}}$ and macroscale L can be systematised if $\sqrt{\overline{u'^2}}/L$ is used as the third parameter which has the dimension of frequency (see Fig, 9.14). It can be seen that for a given value of $\sqrt{\overline{u'^2}}/L$, if Reynolds number is less than 2000, there is one trend of variation; and, if it is greater than 2000, a different trend is indicated. Since for suspended particles Re will be much smaller than 2000, it can be seen that higher the value of $\sqrt{\overline{u'^2}}/L$, for a given Re, smaller will be C_D and, hence, higher will be the fall velocity. The trend is different if Reynolds number exceeds about 2000. For smaller values of $\sqrt{\overline{u'^2}}/L$, C_D is higher and, hence, fall velocity is reduced due to turbulence; however, for higher values of $\sqrt{\overline{u'^2}}/L$, C_D is lower than C_D value for quiescent fluid and, hence, fall velocity is increased. It is further pointed out that the parameter $\sqrt{\overline{u'^2}}/L$ which has the dimensions of frequency can be nondimensionalised by dividing it by natural frequency of the sphere in a given fluid. It needs to be mentioned that the combined effect of sediment concentration and turbulence on the fall velocity is very difficult to predict. Hence, in the design of sedimentation tanks, the designer arbitrarily reduces ω_o by about 50 per cent to account for these two effects.

The second aspect that warrants some discussion concerns the effect of suspended sediment on the turbulence characteristics of the flow field. This aspect has been theoretically investigated among others by Frankle (34*), Barenblatt (34*) and Hino (35). Frankle (34*) between 1953-1960 formulated

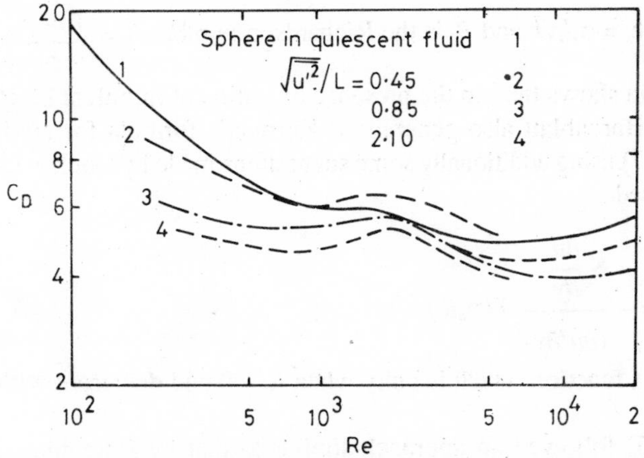

Fig. 9.14 Effect of $\sqrt{u'^2}/L$ on $C_D - R_e$ variation

a system of equations for turbulent flow transporting sediment. These were separate equations for fluid and sediment fluxes, for continuity, momentum and energy for mean and fluctuating motion, and the thermodynamic equations. However, in detailing with the motion in this manner the problem became extremely complicated in that there were 20 equations involving 39 unknowns. According to Vasiliev (34), these equations can form the basis for further detailed investigations.

Barenblatt (34*) between 1953-1956 studied the problem of treating sediment and fluid as a mixture, thereby introducing into the equations density of water sediment complex and mass velocity. That sediment water mixture acts as a continuum has been experimentally verified by Silin et al. (46). Writing equations of continuity, momentum and energy for instantaneous values, averaging procedure was adopted. In further analysis, simplification was done making the following assumptions :

(i) concentration of suspended particles in absolute volume is much smaller than 1.0;

(ii) acceleration of water and sediment particles is much smaller than gravitational acceleration;

(iii) viscosity of mixture is given by Einstein's relation $\mu = \mu_o\,(1 + 2.5\,C_v)$; and

(iv) for small concentration, the fall velocity of particles is unaffected by it.

The resulting equations are given by Vasiliev. While reducing the equations for two dimensional steady uniform flow in open channels, Barenblatt used Kolmogorov's hypothesis (see Chapter V) according to which turbulent kinetic energy per unit mass k, and Kolomogorov's linear scale of turbulence l_k can define dissipation rate of turbulence as well as transfer coefficient. He was, thus, able to show that

$$k = \frac{1}{\sqrt{c_1}}\,u_*^2\left(1 - \frac{y}{y_o}\right)(1 - \alpha_s R_i)^{1/2} \qquad \ldots (9.34)$$

in which c_1 is the constant in the relation for dissipation rate of turbulence,

namely $\varepsilon = c_1 k^{3/2}/l_k$, α_s is the constant in the relation for sediment transfer coefficient $\varepsilon_s = \alpha_s l_k \sqrt{k}$ and R_i is the Richardson number $R_i = \left(\frac{\rho_s - \rho_f}{\rho_f}\right) \frac{g dC/dy}{(du/dy)^2}$.
This relation shows that, in the presence of sediment turbulent kinetic energy decreases. Barenblatt also generalised Karman's formula for mixing length (see Eq. 6.17) using additionally some suggestions made by Obukov and Monin, and proposed

$$l = -\frac{\kappa \dfrac{\partial u}{dy}}{(\partial u/\partial y)^2} F(\alpha_s R_i)$$

where F is a function which is unity when $R_i = 0$ and decreases with increase in R_i.

Hino (35) followed an approach similar to that by Barenblatt, and used energy balance equation and a semi empirical equation for turbulent acceleration balance. On the basis of these equations, he arrived at a number of conclusions regarding effect of suspended sediment on the structure of turbulence, the major ones being the following :

(i) average size of eddies decreases as the sediment concentration increases;
(ii) there is a marginal decrease in turbulence intensity $\overline{u'^2}$ with increase in concentration; however, for neutrally buoyant material, $\overline{u'^2}$ is increased significantly.
(iii) Karman constant decreases with increase in concentration; and
(iv) with the increase in concentration, there is a rapid reduction in average life period of eddies.

Buckley, in 1922, observed that the effect of presence of fine sediment on the Nile river was to reduce the friction factor, thereby increasing velocity for the same depth and slope. Experiments by Vanoni (25) and Ismail (36) in the laboratory channels confirmed this finding. In addition, they showed that in the presence of sediment, Karman constant is reduced, indicating damping of turbulence. Many investigators such as Zagustin and Zagustin (37), Hino (35), Arai and Takahashi (38) produced semi analytical and experimental evidence to show decrease in κ. On the other hand, other investigators such as Kalinske and Hsia (39), Gust (40), Itakura and Kishi (41) and Coleman (42) have concluded from the same and some additional data that κ remains unchanged with increase in the concentration of suspended sediment. For example, Coleman has argued that values of κ obtained by earlier investigators are incorrect since they were obtained by fitting logarithmic law to the velocity distribution over the entire depth, whereas it should have been obtained from velocity data in the wall region. He showed that, when velocity distribution law of the type

$$\frac{u}{u_*} = \left[\frac{2.303}{\kappa} \log \frac{u_* y}{\nu} + A\right] + \frac{\Delta u}{u_*} + \frac{\pi}{\kappa}\, \omega\left(\frac{y}{\delta}\right) \qquad \ldots (9.35)$$

is fitted to the velocity distribution data in open channels with suspended sediment, κ is constant and is not dependent on concentration of suspended sediment. Hence, in spite of a number of investigations, there is still some controversy about variation of κ.

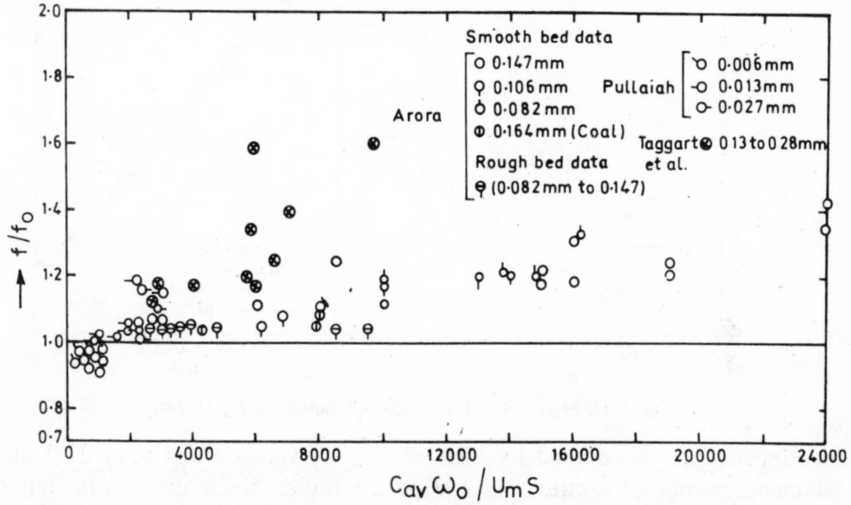

Fig. 9.15 Variation of f/f_o with $C_{av}\omega_o/U_mS$

Another question regarding reduction in resistance to flow in the presence of sediment has been studied by Pullaiah (43) and Arora (44). These studies have revealed that f/f_o depends on the parameter $C_{av}\omega_o/U_mS$ where C_{av} is average concentration in ppm by volume (see Fig. 9.15). If this parameter is less than 1200, friction faction f with suspended sediment is less than clear water friction factor f_o; on the other hand, if this parameter exceeds 1200, f/f_o is greater than unity. It may be mentioned that $C_{av}\omega_o/U_mS$ can be interpreted as the ratio of work done against the density gradient to the energy required to overcome friction, and is, in reality, gross Richardson number.

Gry (45) has presented a heuristic vortex model to give explanation about decrease in frictional resistance in the presence of suspended sediment. According to this model, because of the centrifugal force within the eddy, there is concentration of sediment particles near its core, leading to increase in viscosity. Since, as shown by him, decay period of an eddy is directly proportional to the square of its radius and inversely proportional to the viscosity, because of increase in viscosity the vortex dissociates faster. Effectively, this diffusion process produces an energy transfer from small eddies to larger ones, thereby reducing dissipation of energy of small eddies. Since energy dissipation in turbulent flow is proportional to $\overline{u'^2}/\lambda^2$, it would mean increase in the average size of small eddies and decrease in turbulence level.

In order to verify Gry's hypothesis, Müller (46) conducted experiments in a rough vertical pipe of 200 mm diameter. According to Gry's theory $R_E(\tau)$ the Eulerian correlation coefficient for a fixed time should increase with increase in concentration. This was indeed observed for small times indicating increase in the size of small eddies (see Fig. 9.16). It can also be seen that the effect is more pronounced near the centre of pipe.

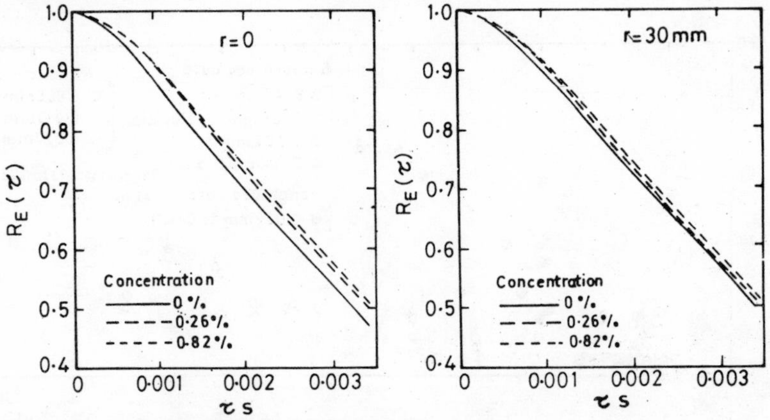

Fig. 9.16 Effect of sediment concentration on $R_E(\tau)$, (46)

Experiments conducted by Silin et al. (46) showed that upto 2.00 mm diameter particles the ratio of *rms* value of turbulent fluctuations of the liquid phase and solid phase was close to unity. Hence, such a fluid can be treated as continuum with increase in mass density due to presence of sediment. Pechenkin and Vedeneev (47) also obtained reduction in turbulent fluctuations due to presence of suspended sediment. It may be mentioned that experimental work done by Silin et al., and Pechenkin and Vedeneev was carried out in pipes.

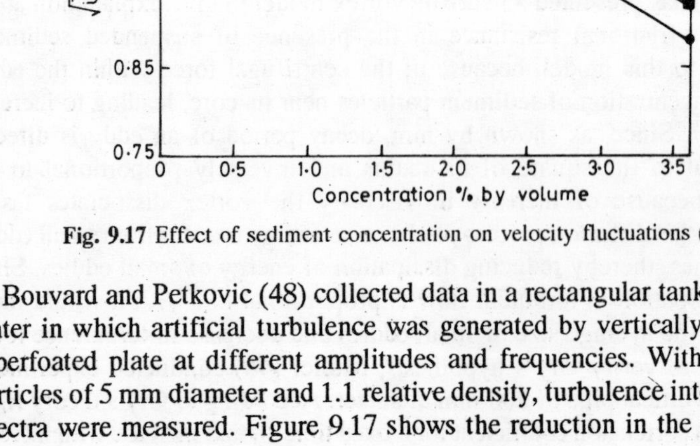

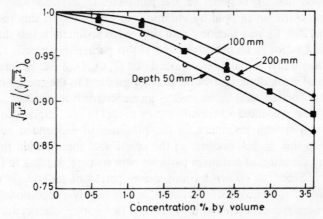

Fig. 9.17 Effect of sediment concentration on velocity fluctuations (48)

Bouvard and Petkovic (48) collected data in a rectangular tank filled with water in which artificial turbulence was generated by vertically oscillating a perfoated plate at different amplitudes and frequencies. With suspended particles of 5 mm diameter and 1.1 relative density, turbulence intensities and spectra were measured. Figure 9.17 shows the reduction in the intensity of turbulence as the concentration in increased. It can be seen that there is about 15 per cent reduction at a concentration of 3.6%. Figure 9.18 shows the energy spectrum. It is clear that in the presence of sediment low frequency turbulence (i.e. larger eddies) have relatively less energy whereas higher frequency components (i.e. smaller eddies) have marginally more energy. It can, thus, be seen that presence of sediment in the flow does affect the structure of turbulence and resistance to flow. However, because of the difficulties of measuring turbulence in water in the presence of sediment, adequate techniques are not available to quantitatively assess these effects.

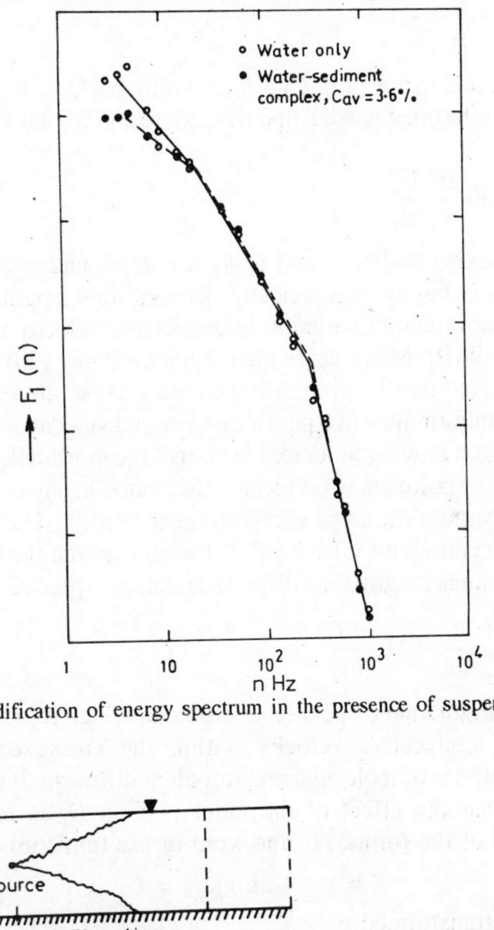

Fig. 9.18 Modification of energy spectrum in the presence of suspended sediment (48)

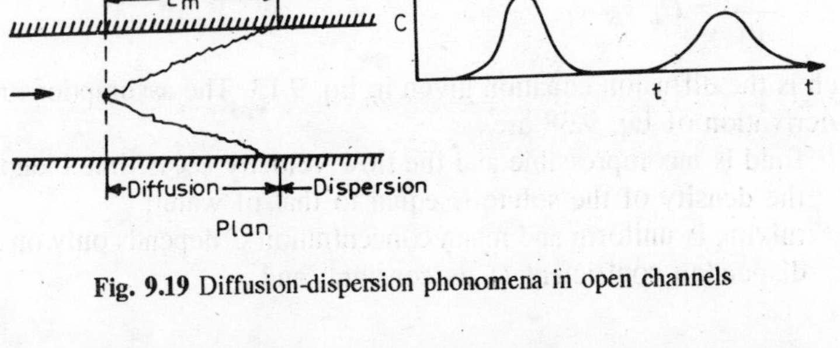

Fig. 9.19 Diffusion-dispersion phonomena in open channels

9.5 LONGITUDINAL DISPERSION IN OPEN CHANNELS

Earlier dispersion-diffusion equation has been obtained for turbulent flow in open channels using the Cartesian co-ordinate system, see Eq. 9.15. Then it was simplified to Eq. 9.16 on the premise that molecular diffusion coefficient D_m being much smaller than ε_x, ε_y, ε_z, the term $D_m \nabla^2 \overline{C}$ can be omitted. If a dye or chemical pollutant is introduced in a stream as a point or line source, it will initially diffuse in the vertical, lateral and longitudinal directions and after a certain distance the dye will be uniformly distributed at any section at a given time. This distance, known as the mixing length L_m, is given by

$$L_m = C_1 \frac{B^2}{R} \frac{U_m}{u_*} \qquad \ldots (9.36)$$

where according to Fischer (49), $C_1 = 0.45$, while McQuivey and Keefer (50) give $C = 0.30$. On the other hand, Kilpatrick, Sayre and Richardson (51) propose the formula

$$L_m = 0.40 \frac{B^2}{y_o} \frac{U_m}{u_*} \qquad \ldots (9.37)$$

Here B is the channel width, y_o and R are the depth and hydraulic radius of the flow and U_m is the average velocity. Beyond this length, the dispersion takes place primarily due to variation in convective velocity within the cross section (see Fig. 9.19). Many years after Taylor's paper (20) was published, Batchelor pointed out that Taylor's theory would apply to dispersion of marked fluid particles in uniform flow in a pipe if distance and velocity were measured in a frame of reference moving at a cross-sectional mean velocity U_m. In such a case, when solute or pollutant is uniformly distributed in any cross-section at a given time, one can use the cross sectional mean velocity U_m and cross sectional average concentration C. Taylor (52) has shown that the mixing process called one dimensional longitudinal dispersion can be expressed by the equation

$$\frac{\partial C}{\partial t} + U_m \frac{\partial C}{\partial x} = D_L \frac{\partial^2 C}{\partial x^2} \qquad \ldots (9.38)$$

Here D_L is the longitudinal dispersion coefficient, which represents effects of the variation of convective velocity within the cross section and also incorporates the effects of molecular and turbulent diffusion. It must, however, be emphasised that the effect of the latter two on D_L is relatively small compared to that of the former. If one were to use transformed coordinates

$$\xi = x - U_m t \text{ and } \tau = t$$

Eq. 9.38 can be transformed to

$$\frac{\partial C}{\partial \tau} = D_L \frac{\partial^2 C}{\partial \xi^2}$$

which is the diffusion equation given in Eq. 9.13. The assumptions made in the derivation of Eq. 9.38 are

(i) fluid is incompressible and the flow velocity U_m is time independent;
(ii) the density of the solute is equal to that of water;
(iii) mixing is uniform and mean concentration C depends only on x and t.
(iv) dispersion coefficient D_L is constant; and

(v) constant convective velocity $\overline{U}$ of the tracer cloud is equal to U_m.
If slug of the solute is injected as a line source in a stream having the slug mass as M, the following boundary conditions can be specified for Eq. 9.38.

$$C(x,0) = 0, \text{ for all } x>0$$

$$\int_{-\infty}^{\infty} C(x,t)\, A dx = M \text{ for all } t$$

$$C(\infty, t) = 0 \text{ for all } t \geq 0, \text{ and}$$

$$dC/dx = 0 \text{ as } x \to \infty$$

The solution of Eq. 9.38 can be obtained by first writing $\xi = x - U_m t$ and using the transformation $\xi / \sqrt{4D_L t} = \eta$. The solution is

$$C(x,t) = \frac{M}{2A\sqrt{\pi D_L t}} \exp\left\{-\frac{(x - U_m t)^2}{4D_L t}\right\} \qquad \ldots (9.39)$$

where A is the cross sectional area of the channel. This is known as the normal or Gaussian distribution which is symmetrical in nature. If a tracer is introduced, and at sufficiently long distances C vs t curves are obtained at two stations 1 and 2, one can determine the constant value of D_L by the formula

$$D_L = \frac{\overline{U}^3(\sigma_{t2}^2 - \sigma_{t1}^2)}{2(x_2 - x_1)} \qquad \ldots (9.40)$$

where σ_t^2 is the variance of C with respect to t, and $\overline{U}$ is velocity of cloud which can be taken as U_m, the average velocity of flow, and the variance σ_t^2 is obtained as

$$\sigma_t^2 = \frac{\int_{-\infty}^{\infty} Ct^2\, dt}{\int_{-\infty}^{\infty} C\, dt} - \overline{t} \qquad \ldots (9.41)$$

where $\overline{t}$ is the time to the centroid of tracer cloud at any station and is determined as

$$\overline{t} = \int_{-\infty}^{\infty} Ctdt \Big/ \int_{-\infty}^{\infty} C dt$$

This is known as the moment method of determination of D_L. A large volume of data has been collected from laboratory and field studies on dispersion during the past three decades. Study of these data has revealed several interesting features about the analysis given above. Many investigators have proposed equation for D_L in the form

$$D_L / u_* y_o = \text{constant} \qquad \ldots (9.42)$$

However, the value of the constant is found to vary considerably; thus, Taylor has recommended the constant as 5.93 while, values recommended by others are listed below.

Investigator	**Constant in Eq. 9.42**
Elder	6.30
Sumer	6.23
Krenkel	9.10
Yotsukura and Fiering	9.0 to 13.0
Glover	500
Thackston and Krenker	7.25

Field studies have shown that D_L/u_*y_o depends on relative roughness, irregularities in channel section and its variation along length, slope, width/depth ratio, presence of bends, etc. Singh (53) has found that, when laboratory and field data are analysed together, for a given value of u_*y_o, D_L can vary by as much as a factor of 10^2. He also found that D_L/u_*y_o is a function of B/y_o. Further, D_L is found to increase gradually in the downstream direction. As a result, it is, at present, very difficult to predict D_L for the known flow conditions.

The second observation that emerges from the analysis is that whereas C vs t curves for a given x should be symmetrical following Gaussian pattern, invariably they are skew in the upstream direction with a long tail of low concentration. Hence Nordin and Sobol (54) as well as Day (55), on the basis of analysis of data from natural streams, concluded that one dimensional Fickian theory described above is not suitable for natural streams.

It has also been observed that in natural streams, there is loss in the average solute concentration, and this loss increases with increase in x. This loss is due to photochemical decay and benthic adsorption by vegetation and suspended sediment present in the stream. Also a significant portion of injected dye may be temporarily detained in the dead zones in the bed and banks of the stream, and later released gradually. Further, the velocity of the tracer cloud $\overline{U}$ is not always the same as average velocity of flow U_m. Even though in flume studies the two are nearly equal, the same is not true for field data. This is believed to be due to nonuniformity in the flow cross section in the longitudinal direction, presence of dead zones, and change in discharge in the longitudinal direction.

In order to overcome these deficiencies in the dispersion model, two approaches have been made. In the first approach the Fickian model is suitably modified. Thus, Nordin and Sobol (54) modified the model dividing the depth in two zones, the top zone of constant velocity and lower region of zero velocity representing the dead zone. A particle can spend certain time in the upper layer and remaining time in lower layer. Liu and Cheng (56, 57) have used a time dependent dispersion coefficient. Bansal (58) has incorporated in his model the increasing loss of tracer.

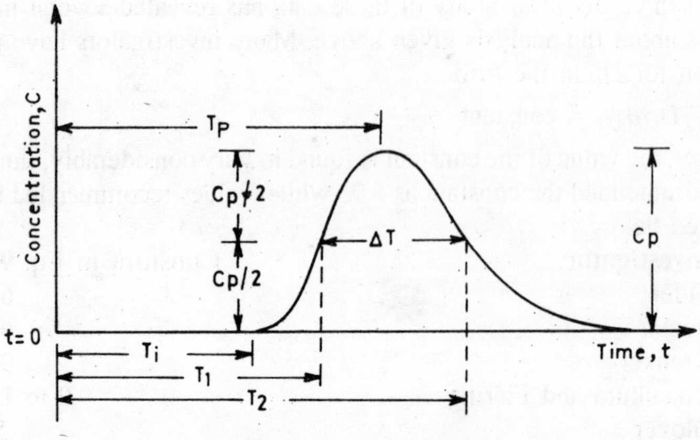

Fig. 9.20 Definition sketch

The second approach is that of similarity solution. This was first adopted by Day and Wood (59) who, after analysing over 700 C vs t curves, found that even though they were unsymmetrical, they were similar. To systematise the C vs t curves, they adopted C_p the peak concentration and time ΔT as the reference parameters. Here $\Delta T = (T_2 - T_1)$ where T_1 and T_2 are the times corresponding to the instants when $C = \frac{C_p}{2}$ (see Fig. 9.20). Thus, C/C_p and $\frac{(t - T_1)}{\Delta T}$ are the two dimensionless parameters used for plotting C vs t curves. Figure 9.21 shows a large volume of data from laboratory and field plotted as C/C_p vs $(t - T_1)/\Delta T$, which clearly validates the concept of similarity. Mean curves of C/C_p vs $\frac{(t - T_1)}{\Delta t}$ obtained by Day and Wood (59), and Singh (59) who has also used this approach, are shown in Fig. 9.22. It can be seen that they are very close to each other. Hence, if one can predict C_p, T_1 and ΔT, one can establish C vs t curve. Day and Wood have suggested the following relationships for these parameters:

$$\left.\begin{aligned} C_p &= \forall C_o/Q\Delta T.I \\ T_1U_m/B &= 0.71\,(x/B) - 0.32 \quad \text{if } x/B < 24.0 \\ T_1U_m/B &= 0.86\,(x/B) - 3.90 \quad \text{if } x/B > 24.0 \\ \frac{\Delta T U_m}{B} &= 0.24\,(x/B) + 0.24 \end{aligned}\right\} \quad \ldots (9.43)$$

where $\forall$ is the volume of tracer injected in the stream, C_o is the initial concentration, I is the area under dimensionless curve in Fig. 9.22 which was found to be 1.12, and B is the channel width. Singh, after analysing his own data as well as those collected by other investigators, found that the prediction

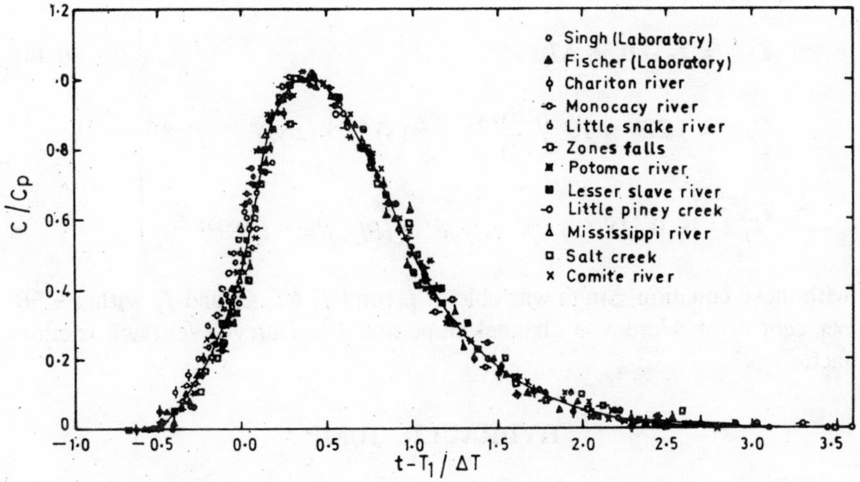

Fig. 9.21 Variation of C/C_p *with* $(t-T_1)/\Delta T$

of C_p, T_1 and ΔT using Day and Wood's relations gave as much as ± 100 per cent error. Hence, on the basis of his own analysis, Singh has proposed the following equations.

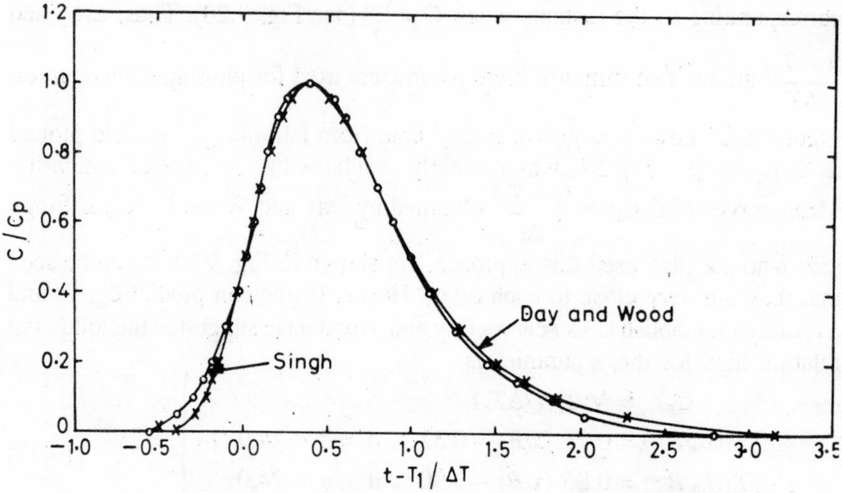

Fig. 9.22. Similarity curves obtained by Day and Wood, and Singh

$$\left.\begin{aligned}
\frac{T_1\sqrt{g}}{\sqrt{x}} &= 0.462\,(S)^{0.423}\,(x/y_o)^{0.597}\,(B/y_o)^{-0.18}\,(f)^{0.421} \\
\frac{\Delta T\sqrt{g}}{\sqrt{x}} &= 0.598\,(S)^{0.362}\,(x/y_o)^{0.349}\,(B/y_o)^{-0.0894}\,(f)^{0.528} \\
T_1 &= T_p - 0.38\,\Delta T \\
C_p &= 1.229 \times 10^6 \left(\frac{\rho B y_o x}{M}\right)^{-0.679} (S)^{-0.064}\,(x/y_o)^{0.138}\,(f)^{-.008} \\
\frac{T_p\sqrt{g}}{\sqrt{x}} &= 0.518\,(S)^{-0.405}\,(x/y_o)^{0.606}\,(B/y_o)^{-0.186}\,(f)^{0.392}
\end{aligned}\right\} \quad \ldots (9.44)$$

with these equation, Singh was able to predict T, ΔT, C_p and T_p within ± 30 per cent error. Here S is channel slope and f is Darcy–Weisbach friction factor.

HYDRAULIC JUMP

Hydraulic jumps are most often used as an energy dissipating device in hydraulic structures. The usual literature available on hydraulic jump deals with derivations regarding ratio of depth after the jump to that prior to the jump, energy loss, and effect of channel shape, which use the continuity, momentum and energy equations for the mean flow, neglecting terms involving turbulence quantities.

Basic elements of hydraulic jump are shown in Fig. 4.1. The conventional treatment of hydraulic jump is, therefore, aimed at obtaining the expressions for y_2/y_1, h_L/y_1 and L_j/y_1 in the form

$$\frac{y_2}{y_1}, \frac{h_L}{y_1}, \frac{L_j}{y_1} = f\,(F_{r1}, \text{ channel shape, slope, etc.})$$

Here y_1 and y_2 are depths before and after the jump, h_L is the energy loss in the jump and L_j the length of the jump; $F_{r1} = U_1/\sqrt{gy_1}$ where $U_1 y_1 = q$, the discharge per unit width.

In doing so, such discussion fails to provide greater insight into the flow phenomenon, and does not provide information on effect of boundary friction, turbulence intensity and nonuniform velocity distribution on the above parameters. Further, from the point of view of realistic design of floor, one needs to know about pressure fluctuations on the floor. Also, it would be of interest to know how turbulence generated in the jump dies out. These aspects are discussed below.

9.6 MOMENTUM AND ENERGY EQUATIONS

Rouse et al. (60) have reduced the differential and integral forms of momentum and energy equations for the case of hydraulic jump in open channels. If it is assumed (see Chapter IV) that

(i) channel is wide and horizontal;
(ii) prior to jump the turbulence is negligible;
(iii) pressure distribution in the vertical is hydrostatic.

Momentum equation between section 1 and section 2 at a distance x downstream, takes the form

$$\int_0^y \rho\bar{u}^2 dy - \int_0^{y_1} \rho\bar{u}^2 dy + \int_0^y \rho\overline{u'^2} dy = \frac{\gamma}{2}(y_1^2 - y^2) - \int_0^x \mu \left(\frac{d\bar{u}}{dy}\right)_{y=0} dx \quad \ldots (9.45)$$

and if momentum equation is applied between sections 1 and 2 at the end of hydraulic jump, one gets

$$\int_o^{y_2} \rho\bar{u}^2 dy - \int_o^{y_1} \rho\bar{u}^2 dy + \int_o^{y_2} \rho\overline{u'^2} dy = \frac{\gamma}{2}(y_1^2 - y_2^2) - \int_o^{L_j} \mu \left(\frac{d\bar{u}}{dy}\right)_{y=0} dx \quad \ldots (9.46)$$

In order to use this equation, one must know how $\bar{u}$ *and* $\overline{u'^2}$ vary with y, and the boundary shear with x. It may be mentioned that if boundary shear is neglected, $\bar{u}$ is assumed to be constant in the vertical at both the sections, and $\overline{u'^2}$ is zero at section 2, Eq. (9.46) reduces to conventional momentum equation found in standard Open Channel Flow texts.

With the same assumptions mentioned earlier, energy equations for the mean and turbulent motions take the following forms :

Mean Motion:

$$\left.\begin{aligned}&\int_0^{y_2}\frac{\overline{V}^2}{2g}\bar{u}\,dy-\int_0^{y_1}\frac{\overline{V}^3}{2g}\,dy+\int_0^{y}\frac{(\bar{u}\,\overline{u'^2}+\bar{v}\,\overline{u'v'})}{g}\,dy\\&-\int_0^y\!\!\int_0^x\left[\frac{\overline{u'v'}}{g}\left(\frac{\partial\bar{u}}{\partial y}+\frac{\partial\bar{v}}{\partial x}\right)+\left(\frac{\overline{u'^2}-\overline{v'^2}}{g}\right)\frac{\partial\bar{u}}{\partial x}\right]dydx\\&=qy_1-qy+\int_0^y\frac{\mu}{\gamma}\left[2\bar{u}\frac{\partial\bar{u}}{\partial x}+\bar{v}\left(\frac{\partial\bar{u}}{\partial y}+\frac{\partial\bar{v}}{\partial x}\right)\right]dy\\&-\int_0^y\!\!\int_0^x\frac{\mu}{\gamma}\left(4\left(\frac{\partial\bar{u}}{\partial x}\right)^2+\left(\frac{\partial\bar{u}}{\partial y}+\frac{\partial\bar{v}}{\partial x}\right)^2\right)dydx.\end{aligned}\right\}\quad\ldots(9.47)$$

Turbulent Motion:

$$\left.\begin{aligned}&\int_0^{y}\bar{u}\frac{\overline{V'^2}}{2g}dy-\int_0^{y_1}\frac{\overline{V'^2u'}}{2g}\,dy+\int_0^y\!\!\int_0^x\left[\left(\frac{\overline{u'v'}}{g}\right)\left(\frac{\partial\bar{u}}{\partial y}+\frac{\partial\bar{v}}{\partial x}\right)+\left(\frac{\overline{u'^2}-\overline{v'^2}}{g}\right)\frac{\partial\bar{u}}{\partial x}\right]dydx\\&=-\int_0^y\frac{\overline{p'u'}}{\gamma}\,dy+\int_0^y\frac{\mu}{\gamma}\left[\frac{\partial}{\partial x}\left(\frac{\overline{V'^2}}{2}+\overline{u'^2}\right)+\frac{\partial\,\overline{u'v'}}{\partial y}\right]dy\\&\int_0^y\!\!\int_0^x K\frac{\mu}{\gamma}\left(\frac{\partial u'}{\partial x}\right)^2dydx\end{aligned}\right\}\quad\ldots(9.48)$$

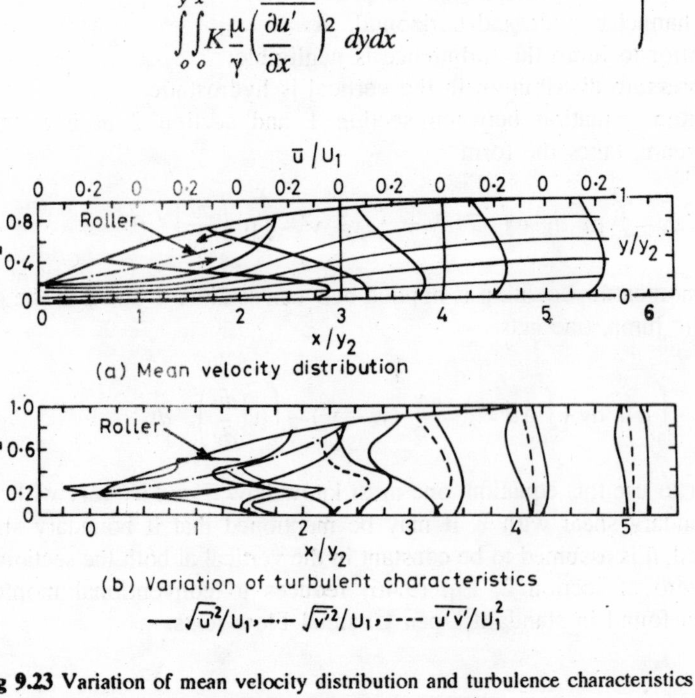

Fig 9.23 **Variation of mean velocity distribution and turbulence characteristics in hydraulic jump at F_{r_1} = 4, (60)**

where $\overline{V}^2 = (\overline{u}^2 + \overline{v}^2 + \overline{w}^2)$, $\overline{V'^2} = (\overline{u'^2} + \overline{v'^2} + \overline{w'^2})$ and K is known as the dissipation constant and assumes a value of 15 when small scale turbulence is isotropic (see Chapter V).

Momentum equation (Eq. 9.45) can be nondimensionalised by dividing each term by $\rho U_1^2 y_1$ yielding the following expression

$$\underset{①}{\frac{1}{U_1^2 y_1}\int_0^y \overline{u}^2 dy} + \underset{②}{\frac{1}{U_1^2 y_1}\int_0^{y_1} \overline{u'^2} dy} + \underset{③}{\frac{1}{2F_{r1}^2}\left(\frac{y}{y_1}\right)^2} + \underset{④}{\frac{1}{Re_1 U_1}\int_0^x \left(\frac{\partial \overline{u}}{\partial y}\right)_{y=0} dx} \quad \ldots (9.49)$$

which should remain constant at each section. Here $Re_1 = \frac{U_1 y_1}{\nu}$. The four terms in the above expression have the following interpretations :

1. Mean momentum flux *MM*
2. Turbulent momentum flux *TM*
3. Pressure term *P*
4. Shear *S*

Experimental data can be used to verify the constancy of sum of the four terms in Eq. (9.49). In a similar manner one can start with energy equation for mean motion, viz. Eq. (9.47) and nondimensionalise it by dividing each term by $y_1 U_1^3/2g$. Further, terms involving viscous shear can be neglected, since they are found to be small experimentally. This would lead one to the conclusion that the sum

$$\left.\begin{aligned} &\underset{①}{\frac{1}{U_1^3 y_1}\int_0^y \overline{V}^2 \overline{u} dy} + \underset{②}{\frac{2}{F_{r1}^2}\frac{y}{y_1}} + \underset{③}{\frac{2}{U_1^3 y_1}\int_0^y (\overline{u}\,\overline{u'^2} + \overline{v}\,\overline{u'v'})\, dy} \\ &- \underset{④}{\frac{2}{U_1^3 y_1}\int_0^y\int_0^x \left[\overline{u'v'}\left(\frac{\partial \overline{u}}{\partial y} + \frac{\partial \overline{v}}{\partial x}\right) + \left(\overline{u'^2} - \overline{v'^2}\right)\frac{\partial \overline{u}}{\partial x}\right] dy dx} \end{aligned}\right\} \quad \ldots (9.50)$$

should remain constant along the length. The four terms appearing in the above expression have the following interpretations

1. Flux of kinetic energy *FKE*
2. Work done by pressure *WP*
3. Work performed by Reynolds stresses *WRS*
4. Production of turbulence *PT*

9.7 EXPERIMENTAL VERIFICATION OF MOMENTUM AND ENERGY BALANCE

Since measurement of turbulence in water is difficult, and it is more so in the hydraulic jump where air entrainment takes place, Rouse et al. (60) conducted experiments in an air model. For given y_1 and F_{r1}, a hydraulic jump was formed in an open channel and free surface profile was carefully noted. A rectangular duct was then constructed with its bottom horizontal and top surface conforming in its shape to the water surface of hydraulic jump for known F_{r1}. The upstream and downstream depths of the duct were equal to y_1 and y_2, respectively, for the jump. Mean velocity distribution was measured by double pitot formed by two L shaped tubes placed back to back; boundary shear was measured by Preston tube while turbulent fluctuations were measured with hot wire anemometer. The authors have discussed in detail the justification, from fluid mechanics point of view, for considering the air model as adequate representation of hydraulic jump.

Figure 9.23 shows the typical results obtained for F_{r1} equal to four. Figure 9.23(a) shows the distribution of mean velocity in the vertical nondimensionalised with U_1, i.e. $\bar{u}/U_1$, for various sections. It shows the regions of high velocity near the bottom as well as region of reverse flow. Figure 9.23 (b) shows variation of $\sqrt{\overline{u'^2}}/U_1$; $\sqrt{\overline{v'^2}}/U_1$ and $-\overline{u'v'}/U_1^2$ in the vertical at various sections. $\sqrt{\overline{w'^2}}$ was also measured, and it was found that $\sqrt{\overline{w'^2}}$ was not much different from $\overline{v'^2}$. This figure indicates that turbulence intensity is maximum where the mean velocity gradient is maximum, i.e. near the roller edge. The intensity of turbulent shear is, likewise, maximum where the mean velocity gradient is maximum.

The analysis of momentum balance for these data is illustrated in Fig. 9.24 where data are presented for Froude number equal to four. As indicated by expression 9.49, the sum of *MM, TM, P* and *S* must remain constant along the length of the jump. The deviation from this constant value depicted in Fig. 9.24 is evidently due to experimental error. Two major conclusions can be drawn from such momentum analysis. Firstly, contribution of turbulence to the momentum flux is important in the intermediate sections, but is small at the end sections. Secondly, contribution of shear, although small, increases with both distance and Froude number. Conclusion regarding Froude number effect is drawn from study of data for other F_{r1} values. In the similar manner, according to expression 9.50, the sum of *FKE, WP, WRS* and *PT* should remain the same along the length of the jump as far as energy balance for mean motion is concerned. This is verified in Fig. 9.25 for $F_{r1} = 4.0$. Energy analysis has revealed that the locus of points of maximum production of turbulence energy coincides closely with the border of the roller, whereas that for maximum dissipation lies, apparently, higher. Further, even though the maximum rate of production per section occurs ahead of the middle of the roller, the maximum rate of dissipation occurs shortly before its end.

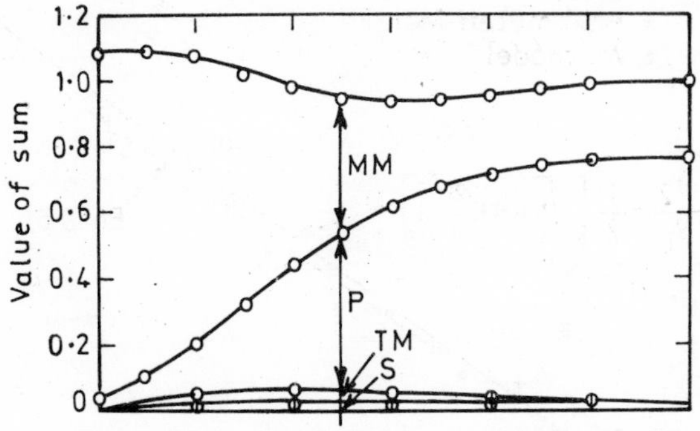

Fig 9.24 Momentum balnce in hydraulic jump at F_{r_1} = 4, (60)

The following general conclusions drawn by Rouse et al. are of relevance. Roller extends about half the length of the jump and half of the total energy of turbulence is produced in half the length of the roller. Further, because of the convective effect of the mean flow and diffusive effect of turbulence, the energy is not dissipated at the point where it is produced, but some distance away, vertically and longitudinally. Lastly, the kinetic energy of turbulence at any section in the jump is relatively small.

Harleman, in his discussion of the paper of Rouse et al. (60) obtained an expression for y_2/y_1, for the jump by using Eq. 9.46. This is

$$F_{r1}^2 = \frac{\frac{J}{2}\left[(J+1)(J-1)+S_1\right]}{\beta_1 J - (\beta_2 + I_2)} \qquad \ldots (9.51)$$

in which $J = y_2/y_1$, $S_1 = \dfrac{2F_{r1}^2}{U_1 R_{e1}} \displaystyle\int_o^{L_j} \left(\frac{\partial \bar{u}}{\partial y}\right)_{y=0} dx$

$I_2 = \dfrac{\int_o^{y_2} \overline{u'^2}\,dy}{U_2^2 y_2}$ and β_1 and β_2 are the momentum correction coefficients at section 1 and 2, respectively. The two integrals S_1 and I_2 were evaluated by

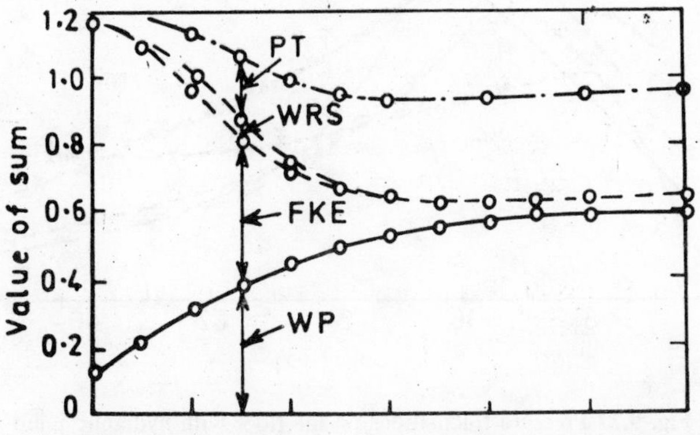

Fig. 9.25 Energy balance in hydraulic at jump F_{r_1} = 4, (60)

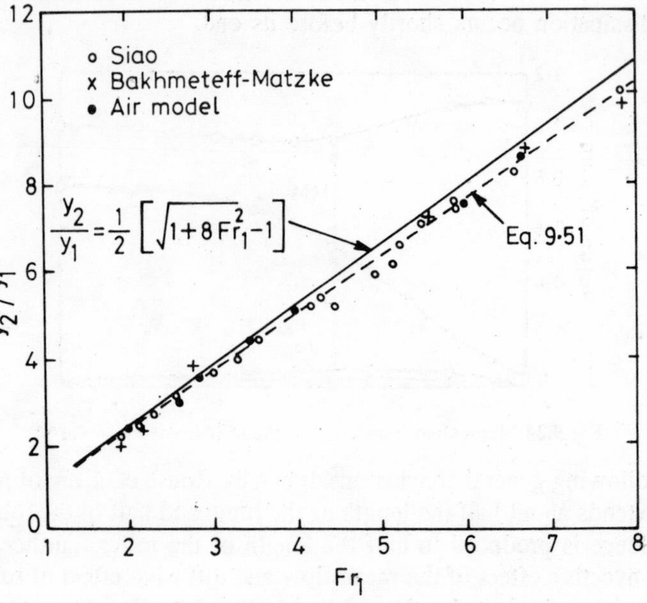

Fig. 9.26. Variation of y_2/y_1 with F_{r1} for hydraulic jump

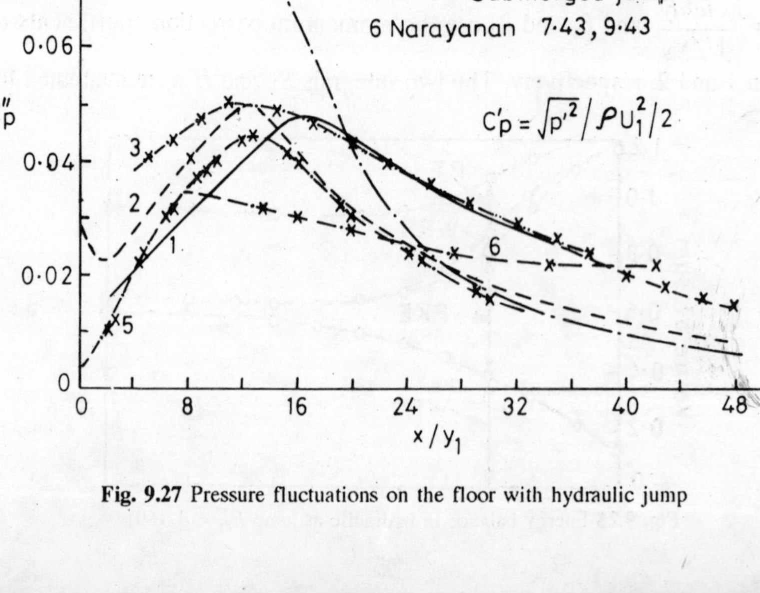

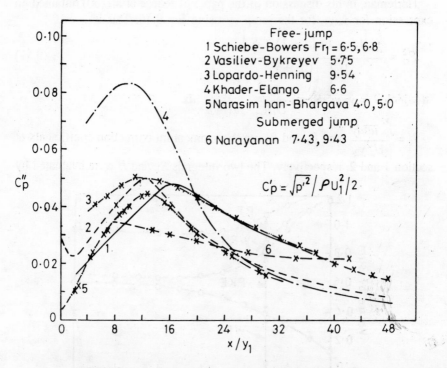

Fig. 9.27 Pressure fluctuations on the floor with hydraulic jump

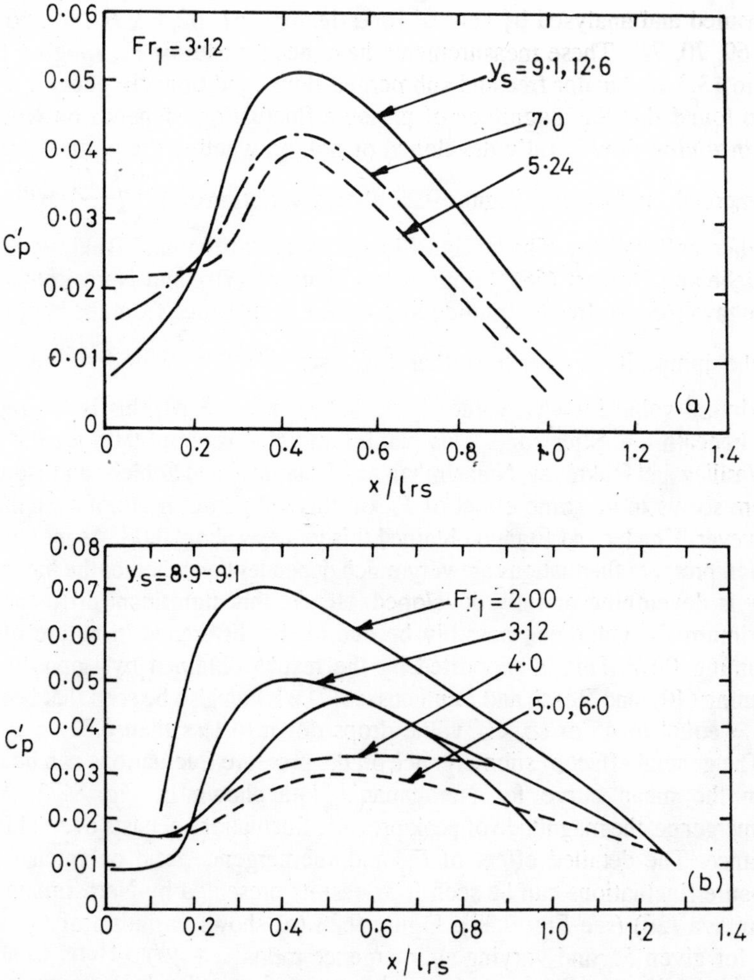

Fig. 9.28 Variation of C'_p with x/l_{rs} and submergence in a hydraulic jump

Harleman for the data pressented by Rouse et al. for various F_{r_1} values. It was found that values of y_2/y_1 obtained from Eq. 9.51 were consistently lower than those obtained by simplified analysis by about three per cent and agreeing well with observed ones (see Fig. 9.26).

9.8 PRESSURE FLUCTUATIONS ON THE FLOOR

Because of the highly turbulent flow conditions in the hydraulic jump and the anisotropic and nonhomogeneous structure of turbulence, significant pressure fluctuations occur on the floor of the stilling basins. These rapid fluctuations in pressure can lead to the damage to the floor due to fatigue, structural resonance or cavitation. Pressure fluctuations also play a significant role in the scour if the bed is erodible.

The pressure fluctuations on the floor beneath the hydraulic jump have been measured and analysed by several investigators (61, 62, 63, 64, 65, 66, 67, 68, 69, 70, 71). These measurements have been made for F_{r_1} ranging from 4.7 to 13.3, and under free and submerged flow conditions. In general, it has been found that the magnitude of pressure fluctuations depends on whether the incoming flow is fully developed or not, on whether the jump is free or submerged, and on F_{r_1}. Figure 9.27 shows variation of $\sqrt{\overline{p_w'^2}}\Big/\frac{\rho U_1^2}{2}$ with x/y_1 for data collected by Khader and Elango (67), Vasiliev and Bukreyev (62), Schiebe and Bowers (66), Lopardo and Henning (70), and Narasimhan and Bhargava (68) for free hydraulic jump; here x is measured from the beginning of the jump. It can be seen that $\sqrt{\overline{p_w'^2}}\Big/\frac{\rho U_1^2}{2}$ designated as C_p' reaches a maximum value for x/y_1, values lying between 8 and 18; this is the region just beneath the roller zone. This maximum value is about 0.05 for the data of Vasiliev and Bukreyev, Narasimhan and Bhargava, and Schiebe and Bowers. There seems to be some effect of F_{r_1} on this value, but it is not systematic. However, Khader and Elango obtained this value as about 0.08. As mentioned earlier, pressure fluctuations are very much dependent on whether the incoming flow is developing or fully developed. Hence, this significant difference in maximum C_p' value can possibly be due to the difference in nature of the incoming flow. This is supported by the results obtained by Lopardo and Henning (40), and Resch and Leutheusser (71). It can also be seen that beyond x/y_1 = equal to 45 or so, C_p' value drops down to less than 0.01.

The general effect of submergence on the pressure fluctuations can be seen from the mean curve for Narayanan's data shown in Fig. 9.27. With submergence, the magnitude of peak pressure fluctuations persists over a longer distance. The detailed effect of F_{r_1} and submergence ratio on intensity of pressure fluctuations can be seen from results presented by Narasimhan and Bhargava (28) (see Fig. 9.28). Figure 9.28 (a) shows variation of C_p' with x/l_{rs} for given F_{r_1} and varying submergence ratios $y_s = y_t/y_1$. Here, l_{rs} is the length of surface roller in submerged jump, and y_t is the downstream water depth. It can be seen that as y_t/y_1 increases, C_p' at any x/l_{rs} value increases; however, beyond y_t/y_1 equal to approximately 9.0, furthur increase in y_t/y_1 has no effect on C_p'. This is probably due to the fact that the roller dimensions vary significantly only upto a certain rise in tail water depth specific for a given F_{r_1}. Any further rise in tail water level has a marginal effect on the roller length along with $\sqrt{\overline{p_w'^2}}\Big/\frac{\rho U_1^2}{2}$ and turbulent charcteristics. Figure 9.28(b) shows effect of F_{r_1} on C_p' for a constant submergence ratio. It can be seen that as F_{r_1} incereases, C_p' decreases. This decreases is because rate of increase of dynamic pressure $\rho U^2/2$ is greater than that of $\sqrt{\overline{p_w'^2}}$ as F_{r_1} increases. Increase in F_{r_1} above 4.0 has no significant effect on the maximum value of C_p'. In the investigations of Narasimhan and Bhargava, the maximum value of C_p' for submerged jump varied from 0.028 to 0.083 within submergence ratio values of 5.28 to 16.8 and F_{r1} values of 2 to 6.

Khadar and Elango (67) have found that at the place of maximum pressure under the roller, the probability distribution function of pressure fluctuations is positively skewed. Thus, at this section, the positive fluctuations are more frequent than the negative ones.

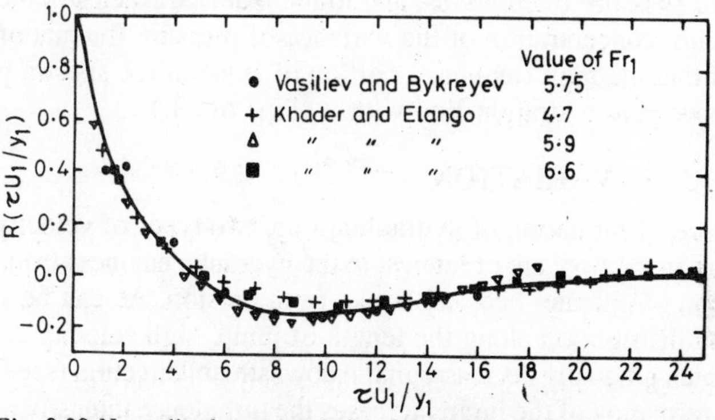

Fig. 9.29 Normalised autocorrelation function for different F_{r1} values

Evidently the pressure fluctuations are not normally distributed. This was also observed by Vasiliev and Bukreyev (62) who found that for x/y_1 values upto 15 the 3rd and 4th moments were greater than zero and four, respectively. It may be mentioned that if pressure fluctuations were normally distributed, the corresponding values of 3rd and 4th moment would have been zero and three, respectively. For larger values of x/y_1, the distribution approached Gaussian or normal law. Normalised autocorrelation function values obtained by Khadar and Elango, and Vasiliev and Bukreyev, at the location of maximum pressure fluctuations for F_{r1} value ranging from 4.7 to 6.6 are plotted in Fig. 9.29. It can be seen that the function attains the first zero value at $\tau U_1/y_1 = 5$,

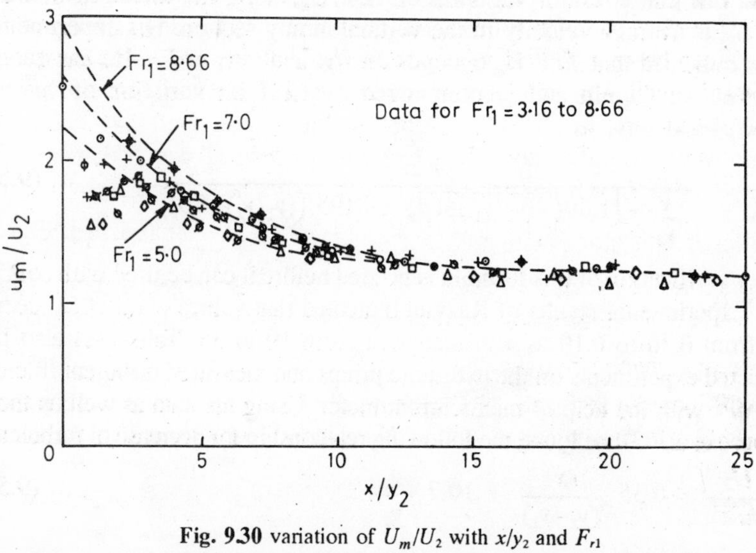

Fig. 9.30 variation of U_m/U_2 with x/y_2 and F_{r1}

and remains negative between $\tau U_1/y_1$ values of 5.0 and 20.0; here τ is the time interval. This implies that macroturbulent pulsations of fairly strong intensity are present at the point of maximum fluctuations. The correlation remains positive for longer durations with distance downstream, indicating that large sized eddies become more and more predominant in the creation of pressure fluctuations. Khader and Elango also studied the spectral density functions of pressure fluctuations, and found that frequencies in the range of 1 to 5 contain concentration of the variance of pressure fluctuations. It was also found that the high frequency portion of some of the spectra plotted on log-log paper give a straight line with a slope of –1/7.

9.9 VELOCITY VARIATION

As a result of formation of hydraulic jump, two types of velocity changes which occur in the flow are of interest to the hydraulic engineer from the point of protection of channel bed and banks from erosion. As can be seen from the velocity distribution along the length of jump, high velocity occurs near the bed, which gradually decreases in the downstream direction (see Fig. 9.23). Secondly, formation of the jump increases the turbulence intensity in the flow which gradually decreases in the downstream direction.

Lipay and Pustovoit (72*) have measured maximum instantaneous velocity u_m at 4 mm to 5 mm from the bed in free hydraulic jump at various sections along the length. On the basis of experimental data, they found that u_m/U_2 varies with x/y_2 and F_{r1}, and this relationship can be expressed as

$$\frac{u_m}{U_2} = 1.2 + \frac{0.2 F_{r1}}{1 + 0.07\,(x/y_2)^2} \qquad \ldots (9.52)$$

(see Fig. 9.30). This is based on experimental data with F_{r1} values between 3.16 and 8.66

Razvan (63) found that, as an approximation, one can assume turbulence to be isotropic and that $\sqrt{\overline{u'^2}}/\bar{u}$ does not vary appreciably along the vertical. Hence, one can consider variation of $\sqrt{\overline{u'^2}}/U_m$ along the direction of flow. Here U_m is average velocity in the vertical at any section. His experimental results indicated that $\sqrt{\overline{u'^2}}/U_m$ depends on x/y_2 and y_2/y_1 or F_{r1}. He has quoted a formula of Cumin and recommended by Levi for variation of $\sqrt{\overline{u'^2}}/U_m$ with x/y_2 and y_2/y_1 as

$$\frac{\sqrt{\overline{u'^2}}}{U_m} = \frac{1.52}{\dfrac{x}{y_2} - \left[1.69\left(\sqrt{(y_2/y_1} - 4.0\right) - 0.195\left((y_2/y_1) - 4.0\right)\right]} \qquad \ldots (9.53)$$

Further verification of this formula is needed before it can be used with confidence. Experimental results of Razvan indicated that values of $\sqrt{\overline{u'^2}}/U_m$ decreased from 0.30 to 0.10 as x/y_2 increased from 10 to 35. Talis (72) also has conducted experiments on the hydraulic jumps and measured turbulent fluctuations $\sqrt{\overline{u'^2}}$ with the help of microcurrent meter. Using his data as well as those of Rouse et al. (60) he found the following relationship for decrease of turbulence

$$\frac{U_2}{\sqrt{\overline{u'^2}}} = 0.35\,\frac{x}{(y_2 - y_1)} + 10.7\,\frac{y_1}{y_2} \qquad \ldots (9.54)$$

Thus, if one prescribes the value of $\sqrt{\overline{u'^2}}/U_2$ for given values of y_1 and y_2, one can determine the value of x required to attain this intensity. Needless to say, this intensity specified should be what would naturally occur in downstream channel.

REFERENCES

1. Blinco, P.H. and E. Partheniades. Turbulence Characteristics in Free Surface Flows Over Smooth and Rough Boundaries. JHR, IAHR, Vol. 9, No. 1, 1971.
2. Ljatkher, V.M. Calculation of Spectra of Turbulent Pulsations in Uniform Flow. Proc. 12th Congress of IAHR, Vol. 2, Fort Collins (USA), 1967.
3. Kemp P.H. and A.J. Grass. The Measurement of Turbulent Velocity Fluctuations Close to a Boundary in Open Channel Flow. Proc. 12th Congress of IAHR, Vol. 2, Fort Collins (USA), 1967.
4. Nalluri, C. and P. Novak. Turbulence Characteristics in Smooth Open Channel of Circular Cross Section. JHR, IAHR, Vol. 11, No. 4, 1973.
5. Richardson, E.V. and R.S. McQuivey. Measurement of Turbulence in Water. JHD, Proc. ASCE, Vol. 94, No. HY-2, March, 1968.
6. McQuivey, R.S. and E.V. Richardson. Some Turbulence Measurements in Open Channel Flow. JHD, Proc. ASCE, Vol. 95, No. HY-1, Jan. 1969.
7. Rao, M.V. A Study of Structure of Shear Turbulence in Free Surface Flows. Ph. D. Thesis, Utah State University, Logan (USA) 1965.

7a. Sushil Kumar, P.K. Pande and R.J. Garde. Turbulence Characteristics of Rough Open Channels. Proc. 44th Annual Research Session of CBIP, Vol. 1, 1975.

8. Li, R.M., J.D. Schall and D.B. Simons. Turbulence Prediction in Open Channel Flow. JHD, Proc. ASCE, Vol. 106, No. HY-4, April, 1980.
9. Imamoto H. Universal Representation of Turbulence Characteristics in Free Surface Flow. Proc. 16th Congress of IAHR, Vol. 5, Sao Paulo (Brazil), 1975.
10. Ishihara, Y. and S. Yokosi. The Spectra of Turbulence in River Flows. Proc. of 12th Congress of IAHR, Vol. 2, Fort Collins (USA), 1967.
11. Komura, S. and M. Kubota. Vorticity Intensities of Turbulent Flow Over Bar Roughness. Proc. 16th Congress of IAHR, Vol. 5, Sao Paulo (Brazil), 1975.
12. Aki, S. Dynamic Characteristics of the Forces Acting On the Spillway Chute. Proc. 12th Congress of IAHR, Vol. 2, Fort Collins (USA), 1967.
13. Hayashi, T., M. Ohashi and Y. Kotani. River Flow Turbulence and Longitudinal Vortices. Published in "Recent Studies on Turbulent Phenomena" Edited by Totsumi et al. Association for Science Documentation Information, Japan, 1985.
14. McQuivey R.S. Large Scale Turbulence in Open Channel Flows. Proc. 16th Congress of IAHR, Vol. 5, Sao Paulo (Brazil) 1975.
15. Vanoni, V.A. Transportation of Suspended Sediment by Water. Trans. ASCE, Vol. 111, 1946.
16. Hayashi, T. Open Channel Flow Turbulence and Large Scale Coherent Structures. Published in "Megatrends in Hydraulic Engineering" Edited by M.L. Albertson and C.N. Papakadis. Colorado State University (USA), 1986.
17. Tominaga, A., I. Nezu, K. Izaki and H. Nakagawa. Three Dimensional Turbulent Structure in Straight Open Channel Flows. JHR, IAHR, Vol. 27, No. 1, 1989.
18. Sutherland, A.J. Proposed Mechanism of Sediment Entrainment by Turbulent Flows. Jour. Geophysical Research, Vol. 72, No. 24, Dec. 1967.
19. Sumer, B.M. and B. Ogzu. Particle Motion Near the Bottom in Turbulent Open Channel Flow. JFM, Vol. 86, Pt. II, 1978.
20. Taylor, G.I. Diffusion by Continuous Movements. Proc. London Mathematical Society, Vol. 20, 1921.

21. Hay, J.S. and F. Pasquil. Diffusion from a Fixed Source at a Height of a Few Hundred Feet in the Atmosphere. JFM, Vol. 2, 1967.
22. Miller, A.C. and E.V. Richardson. Diffusion and Dispersion in Open Channel Flow. JHD, Proc. ASCE, Vol. 100, No. HY-1, Jan. 1974.
23. Engelund, F. Four Papers on Surface Turbulence and Diffusion. Institute of Hydrodynamics and Hyd. Engg., Tech. Univ. Denmark, 1972.
24. Fischer, H.B. Longitudinal Dispersion and Turbulent Mixing in Open Channel Flow. Annual Review of Fluid Mechanics, 1973.
25. Schmidt, W. Der Massenaustausch in Freier Luft und Verwandte Erscheinungen. In "Probleme der Kosmischen Physik" Bd. 7, Hamburg, 1925.
26. Garde R. J. and K.G. Ranga Raju. *Mechanics of Sediment Transportation and Alluvial Stream Problems*. Wiley Eastern Limited, New Delhi, 2nd Edition, 1985. Chapt. 7.
27. McNown, J.S., H.M.Lee, M.B. McPherson and S.M. Engez. Influence of Boundary Proximity on the Drag of Spheres. Proc. Int. Cong. on App. Mech. 7th Congress, London, 1948.
28. McNown J.S. and P.N. Lin. Sediment Concentration and Fall Velocity. Proc. 2nd Midwestern Conference in Fluid Mechanics, Ohio State Univ. (USA) 1952.
29. Maude, A.D. and R.L. Whitmore. A Generalised Theory of Sedimentation. British Journal of App. Physics, Vol. 9, Dec. 1958.
30. Bechteler, W., K. Farber and W. Schrimpf. Settling Velocity Measurements in Quiescent and Turbulent Water. Proc. of 2nd Int. Sym. on River Sedimentation, Nanjing (China), 1983.
31. Field, W.G. Effect of Density Ratio on Sedimentary Similitude. JHD, Proc. ASCE, Vol. 94 No. Hy-3, May 1968.
32. Jobson, H.E. and W.W. Sayre. An Experimental Investigation of Vertical Mass Transfer of Suspended Sediment. Proc. of 13th Congress of IAHR, Vol.II, Kyoto 1969.
33. Boillat, J.L. and W.H. Graf. Settling Velocity of Spherical Particle in Turbulent Media. JHR of IAHR, Vol. 20, No. 5, 1982.
34. Vasiliev O.F. Problems of Two Phase Flow Theory. General Lecture II, Proc. 13th Congress of IAHR, vol. 5.3 Kyoto (Japan), 1969.
35. Hino, M. Turbulent Flow with Suspended Particles. JHD, Proc. ASCE, Vol. 89, No. HY-4, July, 1963.
36. Ismail H.M. Turbulent Transfer Mechanism and Suspended Sediment in Closed Channels. Trans. ASCE, Vol. 117, 1952.
37. Zagustin, A. and K. Zagustin. Mechanism of Turbulent Flow in Sediment Laden Stream. Proc. of 13th Congress of IAHR, Vol. 2, Kyoto (Japan), 1969.
38. Arai, M. and T. Takahashi. The Karman Constant of Flow Laden with High Sediment. 3rd Int. Symp. on River Sedimentation. University of Mississippi (USA), 1986.
39. Kalinske, A.A. and C.S. Hsia. Study of Transportation of Fine Sediments by Flowing Water. Studies in Hydraulic Engineering, Bull. No. 29, Univ. of Iowa (USA), 1945.
40. Gust, G. Observation on Turbulent Drag Reduction in Dilute Suspension of Clay in Sea Water. JFM, Vol. 75, Pt. 1,1976.
41. Itakura, T. and T. Kishi. Open Channel Flow with Suspended Sediments. JHD, Proc. ASCE, Vol. 106, No.Hy-8, Aug. 1980.
42. Coleman, N.L. Velocity Profiles with Suspended Sediment. JHR, IAHR, Vol. 19, No. 3, 1981.
43. Pullaiah, V. Transport of Fine Suspended Sediment In Rigid Bed Channels. Ph.D. Thesis, University of Roorkee (India), 1978.
44. Arora, A.K. Velocity Distribution and Sediment Transport in Rigid-Bed Open Channels. Ph.D. Thesis, University of Roorkee, (India), 1983.
45. Gry, A. The Behaviour of the Turbulent Flow in a 2-Dimensional Open Channel in the Presence of Suspended Particles. Proc. of 12th Congress of IAHR, Fort Collins (U.S.A.), Vol. 2, 1967.

46. Müller, A. Measurement of the Influence of Suspended Particles on the Size of Vortices. Proc. of 12th Congress of IAHR, Vol.4, Fort Collins (U.S.A.), 1967.
47. Silin, N.A., Y.K. Vitoshkin, V.M. Karasik and C.F. Ocherekto. Research on Solid-Liquid Flows with High Consistance. Proc. of 13th Congress of IAHR, Vol. 2, Kyoto (Japan), 1969.

47a. Pechenkin, M.V. and B.E. Vedeneev. Experimental Studies of Flows with High Particle Concentrations. Proc. of 13th Congress of IAHR, Vol. 2, Kyoto (Japan), 1969.

48. Bouvard, M. and S. Petkovic. Modification des Characterestiques d'une Turbulence Sous L'influence des Particules Solides en Suspension. La Houille Blanche, No. 1, 1973.
49. Fischer, H.B. The Mechanics of Dispersion in Natural Streams. JHD, Proc. ASCE, Vol. 93, No. Hy-6, Nov. 1967.
50. McQuivey, R.S. and T.N. Keefer. Simple Method for Predicting Dispersion in Streams. JEED, Proc. ASCE, Vol. 100, No. EE-4, July 1967.
51. Kilpatrick, F.A., W.W. Sayre and E.V. Richardson. Discussion of the paper by Replogle J.A. et al., JHD, Proc. ASCE. Vol. 93, No. Hy-4, July 1967.
52. Taylor, G.I. The Dispersion of Matter in Turbulent Flow Through a Pipe. Proc. Royal Soc. London, A-223, May 1954.
53. Singh, U.P. Dispersion of Conservative Pollutants in Clear Water and Sediment Laden Flows in Open Channels. Ph.D. Thesis, University of Roorkee (India), 1987.
54. Nordin, C.F. and G.V. Sabol. Emiprical Data on Longitudinal Dispersion in Rivers. USGS, Water Resources Investigations 20-74, Washington D.C. 1974.
55. Day T.J. Longitudinal Dispersion in Natural Channels. Water Resources Research, Vol. 11, No. 6, Dec. 1975.
56. Liu, H. and A.H.D. Cheng. Modified Fickian Model for Predicting Dispersion. JHD, Proc. ASCE, Vol. 106, No. Hy-6, June 1980.
57. Liu, H. and A.H.D. Cheng. Closure of Discussion of Ref. 56, JHD, Proc. ASCE, Vol. 108, No. Hy-1, Jan. 1982.
58. Bansal, M.K. Dispersion in Natural Streams, JHD, Proc. ASCE, Vol. 97, No. Hy-11, Nov. 1971.
59. Day, T.J. and I.R. Wood. Similarity of the Mean Motion of Fluid Particles Dispersing in a Natural Channel. Water Resources Research, Vol. 12, No. 4, August, 1976.
60. Rouse, H., T.T. Siao and S. Nagaratnam. Turbulence Characteristics of the Hydraulic Jump. Trans. ASCE, Vol. 124, 1959. Also see discussion by Harleman.
61. Wisner P. On the Bottom Pressure Pulsations of the Closed Conduit and Open Channel Hydraulic Jumps. Proc. of 12th Congress of IAHR, Vol. 2, Fort Collins (U.S.A.), 1967.
62. Vasiliev, O.F. and V.I. Bukreyev. Statistical Characteristics of Pressure Fluctuations in the Region of Hydraulic Jump. Proc. of 12th Congress of IAHR, Fort Collins (U.S.A.), Vol. 2, 1967.
63. Razvan, E. Resultats de l'Etude du Movement Macroturbulent en Aval du Ressaut Hydraulique. Proc. of 12th Congress of IAHR, Vol. 2, Fort Collins (U.S.A.), 1967.
64. Voinitch Sianozhentskij, T.G., V.G. Lomtatidze and G.N. Guazava. Macroturbulence of Bottom Hydraulic Jump and Its Influence on the Stability of the Lower Tail Water. Proc. of 12th Congress of IAHR, Vol. 2, Fort Collins (U.S.A.), 1967.
65. Bowers, C.D. and F.Y. Tsai. Fluctuating Pressures in Spillway Stilling Basins. JHD, Proc. ASCE, Vol. 95, No. Hy. 6, Nov. 1969.
66. Schiebe, F.R. and C.E. Bowers. Boundary Pressure Fluctuations due to Macroturbulence in Hydraulic Jumps. Proc. of Symposium on Liquid Turbulence, Univ. of Missouri-Rolla, USA, 1971.
67. Abdul Khader M.H. and K. Elango. Turbulent Pressure Field Beneath a Hydraulic Jump. JHR, IAHR, Vol. 12, No. 4, 1974.
68. Narasimhan, S. and V.P. Bhargava. Pressure Fluctuations in Submerged Jump. JHD, Proc. ASCE. Vol. 102, No. Hy-3, Mar. 1976.

69. Narayanan R. Pressure Fluctuations Beneath Submerged Jump. JHD, Proc. ASCE, Vol. 104, No. Hy-9, Sept. 1978.
70. Lopardo, R.A. and R.E. Henning. Experimental Advances on Pressure Fluctuations Beneath Hydraulic Jumps. Proc. of 21st Congress of IAHR, Vol. 3, Melborne (Australia), 1985.
71. Resch F.J. and H.J. Leutheusser. Reynolds Stress Measurements in Hydraulic Jumps. JHR, IAHR, Vol. 10, No.8, 1972.
72. Talis, J. Diminution de la Turbulence Derriere de Ressaut. Proc. of 9th Congress of IAHR, Dubrovnik, 1961.

CHAPTER X

Description of Turbulent Flows-III: Free Turbulence Shear Flows

10.1 INTRODUCTION

In the previous chapter, we have discussed categories of flows which deal with flow patterns in the neighbourhood of solid boundaries resulting from the transmission of shear progressively outwards from the boundary. Boundary layer flows, flow in pipes and open channels are a few examples. Yet, there is another class of problems in which zone of shear occurs far away from the fixed boundary. Such a flow is known as Free Turbulence Shear (FTS) flow. These flows, along with the topic of forces on immersed bodies, are discussed in this chapter. Some examples of free turbulence shear flow are :

(i) Two neighbouring fluids brought into relative motion parallel to their common interface;
(ii) two-dimensional or a circular jet issuing in an infinite fluid, as well as jets which are restricted by boundary; and
(iii) wakes behind two dimensional or three dimensional bodies.

Thus, free turbulence shear flow is encountered where there is no direct effect of any fixed boundary on turbulence in the flow. Free turbulence shear flow is nonisotropic in its character; as a result, time averaged values of shear occur in the flow field. Consequently, there is not only dissipation but also production of energy of turbulence, and its convection and diffusion. Theoretically and experimentally, some of the free turbulene shear flows studied include half jets, plane and round jets plane wake behind a cylindrical rod, round wake behind a sphere and plane wake flow behind a row of cylindrical rods. Figures 10.1 and 10.6 show details of flow in the case of diffusion of a jet and wake behind a two dimensional body. Enough experimental evidence is available about the similarity of mean velocity profiles at successive sections in jets and wakes at sufficiently large distances in the downstream direction.

For each type of flow nearly bell shaped velocity profiles are obtained, which are geometrically similar when velocity at each section is nondimensionalised by maximum mean velocity difference and lateral distances are made nondimensional with local width of turbulent region as a length scale or "half value" distance. Half value distance is the distance from the axis of symmetry at which mean velocity is half the maximum mean velocity difference. These velocity and length scales are functions of distance from some apparent origin where the FTS flow seems to originate. Since production of turbulence is determined by the gradient of mean velocity distribution, which in turn depends on turbulence generated upstream and its transport downstream by convection and diffusion, similarity in turbluence distribution pattern is also expected in such flows.

10.2 APPROXIMATIONS IN ANALYSIS (1)

In the analysis of jets and wakes, it is possible to distinguish one main direction in which flow velocity is much greater than in other two directions. On the other hand, the transverse extent of the flow is much smaller than the longitudinal extent. Hence, $\frac{\partial}{\partial x}$ will be much smaller than either $\frac{\partial}{\partial y}$ or $\frac{\partial}{\partial z}$.

Further, the lateral dimension being relatively small, the pressure across the jet or wake can be assumed to be constant and equal to outside pressure as in the case of boundary layers. In addition, in most of the problems in this category, pressure in the longitudinal direction is constant; hence, it is assumed that pressure is constant everywhere in the flow field. With high values of Reynolds number and intense mixing, it is safe to assume that viscous stresses can be neglected in preference to turbulent stresses. It can also be assumed that, if U_o is the constant velocity of jet or free stream velocity in case of wake and $\bar{u}$ is velocity in the jet or wake, then $U_o/\bar{u}$ is of the order of unity in the case of jet, and much less than unity in the case of wake.

With these approximations, pertinent differential equations for steady two dimensional turbulent flow are

$$\left.\begin{aligned} \bar{u}\frac{\partial \bar{u}}{\partial x} + \bar{v}\frac{\partial \bar{u}}{\partial y} &= \frac{1}{\rho}\frac{\partial \tau_t}{\partial y} = \frac{1}{\rho}\frac{\partial}{\partial y}(-\rho\overline{u'v'}) \\ \frac{\partial \bar{u}}{\partial x} + \frac{\partial \bar{v}}{\partial y} &= 0 \end{aligned}\right\} \quad \ldots (10.1)$$

where τ_t is the turbulent shear stress. Corresponding equations for axisymmetric flow are

$$\left.\begin{aligned} \bar{u}\frac{\partial \bar{u}}{\partial x} + \bar{v}\frac{\partial \bar{u}}{\partial r} &= \frac{1}{\rho}\frac{1}{r}\frac{\partial}{\partial r}(-\rho\overline{u'v'}) \\ \frac{\partial}{\partial x}(\bar{u}r) + \frac{\partial}{\partial r}(r\bar{v}) &= 0 \end{aligned}\right\} \quad \ldots (10.2)$$

in which $\overline{V_x}$ is replaced by $\bar{u}$ and $\overline{V_r}$ replaced by $\bar{v}$ for convenience.

TURBULENT JETS

10.3 DESCRIPTION OF FLOW IN A JET

Consider either a two dimensional or a round jet of constant high velocity issuing in an infinite fluid at rest, both fluids being of constant density. Because of very large velocity difference at the surface of discontinuity, large eddies are formed, which cause intense lateral mixing. Mixing process continues in the downstream direction, both inward and outward, as a result of which the fluid within the jet is decelerated while the fluid from the surrounding region is accelerated or entrained. Consequently, the width of the jet, as well as the rate of flow gradually increase in the downstream direction. Here, core region in the jet has a constant velocity and very little turbulence, whereas outside the core region, the mean velocity decreases outward and flow is highly turbulent. At some distance downstream from the efflux, deceleration reaches the centreline of the jet. The region between jet efflux and this section is known as the zone of flow establishment or development as shown in Figure 10.1. In this zone production, growth, interaction and eventual destruction of large scale turbulent structures are the primary mechanisms that are responsible for deceleration.

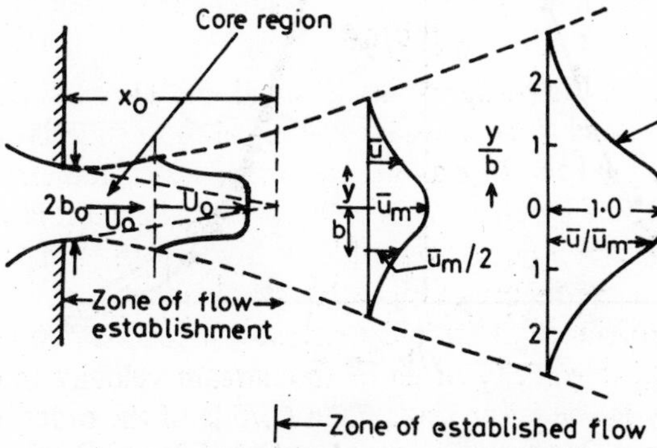

Fig. 10.1 Diffusion of a submerged jet

Downstream of this zone, the entire jet becomes turbulent, and is continuously decelerated. Centreline velocity is less than U_o, and it continuously decreases in the downstream direction with increase in x.

It is in the downstream of the zone of flow establishment that the mean velocity distribution at various sections is similar. This zone is known as the zone of established flow, or fully developed flow. Ultimately, far away from the efflux section, the centreline velocity becomes negligibly small. The precise section of demarcation between zone of establishment and zone of established flow is poorly defined and has to be determined from experiments. The diffusion of high velocity jet in this manner is a common phenomenon in several problems in hydraulic and environmental engineering. "Exit loss" in pipe outlets, efflux of heated or cooled air from ventilation ducts in the room, issuing of a submerged jet from a sluice into the downstream channel, and discharging of sewage from pipe outlets into sea are a few examples (2). If the jet is restricted on one side by a wall, it is known as wall jet.

10.4 ANALYSIS OF 2-D JETS (3,4)

Several characteristics of two or three dimensional jet diffusing in an infinite fluid can be obtained by integral analysis if the velocity distribution is assumed to be similar, i.e. $\bar{u}/\bar{u}_m = f(\eta)$ where $\eta = \frac{y}{b}$ or $\frac{r}{b}$ for two dimensional or axisymmetric jet, respectively. Here b can be taken as the half width of the jet or distance from the axis of symmetry where $\bar{u} = \bar{u}_m/2$, see Fig. 10.1, and $\bar{u}_m$ is the centreline velocity.

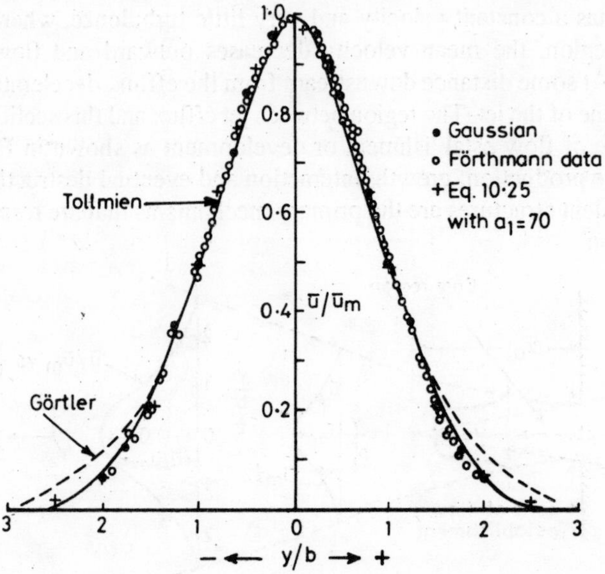

Fig. 10.2 Velocity distribution in zone of established flow (2-D jet)

First, let us consider a two dimensional jet of half width b_o, and apply momentum equation to the mean flow. Since pressure is assumed to be constant, and there is no external force acting on the fluid, momentum flux M from section to section must remain constant. Hence, in the zone of established flow $\frac{\partial M}{\partial x} = 0$. However,

$$M = 2\int_0^\infty \rho \bar{u}^2 dy = 2\rho \bar{u}^2_m b \int_0^\infty f^2(\eta)\, d\eta$$

$\frac{\partial M}{\partial x} = 0$ would imply that $\frac{\partial}{\partial x}\left[2\rho \bar{u}^2_m b \int_0^\infty f^2(\eta)\, d\eta\right] = 0$

Since $\int_0^\infty f^2(\eta)\, d\eta \doteq$ constant for the known velocity distribution law, constancy of M from section to section would imply that

$\frac{\partial}{\partial x}(\bar{u}^2_m b) = 0$ or $\bar{u}^2_m b$ is independent of x; i.e.

$$\bar{u}^2_m b \propto x^o \qquad \ldots (10.3)$$

Assuming further that $\bar{u}_m \propto x^p$ and $b \propto x^q$

$$\bar{u}^2_m b \propto x^{2p+q}$$

or, $$2p + q = 0 \qquad \ldots (10.4)$$

In order to know how half jet width b and $\bar{u}_m$ change with x, we need one more equation. This can be obtained from Equation 10.1, viz.

$$\bar{u}\frac{\partial \bar{u}}{\partial x} + \bar{v}\frac{\partial \bar{u}}{\partial y} = \frac{1}{\rho}\frac{\partial \tau_t}{\partial y}$$

and using the relationships $\bar{u}/\bar{u}_m = f(\eta)$, $\bar{u}_m \propto x^p$, $b \propto x^q$ and $\tau_t/\rho\bar{u}_m^2 = g(\eta)$. Such analysis as given by Rajaratnam (2) yields $q = 1.0$; hence $p = -1/2$. Therefore

$$\begin{aligned} \bar{u}_m &\propto 1/\sqrt{x} \\ b &\propto x \end{aligned} \qquad \ldots (10.5)$$

Rajaratnam has also shown that the same result, viz. $q = 1$ can be obtained from the integrated form of energy equation.

It has already been mentioned that because of the entrainment of the surrounding fluid, the discharge Q in the jet increases in the downstream direction. The nature of dependence of Q as well as b and $\bar{u}_m$ on x can also be determined from dimensional analysis (2). Neglecting the effect of viscosity i.e. for high values of Reynolds number $2b_oU_o/\nu$, one can write

$$b,\ \bar{u}_m,\ Q = f(M_o, \rho, x) \qquad \ldots (10.6)$$

where M_o = Initial momentum flux = $2b_o\rho U_o^2$. Dimensional analysis then yields

$\bar{u}_m/\sqrt{M_o/\rho x}$ = constant C_1

which on substitution of the value of M_o yields

$$\bar{u}_m/U_o = \sqrt{2C_1 b_o/x}; \quad \text{or} \quad \bar{u}_m \propto 1/\sqrt{x} \qquad \ldots (10.7)$$

$$\text{Also } b/x = C_2;\ \text{or } \frac{b}{b_o} = C_2\frac{x}{b_o} \qquad \ldots (10.8)$$

i.e. $b \propto x$.

Lastly $Q/\sqrt{M_o x/\rho} = C_3$, which gives

$$\frac{Q}{Q_o} = C_3\sqrt{\frac{x}{2b_o}} \qquad \ldots (10.9)$$

where $Q_o = 2b_oU_o$ It may be mentinoed that in Eqs. 10.7, 10.8 and 10.9, the constants of proportionality depend on the nature of assumed veloocity distribution law $\bar{u}/\bar{u}_m = f(\eta)$. The velocity distribution in the zone of established flow has been studied by Tollmien (5) and Görtler (6). Tollmein's analysis is given below.

Starting with Eq. 10.1, Tollmien expresses τ_t *as*

$$\tau_t = \rho l^2 \left(\frac{d\bar{u}}{dy}\right)^2 \qquad \ldots (10.10)$$

using Prandtl's mixing length hypothesis; here l is the mixing length. Dimensional considerations indicate that at any section $l = \beta b$ and since $b = C_2 x$, l is expressed as

$$l = \beta C_2 x$$

where β is a constant. Hence $\dfrac{1}{\rho}\dfrac{\partial \tau_t}{\partial y}$ is expressed as

$$\frac{1}{\rho}\frac{\partial \tau_t}{\partial y} = 2(\beta C_2)^2 x^2 \left(\frac{\partial \bar{u}}{\partial y}\right)\left(\frac{\partial^2 \bar{u}}{\partial y^2}\right) \qquad \ldots (10.11)$$

where one can write for convenience $a^3 = 2(\beta C_2)^2$, "a" being a constant. Since velocity distribution can be expressed as

$\bar{u}/\bar{u}_m = f\left(\frac{y}{b}\right)$, it can also be expressed as

$$\bar{u}/\bar{u}_m = f_2(y/ax) = f(\phi) \qquad \ldots (10.12)$$

where $\phi = y/ax$ and a is to be determined later experimentally. Since it has been shown that $\bar{u}_m \propto 1/\sqrt{x}$, one can write

$$\bar{u}_m = \frac{n}{\sqrt{x}} \qquad \ldots (10.13)$$

where n is independent of x. Hence,

$$\bar{u} = \frac{n}{\sqrt{x}} f(\phi) \qquad \ldots (10.14)$$

In order to express $\bar{u}$ and $\bar{v}$ in terms of one variable, it is expedient to introduce a stream function ψ such that

$$\left.\begin{array}{l} \bar{u} = \partial\psi/\partial y \\ \bar{v} = -\,\partial\psi/\partial x \end{array}\right\} \qquad \ldots (10.15)$$

$$\therefore \quad \psi = \int \bar{u}\,dy = \frac{n}{\sqrt{x}} \int f(\phi)\, ax\, d\phi$$

$$\left.\begin{array}{l} \text{or,} \quad \psi = an\sqrt{x}\,F \\ \text{where } F = \int f\, d\phi \end{array}\right\} \qquad \ldots (10.16)$$

Using Eqs. 10.15 and 10.16, one can express $\bar{u}$, $\bar{v}$, $\partial\bar{u}/\partial x$, $\partial\bar{v}/\partial y$ and $\partial\tau_t/\partial y$ in terms of F and its derivatives.

Substitution of these in Eq. 10.1 yields

$$\left.\begin{array}{l} 2F''F''' + FF'' + F'^2 = 0 \\ \text{or } 2F'F'' + \dfrac{d}{d\phi}(FF') = 0 \end{array}\right\} \qquad \ldots (10.17)$$

which on integration yields

$$FF' + F''^2 = C \qquad \ldots (10.18)$$

The boundary conditions for solution of this ordinary nonlinear differential equation (Eq. 10.18) are

(i) $y = 0$, $\quad \bar{u}/\bar{u}_m = 1 \longrightarrow \phi = 0,\ F' = 1.0$ or $F'(0) = 1.0$
(ii) $y = \infty$, $\quad \bar{u}/\bar{u}_m = 0 \longrightarrow \phi = \infty,\ F' = 0$ or $F'(\infty) = 0$
(iii) $y = 0$, $\quad \bar{v} = 0 \longrightarrow \phi = 0,\ F = 0$ or $F(0) = 0$
(iv) $y = 0$, $\quad \tau_t = 0 \longrightarrow \phi = 0,\ F'' = 0$ or $F''(0) = 0$
(v) $y = \infty$, $\quad \tau_t = 0 \longrightarrow \phi = \infty,\ F'' = 0$ or $F''(\infty) = 0$

From these boundary conditions, it can be seen that when $\phi = 0$, Eq. 10.18 gives $C = 0$. Hence, Eq. 10.18 becomes

$$FF' + F''^2 = 0 \qquad \ldots (10.18a)$$

Tollmein has solved this equation numerically and obtained the value of experimental coefficient "a" using data collected by Reichardt.

Görtler (6) assumed that the turbulent shear in Eq. 10.1 can be expressed as $\tau_t = \rho\varepsilon\dfrac{\partial\bar{u}}{\partial y}$, and further, assuming that $\bar{u}/\bar{u}_m = F'\left(\dfrac{\sigma y}{x}\right)$ where σ is a constant, he introduced a stream function ψ such that

$$\psi = \frac{n\sqrt{x}}{\sigma}F'(\xi) \quad \text{where } \xi\ \sigma y/x$$

and obtained values of $\bar{u}$, $\bar{v}$, $\dfrac{\partial\bar{u}}{\partial x}$, $\dfrac{\partial\bar{v}}{\partial y}$ and τ_t. Substitution of these in Eq. 10.1 then yields

$$2FF' + F'' = 0 \qquad \ldots (10.19)$$

which on integration yields

$$F' + F^2 = C \qquad \ldots (10.20)$$

with the boundary conditions

$$F'(0) = 1,\ F'(\infty) = 0,\ F(0) = 0,\ F''(0) = 0,\ F''(\infty) = 0$$

corresponding to five boundary conditions specified earlier by Tollmien. It can be seen from Eq. 10.20 that at $\xi = 0$, $C = 1$. hence, Eq. 10.20 becomes

$$F' + F^2 = 1.0 \qquad \ldots (10.20a)$$

This equation has an exact solution

$$F = \tanh\ \xi = \frac{1 - e^{-2\xi}}{1 + e^{-2\xi}} \qquad \ldots (10.21)$$

which gives

$$\frac{\bar{u}}{\bar{u}_m} = F' = 1 - \tanh^2\xi \qquad \ldots (10.22)$$

The details of Görtler's solution are given by Rajaratnam (2). Here again σ is a constant that needs to be determined from experiments. Vorticity transport hypothesis has also been used to study velocity distribution in a jet. This has been discussed by Goldstein (7). It may be mentioned that turbulence models have been also used to study jets.

10.5 EXPERIMENTAL STUDIES

Many investigators have conducted experiments on plane and round jets to provide useful information to verify the velocity distribution theories and throw light on details of flow. Among these investigators are Förthmann (8), Albertson et al. (4), Zijnen (9) and Heskestad (10).

The experiments have shown that the location of virtual origin from which the jet appears to originate does not concide with the beginning of nozzle. In some cases it has been found to be located behind the nozzle, and in other cases in front of the nozzle. Further, it has been found that the location of virtual origin is sensitive to level of turbulence in the flow in the nozzle. Hence, for practical purposes, origin is taken to coincide with the nozzle.

Albertson et al. (4) have found that the length of zone flow establishment is given by $x_o/b_o = 10.4$. In this zone, they assumed constant velocity in the core region and Gaussian distribution in the mixing zone. For variation of discharge and kinetic energy in the zone of flow establishment they found that

$$\left.\begin{aligned} Q/Q_o &= 1 + 0.04\ x/b_o \\ E/E_o &= 1 - 0.018\ /b_o \end{aligned}\right\} \qquad \ldots (10.23)$$

where $Q_o = 2b_oU_o$ and $E_o = b_o\rho U_o^3$

In the zone of established flow, the constant in the equation $\bar{u}_m/U_o \propto \dfrac{1}{\sqrt{x/b_o}}$ is found to be 3.224 by Albertson et al., while Abramovich (11) found it to be 3.78. Zijnen (9) prefers to express this variation in the form

$$\frac{\bar{u}_m}{U_o} = C_1/\sqrt{(x + C_* \ b_o)/b_o} \qquad \ldots (10.24)$$

where C_1 is found to vary from 3.12 to 3.78 and C_* from 0 to 2.4. Rajaratnam (2) recommends values of C_1 and C_* as 3.50 and zero respectively.

Velocity Distribution

These investigators have also studied the velocity distribution in the zone of established flow. Albertson et al. (4) assumed it to be Gaussian and experimentally found it to be so. Zijnen (9) expressed this distribution in the form

$$\bar{u}/\bar{u}_m = e^{-a_1\lambda^2} \qquad \ldots (10.25)$$

where a_1 varied from 70.7 to 75.0 and $\lambda = y/x$. Substitution of $b = 0.10x$ and $a = 70.7$ gives $\bar{u}/\bar{u}_m = e^{-0.707(y/b)^2}$. Velocity distributions obtained by Tollmien and Görtler can be given in tabular form as variation of $\bar{u}/\bar{u}_m$ with y/b, where $y = b$ when $\bar{u} = 0.5\ \bar{u}_m$. Experiments have shown that value of "a" in Tollmien's analysis varies froom 0.09 to 0.12. If "a" is taken as 0.10, the following table gives values of $\bar{u}/\bar{u}_m$ according to Tollmien.

Table 10.1

Tollmien's solution for velocity distribution in 2-D plane jet

y/b	0.0	0.105	0.209	0.419	0.628	0.838	1.048	1.255
$\bar{u}/\bar{u}_m$	1.00	0.979	0.946	0.842	0.721	0.608	0.474	0.357
y/b	1.465	1.780	1.990	2.200	2.300	2.400	2.500	
$u/\bar{u}$	0.249	0.125	0.067	0.030	0.020	0.009	0.000	

In the same manner the value of σ in Görtler's solution was found to be 7.67 from the measurements conducted by Reichardt (12*). With this value of σ, Görtler's solution has been tabulated below.

Figure 10.2 shows the Gaussian, Tollmien's and Görtler's velocity distributions plotted in non-dimensional form along with Forthmann's data. It can be seen that Gaussian and Tollmien's distributions fit the data very well all over entire range of y/b values. Görtler's agrees well with the data in the central portion for y/b values less than 1.5; however in the outer region it over predicts the velocity. In the zone of established flow the discharge variation and the kinetic energy variation can be obtained analytically. With experimental constants, the relationships as obtained by Albertson et al. are:

Table 10.2
Görtler's solution for velocity distribution in 2-D plane jet

y/b	0.0	0.114	0.227	0.341	0.455	0.568	0.682	0.795	
$\bar{u}/\bar{u}_m$	1.00	0.990	0.961	0.915	0.855	0.788	0.711	0.635	
y/b	0.909	1.022	1.136	1.362	1.590	2.045	2.270	2.500	2.840
$\bar{u}/\bar{u}_m$	0.558	0.486	0.420	0.302	0.218	0.102	0.070	0.048	0.021

$$\left.\begin{aligned} Q/Q_o &= 0.44\sqrt{x/b_o} \\ E/E_o &= 2.64\sqrt{b_o/x} \end{aligned}\right\} \quad \ldots (10.26)$$

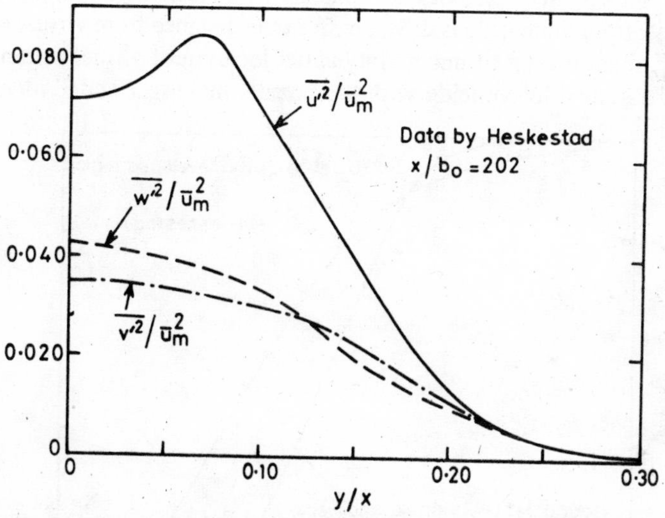

Fig. 10.3 Variation of turbulence intensity across 2-D jet in fully developed region.

Turbulence in 2-D Jets

Some turbulence measurements in two dimensional jet have been made by Zijnen (13), Miller and Comings (14), Heskestad (10), and Gutmark and Wygnanski (15). These studies indicate that distribution of turbulence intensity across the jet is similar beyond $x/b = 85$ to 90. Figure 10.3 shows variation of $\overline{u'^2}/\bar{u}_m^2$, $\overline{v'^2}/\bar{u}_m^2$ and $\overline{w'^2}/\bar{u}_m^2$ *with* y/x as obtained by Heskestad for $x/b_o = 202$.

Gutmark and Wygnanski (15) and Heskestad (10) have also determined the longitudinal and lateral microscales of turbulence λ_f and λ_g; these have values of 6-7 mm and 4-5 mm near the nozzle, and they gradually increase in the downstream direction. Kolmogorov's length scale $l_k = (\nu^3/\varepsilon)^{1/4}$ also increased from 0.08 mm at $x/b_o = 4.0$ to 0.12 at $x/b_o = 240$. For x/b_o greater than approximately 70, these investigators have found that $\sqrt{\overline{u'^2}}/\bar{u}_m = 0.22$ to 0.25, which is slightly smaller than corresponding value for round jet. Figure 10.4 shows variation of intermittency factor γ with y/b for 2-D and round jets.

10.6 ANALYSIS OF ROUND JETS

Analysis of a round jet of diameter D_o diffusing in an infinite fluid is carried out in a manner similar to that of a two dimensional jet. There is a zone of flow establishment immediately downstream of the nozzle in which the turbulence has not reached the axis of the jet. In this region, there is a central core in which velocity is constant and equal to U_o, and a region in which velocity decreases. With increasing distance, the jet dimension increases. This zone is known as the zone of flow establishment. At the end of this zone, the turbulence generated penetrates upto the jet axis from all the directions. Downstream of this section starts the zone of established flow in which velocity distribution profiles are similar; in this zone the centreline velocity decreases with increasing value of x. Experimental investigations show that in the case of round jet, the virtual origin does not coincide with the nozzle, and, therefore, the distance x from the nozzle is different from the distance from virtual origin $\bar{x}$. However, there is a lot of uncertainty about location of virtual origin and, hence, it is assumed to coincide with the nozzle, making x and $\bar{x}$ identical.

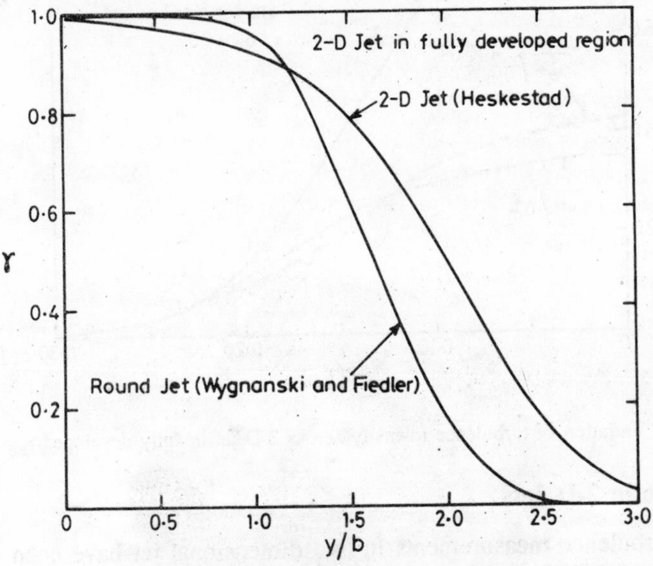

Fig. 10.4 Variation of γ in 2-D and round jets

Since there is no external force acting on the jet, and the pressure can be assumed to be constant everywhere, the momentum flux in the direction of motion must remain the same from section to section. If similar velocity profiles are represented by the equation

$$\bar{u}/\bar{u}_m = f\left(\frac{r}{b}\right) = f(\eta) \qquad \ldots (10.27)$$

where b is the value of r at which $\bar{u} = 0.5\,\bar{u}_m$, and if it is further assumed that $\bar{u}_m \propto x^p$ and $b \propto x^q$, constancy of momentum flux M can be expressed as

$$\frac{dM}{dx} = \frac{d}{dx}\int_0^\infty \bar{u}^2 \rho 2\pi r dr = 0 \qquad \ldots (10.28)$$

which gives $\frac{d}{dx}\left(b^2\rho\bar{u}^2_m \int_0^\infty 2\pi\eta f^2(\eta)d\eta\right) = 0$

Or, since $\int_0^\infty 2\pi\eta f^2(\eta)d\eta$ will be constant for known $f(\eta)$, it implies

$$\left.\begin{aligned} &b^2\bar{u}^2_m \propto x^o \\ \text{or,}\quad &2p + 2q = 0 \\ \text{i.e.}\quad &p + q = 0 \end{aligned}\right\} \qquad \ldots (10.29)$$

One needs another relation between p and q to obtain their values. This is obtained from integral energy equation or from entrainment hypothesis (2). One can also perform dimensional analysis and write

$$\bar{u}_m, b, Q = f(M_o, \rho, x)$$

where $M_o = \pi R_o^2 \rho U_o^2$ and $2R_o = D_o$. Therefore, one gets $\bar{u}_m/\sqrt{M_o/\rho x^2} = C_1$ which then gives

$$\left.\begin{aligned} &\bar{u}_m/U_o = \frac{C_1}{2}\frac{D_o}{x} \\ &\text{Also } b/x = C_2 \text{ and } Q/Q_o = C_3 x/D_o \end{aligned}\right\} \qquad \ldots (10.30)$$

where Q is the discharge at any section and $Q_o = \pi R^2{}_o U_o$.
Therefore, $p = -1$ and $q = +1$.

Velocity Distribution

Again, as in the case of two dimensional jet, we have Tollmien type solution based on Prandtl's mixing length theory. The details of the solution are given by Rajaratnam (2). This solution as tabulated by Abramovich (11) is given in Table 10.3

Table 10.3

Velocity distribution in round jet according to Tollmien and Görtler

Tollmien		*Gortler*	
r/b	$\bar{u}/\bar{u}_m$	r/b	$\bar{u}/\bar{u}_m$
0.000	1.000	0.000	1.000
0.161	0.958	0.166	0.976
0.322	0.884	0.332	0.914
0.484	0.795	0.497	0.826
0.645	0.700	0.663	0.715
0.806	0.605	0.773	0.641
0.967	0.510	0.995	0.505
1.130	0.425	1.105	0.445
1.370	0.300	1.328	0.338
1.531	0.230	1.548	0.254
1.772	0.140	1.770	0.198
2.019	0.075	2.100	0.127
2.340	0.024	2.490	0.079
2.500	0.011	2.762	0.059
2.740	0.000	3.315	0.033

Görtler type solution yields velocity distribution in the form

$$\bar{u}/\bar{u}_m = \frac{1}{1.0 + 0.125\,\xi^2} \qquad \ldots (10.31)$$

and

$$\sigma\bar{v}/\bar{u}_m = \frac{\xi - 0.125\xi^3}{2(1.0 + 0.125\,\xi^2)^2} \qquad \ldots (10.32)$$

where $\xi = \sigma r/x$. The experimentally determined values of "*a*" in Tollmien and σ in Görtler type solutions are 0.066 and 18.5, respectively. Variation of $\bar{u}/\bar{u}_m$ with r/b according to Görtler type solution is also listed in Table 10.3 The velocity distribution can also be represented by the equation

$$\bar{u}/\bar{u}_m = e^{-0.693(r/b)^2} \qquad \ldots (10.33)$$

10.7 EXPERIMENTAL RESULTS

Experiments on flow characteristics of a diffusing round jet have been carried out by Ruden and Kuethe as reported by Goldstein (7), Trupel as reported by Abramovich (11), Reichardt as reported by Schlichting (12), Corrsin (16), Hinze and Zijnen (17), and Albertson et al. (4). These studies have shown that the constant $C_1/2$ in equation for u_m/u_o varies between 5.75 to 6.35. Rajaratnam (2) has recommended a value of 6.30 while List (18) recommends it as 6.20. Hence,

$$\frac{\bar{u}_m}{U_o} = \frac{6.2 \text{ to } 6.3}{x/D_o} \qquad \ldots (10.34)$$

for the zone of established flow. The variation of width of jet is known if C_2 in Eq. 10.30 is known, which is found to vary between 0.082 and 0.097. Taking its value as 0.10, one gets

$$\frac{b}{x} = 0.10 \qquad \ldots (10.35)$$

which is the same as that for plane jet. The length of zone of establishment is found to be

$$x_o = 6.2\,D_o \qquad \ldots (10.36)$$

In the zone of established flow, discharge Q and kinetic energy E vary with x as

$$\left.\begin{aligned} Q/Q_o &= 0.32x/D_o \\ E/E_o &= 4.1\,D_o/x_o \end{aligned}\right\} \qquad \ldots (10.37)$$

as found by Albetson et al. (4).

Turbulence in Round Jets

Turbulence measurements in round jets have been made by Corrsin (16), Wygnanski and Fiedler (19) and by Rodi as reported by List (18). These studies indicate that a round turbulent jet does not attain self preserving state, in the sense of reaching turbulent stress equilibrium, until some 40 jet diameters downstream. Beyond this range, the ratio $\sqrt{\overline{u'^2}}/\bar{u}_m$ on the jet axis is about 0.28 – 0.29, whereas $\sqrt{\overline{v'^2}}/\bar{u}_m$ are $\sqrt{\overline{w'^2}}/u_m$ are nearly equal, and they asymptotically reach a value of 0.23–0.25. Beyond a distance of $x/D_o > 40$ the distribution of turbulent shear stress as well as streamwise turbulence intensities across the jet are similar when nondimensionalised properly (see Fig. 10.5).

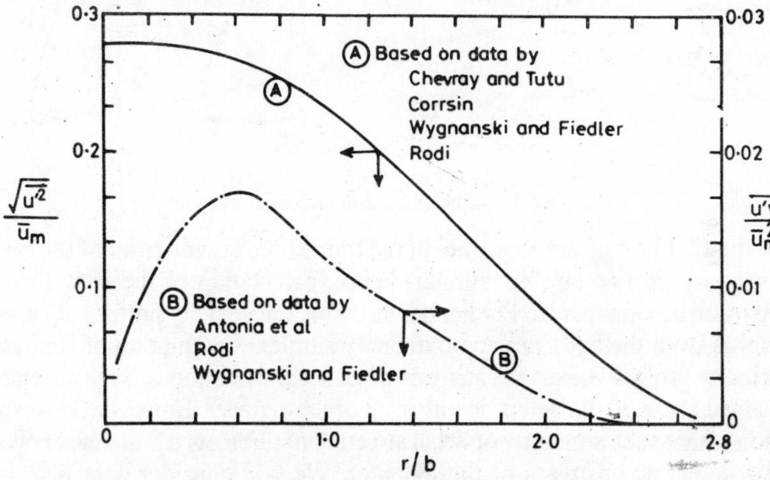

Fig. 10.5 Distribution of turbulent intensity and shear across round jets

Other properties of turbulent jets have also been measured in some detail. These include velocity correlation functions, velocity and shear stress spectra, concentration and *rms* concentration spectra, various micro and macro scales of velocity and concentration as well as time, energy balance for mean and turbulent motion, etc. In this connection, the reader can see publications by Becker et al. (20), Wygnanski and Fiedler (19), Chevray and Tutu (21), and Birch et al. (22).

WAKES

The flow behind a solid body forms a wake. The mean velocity in the wake is less than the free stream velocity in the outside stream. Typical flow pattern in the wake behind a two dimensional body is shown in Fig. 10.6. At very small Reynolds numbers, the flow in the wake is laminar which becomes turbulent when Reynolds number becomes high. In such a case, regular vortices develop behind the body, which finally get dissolved into small irregular eddies to form the mixing region. As these eddies diffuse laterally in the downstream direction, an ever increasing portion of the ambient fluid is retarded, and previously retarded fluid gets acclerated. In the wake region, the boundary layer approximations are applicable.

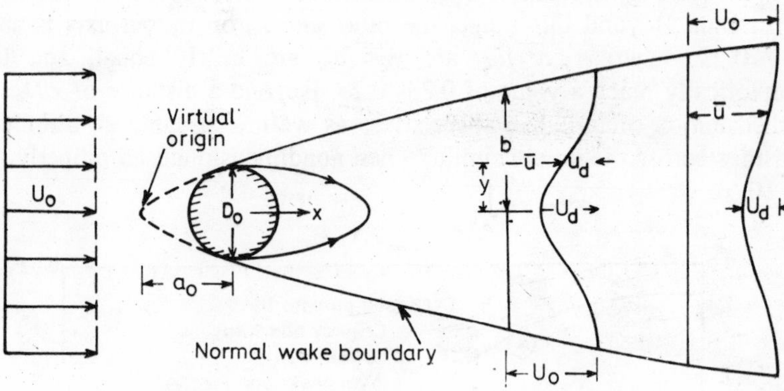

Fig. 10.6 Wake behind a 2-dimensional cylinder

In the first part of the wake, i.e. in the immediate downstream of the body, the velocity profiles are not similar; hence, calculation of the flow field in this portion is complicated. Further downstream, the velocity profiles at various distances from the body are approximately similar. Assumption of similarity of velocity profiles makes the analysis of wakes rather simple. Such similarity is attained about 80 diameters downstream of the cylinder. However, Townsend has found that such similarity of detail structure of turbulence is attained beyond 50 diameters downstream of the cylinder. We will consider here both two dimensional and axisymmetric wakes.

10.8 INTEGRAL RELATIONSHIPS

As in the case of jets and boundary layers, one can assume pressure to be constant in the flow region. Application of momentum relationship then provides one equation between maximum velocity difference U_d at the centreline of the wake and wake width b. Consider wake flow behind an axisymmetric body and substitute $\bar{u}_d = (U_o - \bar{u})$. Since no other external force is acting on the flow, rate of change of momentum will be balanced by the drag force acting on the body. Hence, one can write

$$-C_D \pi R_o^2 \rho \frac{U_o^2}{2} = \text{(Final momentum} - \text{Initial momentum)}$$

$$= \int_0^r 2\pi r \rho \bar{u}^2 dr + \left[\int_0^r 2\pi r \rho U_o^2 dr - \int_0^r 2\pi r \rho U_o \bar{u}\, dr \right] - \int_0^r 2\pi r \rho U_o^2 dr \quad \ldots (10.38)$$

Here the first term on the right hand side represents the momentum outflow through the downstream section; the quantity in the rectangular bracket represents momentum overflow through the cylindrical surface. The third term represents momentum outflow across the upstream section. In writing this equation, fluctuating part of velocity has been neglected. It is also assumed that the upper limit of integration r is greater than the radius of wake. Substitute $\bar{u}_d = (U_o - \bar{u})$, the velocity difference, in Eq. 10.38, which then becomes

$$-C_D \pi R_o^2 \rho \frac{U_o^2}{2} = -2\pi\rho \int_o^{r_1} U_o \bar{u}_d \, r \, dr \qquad \ldots (10.39)$$

In simplifying Eq. 10.38 to Eq. 10.39, it is assumed that $\bar{u}_d$ being small, $\bar{u}_d^2 << U_o \bar{u}_d$ and, hence, can be neglected. As done by Rouse (3), one can write $U_o \bar{u}_d / U_d$ and introduce a dimensionless variable r/b. Then the above equation becomes

$$\frac{1}{4} C_D R_o^2 U_o^2 = b^2 \, U_o U_d \int_o^{\infty} \left(\frac{\bar{u}_d}{U_d}\right) \frac{r}{b} \, d\left(\frac{r}{b}\right) \qquad \ldots (10.40)$$

Here U_d is the maximum value of $\bar{u}_d$ which occurs along the axis of symmetry. Further, the limit of integration is extended to infinity. Since for known body shape, U_o and fluid, C_D is constant, left hand side in Eq. 10.40 assumes a constant value. And, since velocity distribution is similar and known, the quantity under integration sign is constant. Therefore, in the wake region where velocity distribution is similar

$$\left.\begin{aligned} & b^2 U_d = \text{constant} \\ \text{or} \quad & b^2 U_d \propto x^o \end{aligned}\right\} \qquad \ldots (10.41)$$

Another equation is needed to determine how U_d and b vary with x. This can be obtained from the energy equation as shown by Rouse (8). Such an analysis for wakes behind axisymmetric bodies indicates that

$$\left.\begin{aligned} & U_d \propto x^{-2/3} \\ & b \propto x^{1/3} \end{aligned}\right\} \qquad \ldots (10.42)$$

Similar analysis for wakes behind two dimensional bodies yields

$$\begin{aligned} & U_d \propto x^{-1/2} \\ & b \propto x^{1/2} \end{aligned} \qquad \ldots (10.43)$$

10.9 VELOCITY DISTRIBUTION AND STRUCTURE OF TURBULENCE

The velocity distribution in wakes behind 2-D bodies was first investigated by Schlichting (28) using Prandtl's mixing length theory, and also by Görtler, between 1930-1942. With the assumptions listed at the beginning of this chapter, the equation of motion reduces to

$$\bar{u} \frac{\partial \bar{u}}{\partial x} + \bar{v} \frac{\partial \bar{u}}{\partial y} = \frac{1}{\rho} \frac{\partial \tau_t}{\delta y}$$

Substituting $\bar{u}_d = U_o - \bar{u}$, one can evaluate $\frac{\partial \bar{u}}{\partial x}$ and $\frac{\partial \bar{u}}{\partial y}$. Further, since

$\tau_t = \rho l^2 \left(\frac{\partial \bar{u}}{\partial y}\right)^2$, the above equation can be written as

$$\left.\begin{aligned} & -(U_o - \bar{u}_d) \frac{\partial \bar{u}_d}{\partial x} - \bar{v} \frac{\partial \bar{u}_d}{\partial y} = 2 l^2 \left(-\frac{\partial \bar{u}_d}{\partial y}\right) \left(\frac{\partial^2 \bar{u}_d}{\partial y^2}\right) \\ \text{or,} \quad & (U_o - \bar{u}_d) \frac{\partial \bar{u}_d}{\partial x} + \bar{v} \frac{\partial \bar{u}_d}{\partial y} = -2 \, l^2 \frac{\partial \bar{u}_d}{\partial y} \frac{\partial^2 \bar{u}_d}{\partial y^2} \end{aligned}\right\} \qquad \ldots (10.44)$$

Since $U_o >> \bar{u}_d$ and $U_o \dfrac{\partial \bar{u}_d}{\partial x} >> \bar{v} \dfrac{\partial \bar{u}_d}{\partial y}$, Eq. 10.44 can be reduced to

$$U \frac{\partial \bar{u}_d}{\partial x} = -2\, l^2 \frac{\partial \bar{u}_d}{\partial y} \frac{\partial^2 \bar{u}_d}{\partial y^2} \qquad \ldots (10.45)$$

Assuming that mixing length is constant over the section, and is proportional to the width there, one can write $l = \beta b$, where β is a constant. Letting $\eta = y/b$ to be the dimensionless variable, and substituting

$b = B\,(C_D\, D_o\, x)^{1/2}$ where D_o is the diameter of cylinder

and $\bar{u}_d = U_o \left(\dfrac{x}{C_D D_o}\right)^{-1/2} f(\eta)$

Equation 10.45 reduces to

$$\frac{1}{2}(f + \eta f') = \frac{2\beta^2}{B} f' f'' \qquad \ldots (10.46)$$

where B is a constant. Integrating this equation one gets

$$\frac{1}{2}\eta f = \frac{\beta^2}{B} f'^2 + C \qquad \ldots (10.47)$$

with the boundary condition

$\bar{u}_d = 0$ and $\dfrac{\partial \bar{u}_d}{\partial y} = 0$ at $y = b$

i.e. $f = f' = 0$ at $\eta = 1$. Hence, $C = 0$

One more integration yields

$$f = \frac{1}{9} \cdot \frac{B}{2\beta^2} (1 - \eta^{3/2})^2$$

The constant B is determined from application of momentum equation which yields $B = \sqrt{10}\,\beta$. Hence,

$b = \sqrt{10}\,\beta (C_D\, D_o\, x)^{1/2}$

$$\text{and } \frac{\bar{u}_d}{U_o} = \frac{\sqrt{10}}{18\beta} \left(\frac{x}{C_D D_o}\right)^{-1/2} \left(1 - \left(\frac{y}{b}\right)^{3/2}\right)^2 \qquad \ldots (10.48)$$

Experimental value of β obtained from Reichardt's data is 0.18. Görtler (24) started with the assumption that τ_t can be expressed as $\tau_t = \rho\varepsilon \dfrac{\partial \bar{u}_d}{\partial y}$ and that $\varepsilon_o = \kappa b U_d$ where κ is constant, and U_d is the maximum value of $\bar{u}_d$. Using this expression for shear, the velocity distribution is shown to follow Gaussian distribution, viz.

$$\frac{\bar{U}_d}{U_o} = \frac{1}{4\sqrt{\pi}} \sqrt{\frac{U_o C_D D_o}{\varepsilon_o}} \left(\frac{x}{C_D D_o}\right)^{-\frac{1}{2}} (\exp -\eta^{-2/4}) \qquad \ldots (10.49)$$

Experimentally, $U_o C_D D_o/\varepsilon_o$ is found to be 0.0222. With this value the two velocity distribution laws are plotted in Fig. 10.7 along with Reichardt's data. It can be seen that the agreement of these two equations with the measured data is quite good. Similar results have been obtained by Fage and Falkner (7*).

The velocity distribution in the turbulent wake behind a body of revolution has been studied by Swain (25) and also discussed by Goldstein (7). Swain showed that Eq. 10.48 holds good for circular wakes. Figure 10.8 shows variation of $\bar{u}_d/U_d$ with $r/r_{1/2}$ where $r_{1/2}$ is value of r at which $\bar{u}_d/U_d = 0.50$. On this figure are also shown the data by Hall and Hislop. The agreement seems to be fairly good.

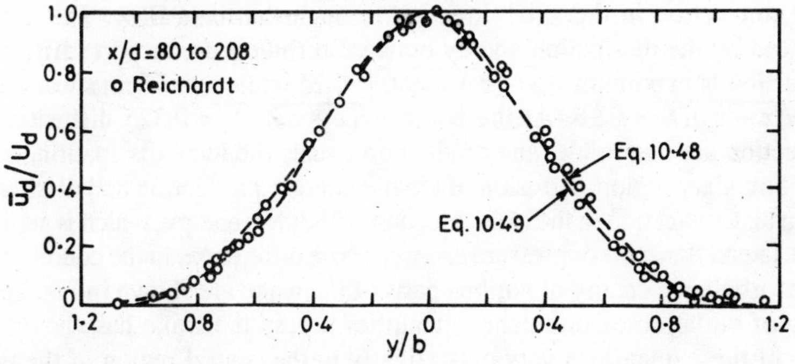

Fig. 10.7 Velocity distribution in a wake behind two dimensional body (12)

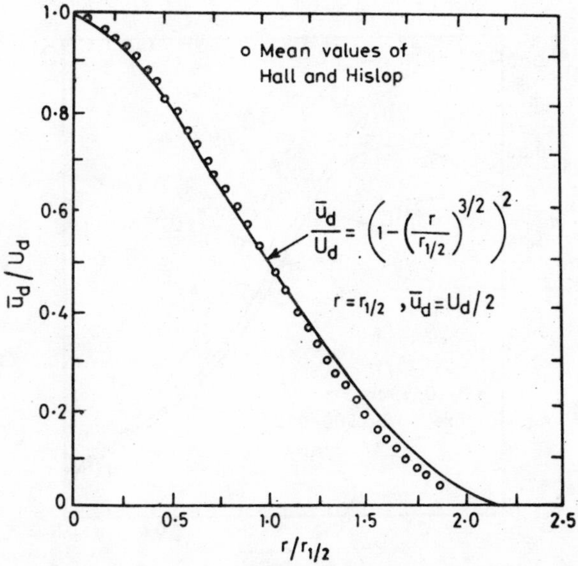

Fig. 10.8 Variation $\bar{u}_d/U_d$ with $r/r_{1/2}$ for wake behind body of revolution

Among others, extensive measurements of turbulence quantities in the wake have been made by Townsend and reported by Hinze (4). These have been made in wake behind a two dimensional circular cylinder of diameter 0.159 cm placed in air at a velocity at 1.28 m/s. Figure 10.9 shows the variation of $\overline{u'^2}/U_d^2$, $\overline{v'^2}/U_d^2$ and $\overline{w'^2}/U_d^2$ with $y/\sqrt{(x + a_o) D_o}$; the mean curves are for data at $(x + a_o)/D_o$ values ranging from 500 to 950. It can be seen that far away from the axis, turbulence tends to be isotropic, however, it is not so near the axis. These data also indicate similarity of distribution of turbulence intensity.

Townsend, and Kobashi (1*) have studied the energy balance of turbulent flow in the wake by measuring various terms appearing in the energy balance equation, viz. convection, diffusion, production and dissipation. This analysis has shown that near the axis, the production is negligible and the main gain is by convection in the axial direction through the main flow. This gain is balanced by the dissipation and by outward turbulence transport (diffusion). Production is maximum at $y/\sqrt{(x + a_o)D_o} = 0.25$ while maximum shear occurs at $y/\sqrt{(x + a_o)D_o} = 0.18$. At the point $y/\sqrt{(x + a_o) D_o} = 0.325$ diffusion and convection are negligible, and production equals the local dissipation. In the outer boundary region, diffusion of kinetic energy, production and dissipation are small. Convection in the axial direction withdraws energy, which is supplied by the lateral transport of pressure energy. These differences in the contributions to the turbulence energy of various parts of the wake are shown in Fig. 10.10. Study of variation of turbulence quantities across the wake has shown that many of these quantities vary only slightly in the central region of the wake and decrease more ripidly near the outer region. The micro and macro scales of turbulence also show slight variation across the wake.

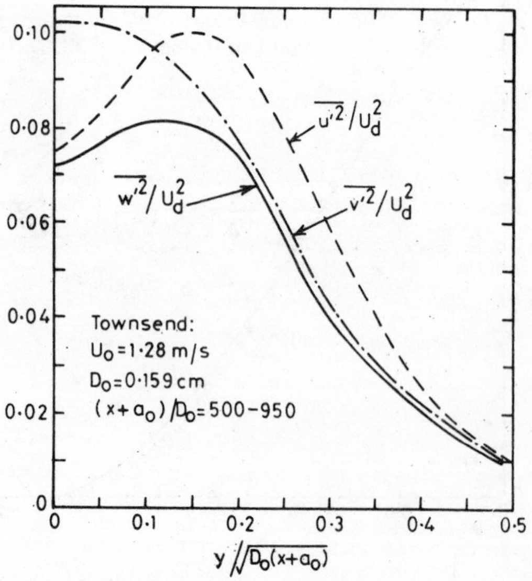

Fig. 10.9 Turbulence characteristics in wake behind a 2-D circular cylinder

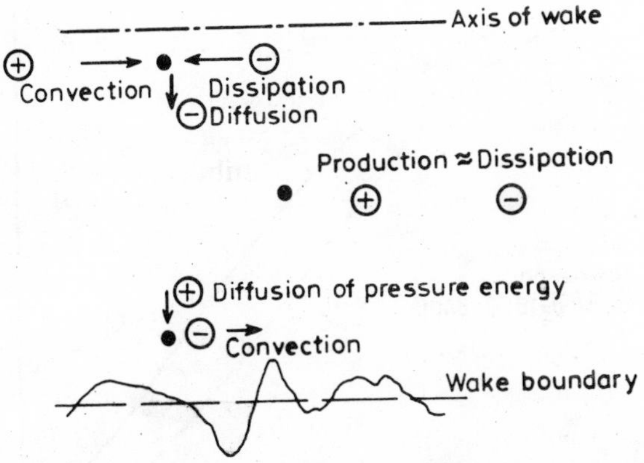

Fig. 10.10 Contribution to energy balance in different parts of the wake

As in the case of a turbulent boundary layer and a jet, the boundary which separates a wake from the nonturbulent surrounding fluid is quite irregular. In the central portion, the turbulence is practically continuous with respect to time, and it becomes more and more intermittent near the boundary of the wake (see Fig 10.11); the instantaneous picture of the wake boundary is shown in this figure. The intermittency factor γ has been earlier defined as ratio of time during which turbulence occurs and total time. Figure 10.12 shows variation of γ with $y/\sqrt{(x + a_o)D_o}$ for wake flow behind 2-D circular cylinder. It can be seen that for large values of $(x + a_o)/D_o$, there is similarity in variation of γ with $y/\sqrt{(x + a_o)D_o}$.

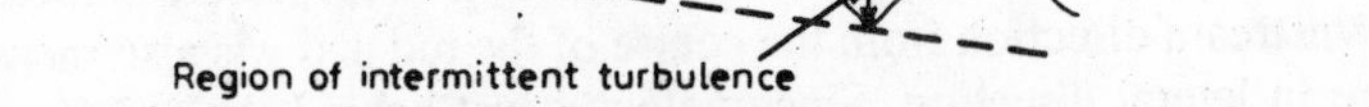

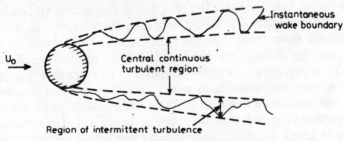

Fig. 10.11 Instantaneous picture of wake

As in the case of other turbulent flows, turbulence modelling has been used to predict the mean velocity distribution and turbulent kinetic energy across the jet (25). After assuming that the integral scale of turbulence is proportional to the wake width b at any section, Adachi and Yoshida (25), have used turbulence modelling, using kinetic energy of turbulence to be a transferable quantity. Good conformity between experimental and calculated velocity distribution, and distribution of kinetic energy have been reported.

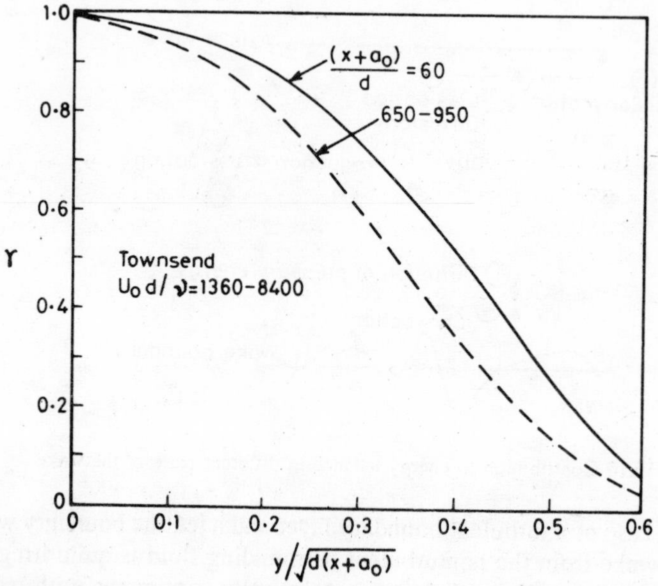

Fig. 10.12 Variation of intermittency factor in the wake of 2-D circular cylinder

10.10 WAKE BEHIND A ROW OF PARALLEL RODS

Characteristics of the wake behind a row of parallel rods have been studied theoretically and experimentally by Olsson (12*) and Anderlik (7*). Such rods at close spacing are often used in the upstream of wind tunnels to obtain locally uniform velocity. Let M be the centre to centre spacing between the rods kept in a free stream velocity U_o. Wakes will form behind each rod, and at a certain distance downstream from the rods, the wake width b will be equal to M. Beyond this distance, the wakes of individual rods will interfere and the velocity at any distance x from the rods will be a periodic function of y. If this velocity distribution is to satisfy Eq. 10.45, one can assume that $u_d \sim x^p f(y)$ and substitute it in Eq. 10.45; this gives $p = -1$. Here x is measured in the downstream direction from the centre of the rod and y is also measured from there in lateral direction. Since velocity just behind the rod for any x value will be the least, $\bar{u}_d$ will be maximum there; when $y/M = 0.5$, the velocity will be maximum and $\bar{u}_d$ will be minimum. Such a periodic velocity distribution can be expressed by the equation

$$\bar{u}_d/U_o = A(x/M)^{-1} \cos (2\pi y/M)$$

where A is a constant. Substitution of the above equation in Eq. 10.45 yields $A = (M/l)^2/8\pi^3$ where l is the mixing length. Comparison of this equation with experimental data by Olsson indicated that $\frac{l}{M} = 0.103$. Hence $\bar{u}_d/U_o$ is given by

$$\bar{u}_d/U_o = \frac{(M/l)^2}{8\pi^3}\left(\frac{x}{M}\right)^{-1} \cos(2\pi y/M)$$

with $\frac{l}{M} = 0.103$. This is valid for x/M greater than 4.0.

FORCES ON IMMERSED BODIES

Whenever there is a relative motion between a body and a real fluid, a resultant force acts on the body which can be resolved into two components : one in the direction of motion known as the drag force and the other in the direction perpendicular to motion known as the lift force. The magnitudes of these two components are usually related to the flow, fluid and geometric parameters through the relations

Drag Force $F_D = C_D A \rho U_o^2/2$

Lift Force $F_L = C_L A \rho U_o^2/2$

where C_D and C_L are known as the drag and lift coefficient, respectively; A is the cross-sectional area perpendicular to the flow and U_o is the characteristic velocity. For completely submerged bodies and at velocities with Mach number much less than unity, C_D and C_L are usually functions of Reynolds number of the flow and geometric parameters related to the body. Figure 10.13 shows variation of C_D with Reynolds number for a variety of body shapes, such as sphere, two dimensional circular cylinder, and two dimensional flat plate and a circular disc held perpendicular to the flow. It can be seen from this figure that in general C_D decreases with increase in Re when Re is small. Beyond Reynolds number of 10^3, C_D remains constant for 2-D flat plate and circular disc held perpendicular to flow. This is due to the fact that with such sharp edged bodies the location of the point of flow separation is fixed. On the other hand, in the case of circular cylinder and sphere, C_D slightly increases in the Reynolds number range of 10^3 and 10^5 and

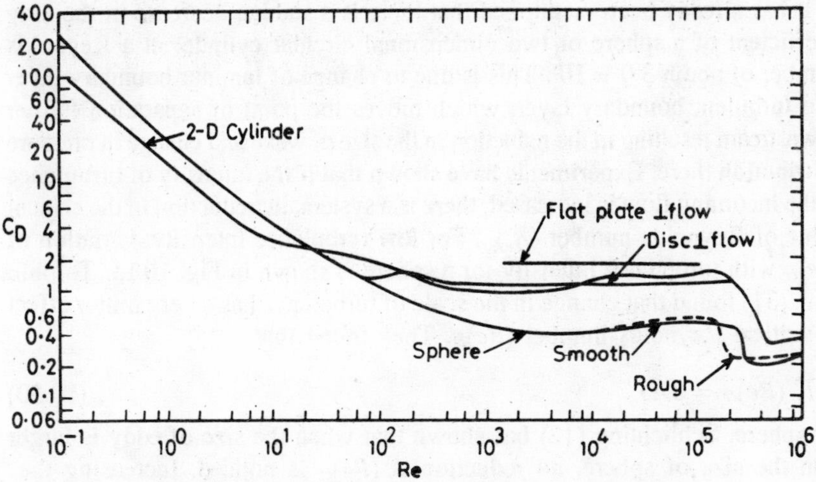

Fig. 10.13 Variation of C_D with Re for various body shapes

then suddenly decreases at Reynolds number of about 3×10^5. The sudden decrease in C_D is due to change in the boundary layer from laminar to turbulent, which drastically reduces the size of wake and alters the pressure distribution there resulting in reduction of drag force and C_D.

In the Fluid Mechanics text books, normally a chapter is devoted to study variation of C_D and C_L for certain body shapes. However, in such a discussion no mention is usually made of fluctuating nature of F_D, F_L, C_D and C_L when the flow field is turbulent. Figure 10.14 shows fluctuating drag force on a cube held normal to turbulent air stream as obtained by Satapathy (26). Similar measurements have been made by Einstein and El-Samni (27), Ko and Graf (28,29), Cheng et al. (30) and others on drag and lift force fluctuations. From these studies, it has been found that the pressure as well as the drag and lift force fluctuations follow nearly Gaussian distribution. From the design point of view, one would like to know more about the fluctuating nature of these forces and their relation to turbulence characteristics of the ambient flow. As regards the force exerted by turbulent flow on a body, the following two aspects are of relevance, namely, effect of characteristics of turbulence on the drag, and Karman vortex trail and its effects.

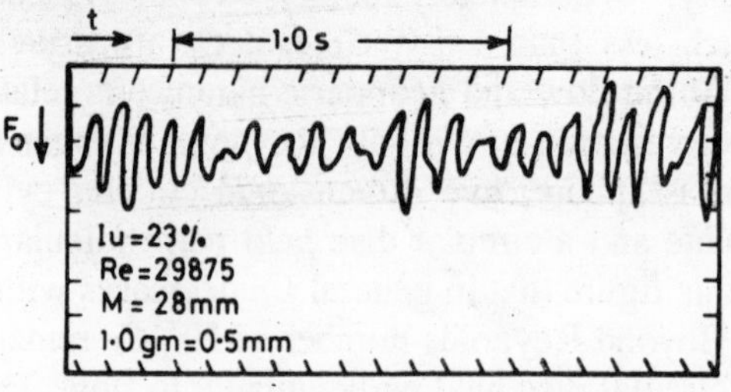

Fig. 10.14 Fluctuating drag force on a cube

10.11 EFFECT OF TURBULENCE CHARACTERISTICS ON DRAG

It has already been mentioned that there is a sudden decrease in the drag coefficient of a sphere or two dimensional circular cylinder at a Reynolds number of about 3.0×10^5. This is due to change of laminar boundary layer into turbulent boundary layer, which moves the point of separation farther downstream resulting in the reduction in the size of wake and change in pressure distribution there. Experiments have shown that if the intensity of turbulence in the incoming flow is increased, there is a systematic reduction in the critical value of Reynolds number $(R_e)_{cr}$. For low turbulence intensity, variation of $(Re)_{cr}$ with turbulence intensity for a sphere is shown in Fig. 10.15. Torobin et al. (31) found that change in the scale of turbulence has a very minor effect on critical Reynolds number $(Re)_{cr}$. They found that

$$I_u^2 \, (Re)_{cr} = 0.45 \qquad \ldots (10.50)$$

for sphere. Schlichting (12) has shown that when the size of eddy is larger than the size of sphere, no reduction is $(Re)_{cr}$ is noticed. Increasing the turbulence intensity causes a moderate increase in drag coefficient in the subcriticial as well as supercritcal regions.

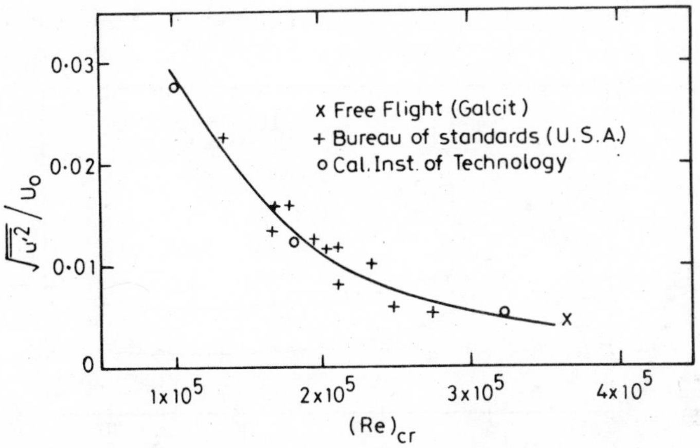

Fig. 10.15 Dependence of $(Re)_{cr}$ on turbulence intensity for a sphere (7)

After carrying out theoretical analysis, Taylor (32) showed that *rms* value of pressure gradient $\sqrt{(\partial p'/\partial x)^2}$ is a function of $\frac{\rho \overline{u'^2}}{L} \sqrt{\frac{\sqrt{\overline{u'^2}}\, L}{\nu}}$ where L is eddy size or integral macroscale of turbulence. This parameter can be written as $I_u\,(D/L)^2$ where $I_u = \sqrt{\overline{u'^2}}/U_o$ and D is characteristic diameter; $I_u\,(D/L)^2$ is many times known as Taylor's parameter or Taylor number. Schubauer (33) has shown that the relative location of the transition point X_t/D on an elliptic cylinder is related to Taylor parameter. Loisyanskii (34) has explained this phenomenon, using the argument that at low values of Taylor's parameter, external disturbances have little effect on the laminar boundary layer and transition takes place due to internal instability of the boundary layer. At higher values of this parameter, the influence of external disturbance predominates and the length of laminar part X_t rapidly decreases.

In general, it has been found that for streamlined bodies $(Re)_{cr}$ and $\overline{C_D}$ decrease with increase in turbulence intensity and scale of turbulence, while for bluff bodies the effect is not conclusive. For bluff bodies $\overline{C_D}$ sometimes increases and sometimes decreases with increased turbulence. This has been found to be dependent on the aspect ratio of the body which controls the reattachment of separated shear layer. Roberson et al. (36) have found that bodies for which reattachment of the flow is not a factor, increase in turbulence intensity increases $\overline{C_D}$. On the other hand bodies, for which reattachment or near attachment of flow occurs due to enhanced momentum transfer in the shear layer and shifting of vortex generating zone due to increase in turbulence, may experience either a decrease or increase in $\overline{C_D}$ with increased turbulence intensity depending on the shape of the body. Vickery (37,38,39) has studied effect of turbulence and aspect ratio of body on $\overline{C_D}$ and $\sqrt{C_D'^2}$ His (39) results for rectangular blocks and cylinders of various lengths kept on the floor snow effect of aspect ratio on $\sqrt{C_D'^2}/\overline{C}_D$ for a given turbulence intensity (see Fig. 10.16).

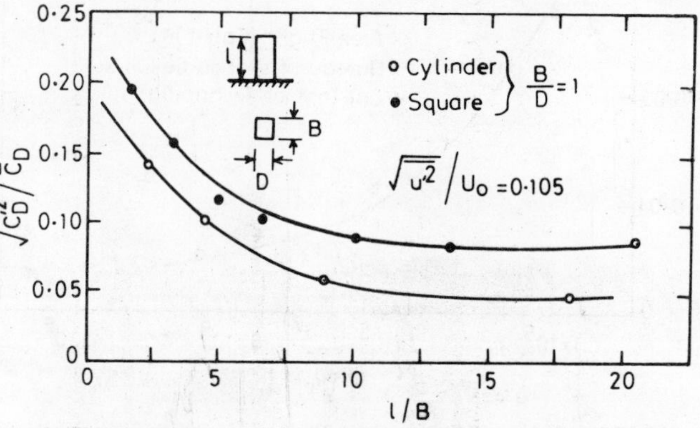

Fig. **10.16** Varition of $\sqrt{C_D'^2}/\overline{C_D}$ with l/B for cylinders

Ko and others (29, 30), as well as Satapathy (27) have studied the effect of free stream turbulence on mean drag coefficient $\overline{C_D}$. For two dimensional circular cylinders, Ko and others found that $I_u^2/\overline{C}_D = f\left(\frac{I_u(D/L)}{\log Re}\right)$. Graf and Mansour (35) found that for cubes as well as square rods $I_u^2/\overline{C_D}$ is a function of $I_u(D/L)^{1/5}/(\log Re)$. Figure 10.17 shows variation of $I_u^2/\overline{C}_D$ with $I_u(D/L^{1/5}/$ (log *Re*) for sphere as obtained by Satpathy. For various body shapes the equations proposed are:

$$\left.\begin{aligned} &\text{Circular cylinders (29,30)} && I_u^2/\overline{C_D} = 0.129\,(B) + 13.652\,B^2 \\ &\text{Cubes (36)} && I_u^2/\overline{C_D} = -\,0.132\,(B) + 10.6\,B^2 \\ &\text{Square rods (36)} && I_u/\overline{C_D} = -\,0.0637\,(B) + 133.6\,B^2 \\ &\text{where} && B = \frac{I_u(D/L)^{1/5}}{\log Re} \end{aligned}\right\} \quad (10.51)$$

It may be mentioned that $I_u^2/\overline{C}_D$ can be interpreted as the ratio (kinetic energy of turbulence per unit volume)/(force exerted by mean flow per unit area).

Satapathy has conducted experiments on spheres, cubes with their face normal to flow, cubes with their face at 45° to flow, and circular and square plates. The Reynolds number varied from 3×10^3 to 8×10^4, turbulence intensity from 16 to 26 per cent and relative scale of turbulence D/L from 0.22 to 8.6. These experiments were conducted in two different sizes of wind tunnels. Major conclusions drawn by him are as follows:

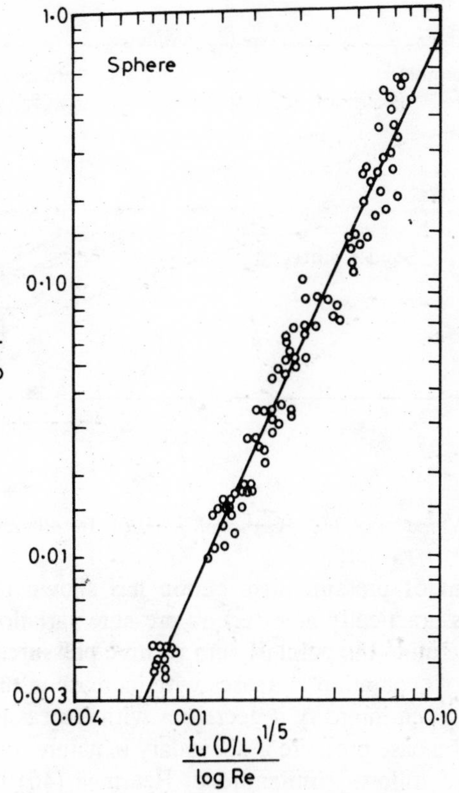

Fig. 10.17 Variation of $\frac{I_u^2}{\overline{C_{DO}}}$ with $\frac{I_u\,(D/L)^{1/5}}{\log Re}$

1. To eliminate effect of blockage, $\overline{C_D}$ and $\sqrt{C_D'}$ need to be corrected for blockage and to obtain corrected values designated as $\overline{C_{DO}}$ and $\sqrt{\overline{C_{DO}^{\prime 2}}}$
2. $\overline{C_{DO}}$ is related to $I_u(D/L)^{1/5}/\log Re$ by the relation

$$a\left[Iu(D/L)^{1/5}/\log Re\right]^b = I_u^2/\overline{C_{DO}} \quad \ldots (10.52)$$

where a and b are dependent on body shape as shown below:

Body Shape	a	b
Sphere	0.57	2.15
Cube (normal)	0.33	1.89
Cube (45°)	2.02	1.89
Plate (normal)	0.22	1.95

See Fig. 10.17 for sphere.

3. $\sqrt{\overline{C_D^{\prime 2}}}/\overline{C_{DO}}$ is related to $I_u(D/L^2)$ as shown in Fig. 10.18. On this figure are shown mean lines for sphere, cube, square plate and circular plate. It can be seen that two envelope lines can be drawn to contain these four curves.

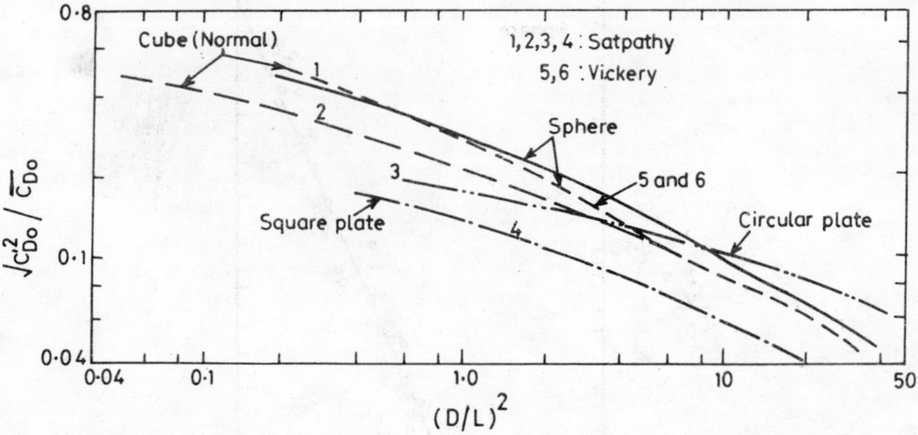

Fig. 10.18 Variation of $\sqrt{\overline{C_D'^2}}/\overline{C_{DO}}$ with $I_u\,(D/L)^2$ for different body shapes

4. Measurement of pressure distribution has shown that the turbulence intensity has practically no effect on pressure variation on the front side of the sphere up to the point of zero relative pressure. The base pressure in the zone of separation increases with increase in turbulence intensity, while for a given intensity it decreases with increasing scale. However, scale effect on base pressure is secondary in nature. Pressure distribution around cubes follows similar trend. Bearman (40) has suggessted the mechanism which brings about changes in pressure distribution or drag in the presence of turbulence. According to him, the introduction of turbulence induces increased entrainment of fluid from the wake in the case of a bluff body, thereby decreasing its size and thickening the shear layer.

10.12 LATERAL OSCILLATIONS

Primary characteristic of all symmetrical 2-D bodies is the lateral pendulation of the separation zone as eddies form alternately on either side of centre line of flow and pass into the fluid. This phenomenon is known as Karman vortex trail, since it was first studied theoretically by Karman. Generation of these vortices proceeds in close accordance with the basic theorem of hydrodynamics which states that the circulation around any closed curve within the fluid must remain constant with time. Alternate release of the vortices gives a flow configuration as shown in Fig. 10.19. With notations shown in the figure Karman found that the configuration is stable when $a/b = \cosh^{-1}\sqrt{2} = 0.281$. The frequency f of alternate shedding of vortices is related to U_o and u_r the velocity of downward movement of vortices by the expression $f = (U_o - u_r)/b$. The circulation Γ induced by vortices is related to u_r by the relation $u_r = \Gamma/\sqrt{8}b$. The dimensionless parameter characterising the alternate shedding of vortices is Strouhal number $S = fD/U_o$ where D is characteristic length, e.g., diameter of cylinder. In general, Strouhal number is a function of Reynolds number and body shape; however since $\overline{C_D}$ is a function of *Re* and body shape, one would expect Strouhal number to be uniquely related to $\overline{C_D}$. That this is indeed the case can be seen from Fig. 10.20 in which *S* is plotted against $\overline{C_D}$ for various two dimensional body shapes. The relationship between the two can be expressed by the equation

$$\overline{S} = 0.21 / \overline{C_D}^{0.75}$$

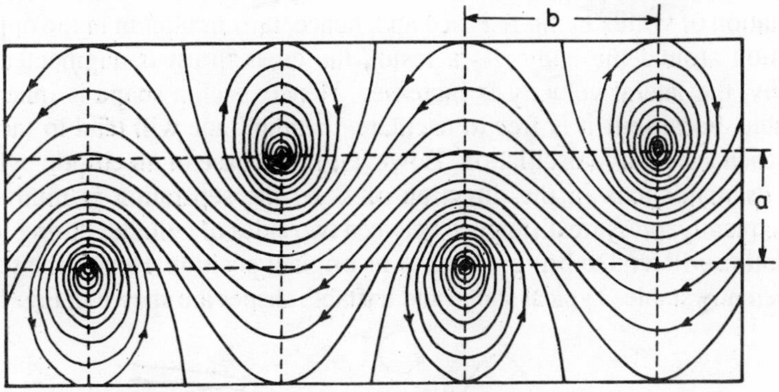

Fig. 10.19 Stream lines of Karman vortex-trail, relative to vortices

Alternate shedding of vortices causes a lateral force acting on the cylindrical body which can cause forced vibration of structures, such as smoke stacks, telephone wires, electrical cables, suspension bridges, submarine periscopes, etc. If natural frequency of these structures is close to the forced frequency, there will be danger of resonance.

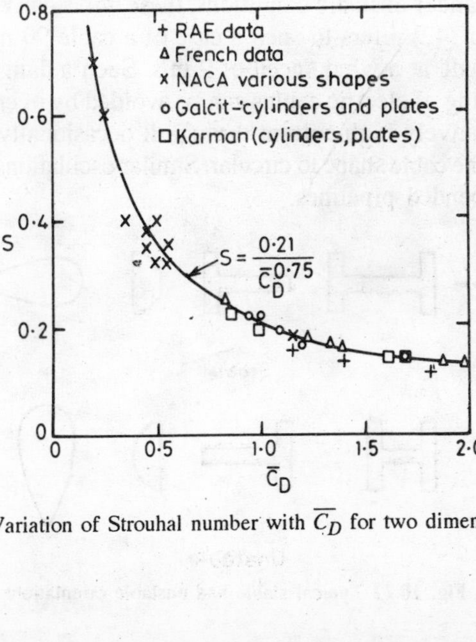

Fig. 10.20 Variation of Strouhal number with $\overline{C_D}$ for two dimensional bodies

The role played by the shape of cylindrical 2-D body in stabilising the condition or enhancing the unstable flow caused by Karman vortex trail has been well discussed by Rouse (41). Consider flow around an elliptic cylinder with its major axis at right angles to the flow. If a momentary cross thrust is applied to it due to release of vortex, it produces an oscillation of the cylinder. The vector sum of the longitudinal velocity of approach and lateral velocity of oscillation yields a resultant velocity somewhat inclined to the axis of symmetry as shown in Fig. 10.21. This condition tends to increase the circulation of vortex being released and, hence, the circulation in the opposite direction around the body. As a result, the cross thrust is augmented and, thereby, the lateral velocity is increased. Hence, such a shape is inherently unstable, because if it is free to oscillate, its amplitude will tend to increase with each successive oscillation. However, in the case of an elliptic cylinder with its major axis in the direction of flow, the separation is minimised. Consequently, pressure difference between two sides of cylinder during lateral oscillation will tend to bring it back to its original position. Hence, the section is inherently stable. Typical stable and unstable shapes are shown in Fig. 10.22.

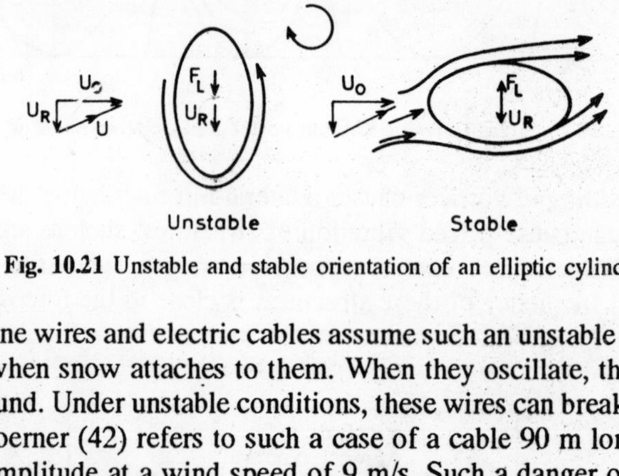

Fig. 10.21 Unstable and stable orientation of an elliptic cylinder

Telephone wires and electric cables assume such an unstable shape in cold countries when snow attaches to them. When they oscillate, they produce a musical sound. Under unstable conditions, these wires can break if amplitude is large. Hoerner (42) refers to such a case of a cable 90 m long oscillating at ± 3 m amplitude at a wind speed of 9 m/s. Such a danger of oscillations leading to breaking of electric cables can be avoided by overheating the cable by passing a relatively high current through it occasionally which melts the ice and restores the cable shape to circular. Similar oscillations have been found to occur in suspended pipelines.

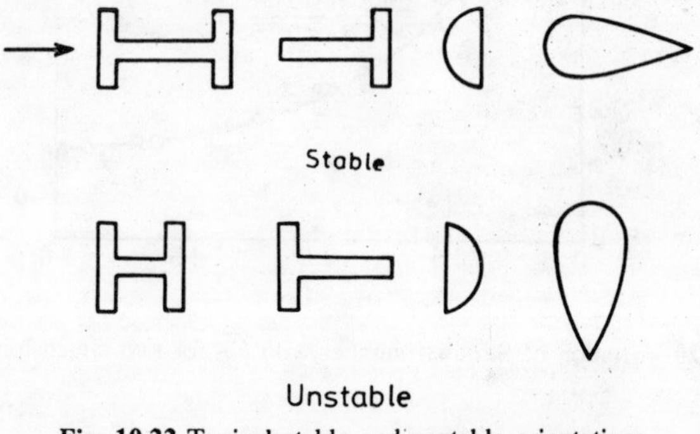

Fig. 10.22 Typical stable and unstable orientations

Steel RCC chimneys and towers are subject to such lateral vibrations at moderate wind velocities. The amplitude of vibration can be as large as one meter at the top if tower is 150-200 m tall. Vibrations over long period can cause fatigue stresses leading to failure. The problem can be more serious if forced frequency is close to the natural frequency of the structure. Some of the methods that are used to control vibrations of such slender structures include:

i. changing natural frequency of structure by structural changes;
ii. changing dynamic and static action of the structure by providing additional support, e.g. guying;
iii. increasing structural damping by concrete lining, filling cavities, etc.
iv. modification of flow regime by providing spoilers. These include splitter plate on the downstream side which stabilises vortices, providing spiral rib around the cylinder, etc.

REFERENCES

1. Hinze, J.O. *Turbulence-An Introduction to its Mechanism and Theory*. McGraw Hill Book Co. Inc. 1955, Chapter VI.
2. Rajaratnam N. *Turbulent Jets*. Development in Water Resources Series 5, Elsevier Publication, Amsterdam, 1984.
3. Rouse, H. (Ed.), *Advanced Mechanics of Fluids*. John Wiley and Sons Inc., N.Y. 1959 Chapter VIII.
4. Albertson, M.L., Y.B. Dai, R.A. Jensen and H. Rouse. Diffusion of Submerged Jets. Trans ASCE, Vol 115, 1950.
5. Tollmien, W. Berechnung Turbulenter Ausbreitungsvorgänge Z.A.A.M. Vol 6, 1926 (English Translation NACA, TM 1085, 1945)
6. Görtler, H. Berechnung Von Aufgaban der Freien Turbulenz auf Grund Eines Neunen Naherun Gsansatzes. Z.A.A.M. Vol 22, 1942.
7. Goldstein, S. *Modern Developments in Fluid Dynamics*. Vol 2, Dover Publications Inc., New York, 1965.
8. Forthmann, E. Turbulent Jet Expansion. English Translation NACA TM-789, 1936. (Original paper in German, 1934, Ing. Archiv. 5).
9. Zijnen, B.G. Van der Hagge. Measurements of the Velocity Distribution in a Plane Turbulent Jet of Air. App. Sci. Res., Sect A, Vol 7, 1958.
10. Heskestad, G. Hot-wire Measurements in a Plane Turbulent Jet. J. Appl. Mech., Trans. ASME, 1965.
11. Abramovich, G.N. *The Theory of Turbulent Jets* (Translation), M.I.T. Press, Massachusetts, USA, 1963.
12. Schlichting, H. *Boundary Layer Theory*. McGraw Hill Book Co., New York, 1968.
13. Zijnen, B.G. Van der Hagge. Measurements of Turbulence in Plane Jet of Air by Diffusion Method by the Hot-wire Method. App. Sci. Res., Sect A, Vol 7, 1958.
14. Miller, D.R. and E.W. Comings. Static Pressure Distribution in Free Turbulent Jet. JFM. Vol 3, 1957.
15. Gutmark, E. and I. Wygnanski. The Planer Turbulent Jet. JFM, Vol 73, 1976.
16. Corrsin, S. Investigation of Flow in an Axially Symmetric Heated Jet of Air. NACA Wartime Report, W-94, 1946.
17. Hinze, J.O. and Zijnen, B.G. Van der Hagge. Transfer of Heat and Matter in the Turbulent Mixing Zone of an Axially Symmetrical Jet. Jour. of App. Sci. Res., Sect A, Vol 1, 1949.
18. List E.J. Mechanics of Turbulent Buoyant Jets and Plumes. In Rodi, W. (ed.). *Turbulent Buoyant Jets and Plumes*. Vol 6, in the Science and Application of Heat and Mass Transfer. Pergamon Press, 1982.

19. Wygnanski, I. and Fiedler, H. Some Measurements in the Self-Preserving Jet, JFM. Vol 38, 1969.
20. Becker, H.A., H.C. Hottel and C.C. William. The Nozzle-Fluid Concentration Field of the Round Turbulent Free Jet. JFM. Vol 38, 1969.
21. Chevray, R. and N.K. Tutu. Intermittency and Preferential Transport of Heat in a Round Jet. JFM, Vol 88, 1978.
22. Birch, A.D., D.K. Brown, M.G. Dodson and J.R. Thomas. Turbulent Concentration Field of a Methane Jet. JFM. Vol 88, 1978.
23. Schlichting, H. Uber das Ebene Windschattenproblem. Thesis, Göttingen University (Germany), 1930.
24. Swain, L.M. On the Turbulent Wake Behind a Body of Revolution. Proc. Royal Society of London. Vol 125A, 1929.
25. Adachi, T. and K. Yoshida. Study of Two-Dimensional Similar Velocity Distribution in the Turbulent Wakes. Bull. JSME. Vol 5, No. 204, June 1982.
26. Satapathy, B. Turbulence Effects on the Drag of Three-Dimensional Bodies in Mid-Stream. Ph.D. Thesis, Univ. of Roorke, Roorkee, 1980.
27. Einstein, H.A. and EL-Samni. Hydrodynamic Forces on a Rough Wall. Rev. of Modern Physics. Vol 21, No. 3, July 1949.
28. Ko, S.C. and W.H. Graf. Drag Coefficient and Turbulence Characteristics. Proc. of 14th Congress of IAHR, Paris (France), Vol 2, 1971.
29. Ko, S.C. and W.H. Graf. Drag Coefficients of Cylinders in Turbulent Flow. JHD, Proc. ASCE, Vol 98, No HY-5, 1972.
30. Cheng, E.D.H. and C.G. Clyde. Instantaneous Hydrodynamic Lift and Drag Forces on Large Roughness Elements in Turbulent Open Channel Flow. *Sedimentation Symposium to Honour Einstein.* Ed. H.W. Shen, Fort Collins (USA), 1972.
31. Torobin, L.B. and W.H. Gauvin. Drag Coefficient of Single Spheres Moving in Steady and Accelerated Motion in a Turbulent Fluid. Jour. of A.I. Ch. E. Vol 7, December 1961.
32. Taylor, G.I. Statistical Theory of Turbulence. Part V, Effect of Turbulence on Boundary Layers. Proc. RSL, Series A, Vol 156, 1936.
33. Schubauer, G.B. The Effect of Turbulence on Transition in the Boundary Layer of an Elliptic Cylinder. Proc. 5th International Conference on Fluid Mechanics, 1937.
34. Loisyanskii, L.G. *Mechanics of Liquids and Gases*. Perganon Press, 1966.
35. Graf W.H. and F.F. Mansour. Turbulent Drag Coefficient of Sharp Edged Objects. JHR, IAHR, Vol 13, No. 2, 1975.
36. Roberson, J.A., C.Y. Lin, G.S. Rutherford and M.D. Stine. Turbulence Effects on Drag of Sharp Edged Bodies. JHD, Proc. ASCE, Vol 98, NO. HY-7, 1972.
37. Vickery, B.J. On the Flow Behind a Coarse Grid and its Use as a Model of Atmospheric Turbulence in Studies Related to Wind Loads on Buildings. NPL Aero. Rep. 1143, 1965.
38. Vickery, B.J. Fluctuating Lift and Drag on a Long Cylinder of Square Cross Section in a Smooth and Turbulent Stream. JFM, Vol 25, Part 3, 1966.
39. Vickery, B.J. Load Fluctuations in Turbulent Flow. JEMD, Proc. ASCE, Vol. 94, EM-1, 1968.
40. Bearmen, P.W. An Investigation of the Forces on Flat Plates Normal to a Turbulent Flow. JFM, Vol. 46, Pt. 1, 1971.
41. Rouse, H. (Ed.) *Engineering Hydraulics*. John Wiley and Sons Inc. New York, 1950, Chapter 1.
42. Hoerner, S.F. *Fluid Dynamic Drag-Practical Information on Aerodynamic Drag and Hydrodynamic Resistance*. Published by Author at New Jersey (USA), 1965.

Author Index

Subject Index